Springer Series in
Surface Sciences

33

Editor: Gerhard Ertl

Springer Series in **Surface Sciences**

1 **Physisorption Kinetics**
By H. J. Kreuzer, Z. W. Gortel

2 **The Structure of Surfaces**
Editors: M. A. Van Hove, S. Y. Tong

3 **Dynamical Phenomena at Surfaces, Interfaces and Superlattices**
Editors: F. Nizzoli, K.-H. Rieder, R. F. Willis

4 **Desorption Induced by Electronic Transitions, DIET II**
Editors: W. Brenig, D. Menzel

5 **Chemistry and Physics of Solid Surfaces VI**
Editors: R. Vanselow, R. Howe

6 **Low-Energy Electron Diffraction**
Experiment, Theory and Surface Structure Determination
By M. A. Van Hove, W. H. Weinberg, C.-M. Chan

7 **Electronic Phenomena in Adsorption and Catalysis**
By V. F. Kiselev, O. V. Krylov

8 **Kinetics of Interface Reactions**
Editors: M. Grunze, H. J. Kreuzer

9 **Adsorption and Catalysis on Transition Metals and Their Oxides**
By V. F. Kiselev, O. V. Krylov

10 **Chemistry and Physics of Solid Surfaces VII**
Editors: R. Vanselow, R. Howe

11 **The Structure of Surfaces II**
Editors: J. F. van der Veen, M. A. Van Hove

12 **Diffusion at Interfaces: Microscopic Concepts**
Editors: M. Grunze, H. J. Kreuzer, J. J. Weimer

13 **Desorption Induced by Electronic Transitions, DIET III**
Editors: R. H. Stulen, M. L. Knotek

14 **Solvay Conference on Surface Science**
Editor: F. W. de Wette

15 **Surfaces and Interfaces of Solids**
By H. Lüth

16 **Atomic and Electronic Structure of Surfaces**
Theoretical Foundations
By M. Lannoo, P. Friedel

17 **Adhesion and Friction**
Editors: M. Grunze, H. J. Kreuzer

18 **Auger Spectroscopy and Electronic Structure**
Editors: G. Cubiotti, G. Mondio, K. Wandelt

19 **Desorption Induced by Electronic Transitions, DIET IV**
Editors: G. Betz, P. Varga

20 **Scanning Tunneling Microscopy I**
General Principles and Applications to Clean and Adsorbate-Covered Surfaces
Editors: H.-J. Güntherodt, R. Wiesendanger

21 **Surface Phonons**
Editors: W. Kress, F. W. de Wette

22 **Chemistry and Physics of Solid Surfaces VIII**
Editors: R. Vanselow, R. Howe

23 **Surface Analysis Methods in Materials Science**
Editors: D. J. O'Connor, B. A. Sexton, R. St. C. Smart

24 **The Structure of Surfaces III**
Editors: S. Y. Tong, M. A. Van Hove, K. Takayanagi, X. D. Xie

25 **NEXAFS Spectroscopy**
By J. Stöhr

26 **Semiconductor Surfaces and Interfaces**
By W. Mönch

27 **Helium Atom Scattering from Surfaces**
Editor: E. Hulpke

28 **Scanning Tunneling Microscopy II**
Further Applications and Related Scanning Techniques
Editors: R. Wiesendanger, H.-J. Güntherodt

29 **Scanning Tunneling Microscopy III**
Theory of STM and Related Scanning Probe Methods
Editors: R. Wiesendanger, H.-J. Güntherodt

30 **Concepts in Surface Physics**
By M. C. Desjonquères, D. Spanjaard

31 **Desorption Induced by Electronic Transitions, DIET V**
Editors: A. R. Burns, E. B. Stechel, D. R. Jennison

32 **Scanning Tunneling Microscopy and Related Techniques**
By Bai Chun-Li

33 **Adsorption on Ordered Surfaces of Ionic Solids and Thin Films**
Editors: H.-J. Freund, E. Umbach

H.-J. Freund E. Umbach (Eds.)

Adsorption on Ordered Surfaces of Ionic Solids and Thin Films

Proceedings of the 106th WE-Heraeus Seminar,
Bad Honnef, Germany, February 15–18, 1993

With 139 Figures

Springer-Verlag
Berlin Heidelberg New York
London Paris Tokyo
Hong Kong Barcelona
Budapest

Professor Dr. Hans-Joachim Freund

Ruhr-Universität Bochum, Physikalische Chemie I,
Universitätsstrasse 150, D-44780 Bochum, Germany

Professor Dr. Eberhard Umbach

Experimentelle Physik II, Universität Würzburg,
Am Hubland, D-97074 Würzburg, Germany

Series Editors

Professor Dr. Gerhard Ertl

Fritz-Haber-Institut der Max-Planck-Gesellschaft, Faradayweg 4–6,
D-14195 Berlin, Germany

Professor Robert Gomer, Ph.D.

The James Franck Institute, The University of Chicago, 5640 Ellis Avenue,
Chicago, IL 60637, USA

Professor Douglas L. Mills, Ph.D.

Department of Physics, University of California,
Irvine, CA 92717, USA

Managing Editor: Dr. Helmut K.V. Lotsch

Springer-Verlag, Tiergartenstrasse 17,
D-69121 Heidelberg, Germany

ISBN-13:978-3-642-78634-1 e-ISBN-13:978-3-642-78632-7
DOI: 10.1007/978-3-642-78632-7

Softcover reprint of the hardcover 1st edition 1993

Typesetting: Camera ready copy from the authors
54/3140 - 5 4 3 2 1 0 - Printed on acid-free paper

Preface

The adsorption on "salt" and "rust" has recently attracted much attention. The reason is at least threefold.

First, little is known about the structure of insulating ionic crystals and, in particular, the adsorption on well-defined surfaces of these crystals. This situation is entirely different from that of metal or semiconductor surfaces for which a broad and detailed knowledge exists based on the extensive research of an enormous number of groups using many different techniques. In contrast, insulator surfaces have played the role of a "Sleeping Beauty" in basic surface science for a long time. Of course, an unexplored field is attractive. Moreover, it turned out that many results, concerning, for example, the chemical bonding and the kinetics of adsorbates, are different for insulators as compared to metal or semiconductor surfaces.

The second reason is that most surface sensitive techniques deal with charged particles and/or ionizing radiation. These cause charging problems and rapid degradation of the sample, and hence result in local gradients of fields, order, stoichiometry, heat, etc. The recent development of new techniques of analysis (e.g. atomic force microscopy) and especially of sample preparation has mostly solved or circumvented these problems. In particular, the successful preparation of epitaxial ultrathin films of halides and oxides on single-crystal surfaces of semiconductors or metals has had a major impact. With these films it became possible to perform detailed and reproducible experiments on macroscopic areas of single-crystalline layers of insulators using nearly the full potential of modern surface techniques.

The third reason is technology's incentive to obtain a detailed knowledge about insulator surfaces. The search for novel materials with tailored properties, the reduction of dimensions (ultrathin films, micro- or nanostructures), and the need for parameter optimization are important scientific and economic challenges which frequently concentrate on, or at least include, surfaces or interfaces of insulators. Numerous examples can be found in the fields of materials research, semiconductor production, thin-film technology, optics, solar-energy conversion, reaction kinetics, electrochemistry, catalysis, microelectronics, micromechanics, sensors, etc.

These factors have inspired a considerable activity in the field of insulator surfaces within the last few years. With this activity the number of new and exciting results rapidly increased. We believe that at present this new field of surface science has reached an extent and maturity that makes it useful to take stock and to give an overview of its present status. Sponsored by the Dr. Wilhelm and Else Heraeus-Stiftung we organized a symposium on this topic in Bad Honnef, Germany, in February 1993. In addition, we asked the invited speakers to contribute short overview articles which should give both new results as well as a review of previous results obtained by their groups. We tried to organize the selection, the arrangement, and the contents of the articles such that a coherent picture of the present status of this field emerges. The book is meant to serve as a basis for the expected rapid development of the understanding of insulator surfaces and the adsorption on them.

As "report" of the present status the book is organized as follows: In Part I halide surfaces are discussed. The preparation and characterization of ultrathin epitaxial halide films on semiconductor crystals as well as the adsorption of atoms and molecules on such films and on halide single crystals are the topic of six articles. In Part II three articles deal with the geometric and electronic structure of clean surfaces of metal oxides, whereas in Part III four authors review the adsorption of atoms and small molecules and two authors the deposition of metal atoms and ultrathin films on well-defined, single-crystalline oxide surfaces. In Part IV, the results from theoretical calculations of ionic surfaces and especially of the adsorption properties on oxides are presented. In Part V and VI six articles cover research aspects that are directed towards applications. The three articles of Part V are concerned with the chemical reactivity of oxide surfaces, whereas those of Part VI explicitly concentrate on catalysis and sensors. The reference lists in most articles were reduced such that they are representative of the work of this particular group rather than of the whole field. All articles were thoroughly refereed.

Finally, we would like to thank all who have contributed to the symposium and to this book, in particular the Dr. Wilhelm and Else Heraeus-Stiftung for the generous sponsorship. We are also very grateful to Dr. V. Schäfer and Mrs. J. Hartmann for the organization, to all referees for doing a careful job, to Springer-Verlag for the good cooperation, and to the coworkers in Bochum for the final layout of the articles, especially Dr. H. Hamann, as "editor-in-chief", and Dr. K. Al-Shamery, B. Baumeister, M. Beckendorf, D. Cappus, I. Hemmerich, U. J. Katter, Dr. H. Kuhlenbeck, H. Schlienz, F. Winkelmann, K. R. Wirth, and S. Wohlrab as the "editorial crew".

Bochum
Würzburg
June 1993

H.-J. Freund
E. Umbach

Contents

Part I **Halide Ultrathin Films and Single Crystals: Structure and Adsorption**

Part IV Adsorption on Oxide Surfaces: Theory

Part V Chemical Reactivity

Part VI Application-Oriented Studies

Part I

Halide Ultrathin Films and Single Crystals: Structure and Adsorption

Epitaxy of CaF_2/Si(111) and LiF/Ge(100)

F. J. Himpsel

IBM T. J. Watson Research Center, P.O. Box 218
Route 134, Yorktown Heights, NY 10598, USA

Abstract. The epitaxial growth of insulators on semiconductors is exemplified by the CaF_2/Si(111) and LiF/Ge(100) systems, which are representative of the fluorite and rocksalt structures, respectively. High quality epitaxial films are grown, and are used to study the bonding at the interface (covalent versus ionic), surface states of the insulator, negative electron affinity, and band dispersions of the monolayer and thick film (band widening due to self-energy, band narrowing in the monolayer, band offset). The regrowth of semiconductors on epitaxial insulators is investigated via adsorption of Si_2H_6 and Si_2Cl_6 on CaF_2/Si(111) activated by irradiation.

1. Introduction

Epitaxial growth of insulators on semiconductors looks promising from a variety of angles [1,2]. On the applied side, there has been a trend towards stacked, three-dimensional device structures, which require insulating spacers between semiconductors and metals that are able to transmit the crystalline perfection of the substrate. On the fundamental side there has been the long-standing question about the electronic structure of localized, highly-correlated bands, e.g. the influence of the self-energy on the band width and other many body effects. The valence bands of wide gap insulators are nice and simple model systems for studying such effects. In addition one can pursue the electronic structure into lower dimensions, e.g. at semiconductor/insulator interfaces and insulator surfaces. Due to the drastic differences in bonding (covalent versus ionic) it is possible to change the band gaps by a factor of two and more at interfaces. The surface of an insulator can be converted into a metal by irradiation, and thus be activated for further epitaxial growth on top of the insulator. Insulator surfaces, such as LiF(100), exhibit the largest known negative electron affinities, giving rise to intense electron emission at the vacuum level and to electron trapping into two-dimensional image states.

This article focuses on two systems which exhibit a variety of these interesting phenomena, i.e. CaF_2/Si(111) and LiF/Ge(100). Together with NaCl/Ge(100) (see the articles by Henzler, Pfnür and Umbach and [3]), they represent the best-studied insulator/semiconductor systems. All of them exhibit excellent lattice match (LiF/Ge(100) by a factor of $\sqrt{2}$ giving rise to a 45° rotated structure). They also expose the lowest energy, non-polar surfaces, which are (111) for the fluorite and (100) for the rocksalt structures. In the following we will start out at the semiconductor/insulator interface, then proceed outwards to bulk properties of epitaxial insulators, to their surfaces and to the growth of overlayers on insulators.

2. Interface Bonding

The best system for studying the bond across the interface is CaF_2/Si(111), since the first layer wets perfectly and gives rise to large changes in the electronic structure [4-23]. There are a few obvious questions to ask about the bonding at the interface: Does Si bond to Ca or to F? Is the bonding covalent, as in Si, or ionic, as in CaF_2? What is the bond geometry? The first two questions can be answered in straightforward fashion by looking at the core level shifts induced at the interface (Fig. 1 [5,6]). Looking at the Si 2p core level we see that the

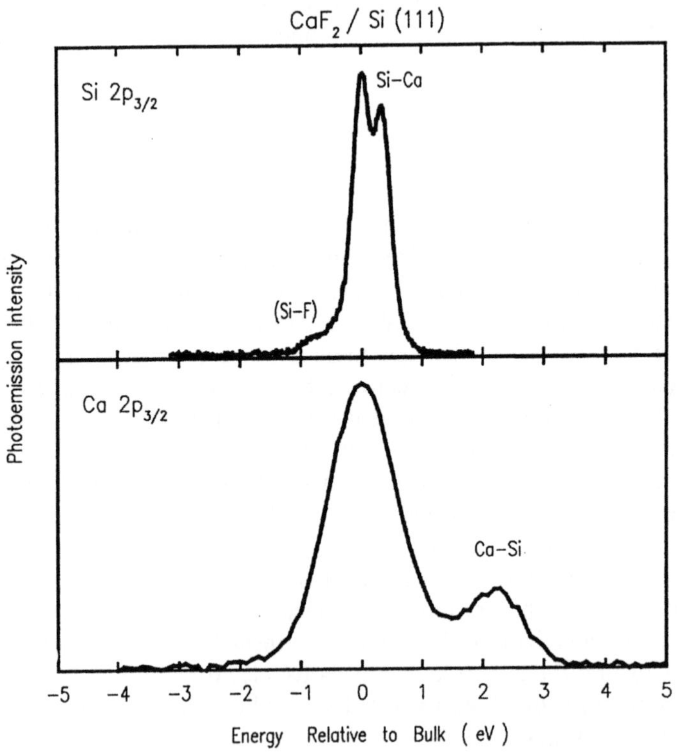

Fig. 1: Core level spectra for epitaxial CaF_2 on Si(111) showing that Si bonds to Ca across the interface, and not to F [5,6]. The small shift of the Si level indicates a covalent bonding on the Si side, while there is ionic bonding on the Ca side with Ca in the unusual 1+ oxidation state.

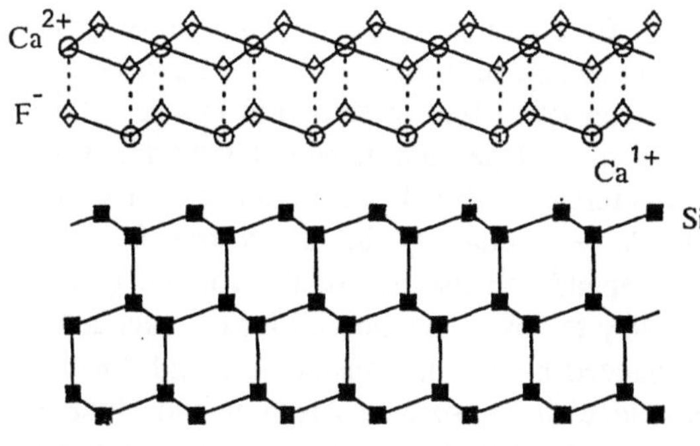

Fig. 2: Structural model for the $CaF_2/Si(111)$ interface grown at high temperature [22].

majority of the interface Si atoms have their core levels shifted up in energy, as expected from a transfer of negative charge from the electropositive Ca towards Si (Si-Ca peak in Fig. 1). Only a few Si atoms have core levels shifted the other way (Si-F shoulder in Fig. 1), and their number is found to become smaller with better-oriented and more perfect interfaces, as long as the growth temperature is high enough. The Si-Ca bonding is confirmed by looking at the Ca 2p core level, which selects the other side of the interface. It exhibits a clear interface peak, labelled Ca-Si in Fig. 1.

To understand the interface bond better we have to look at the bonding orbitals. Taking Si at the interface as covalent and Ca in the 1+ oxidation state we have two half-filled orbitals to work with, i.e. the Si broken bond orbital and the Ca 4s. These form bonding/antibonding combinations. The splitting between the occupied, bonding combination and the unoccupied, antibonding is a measure of the interface bond energy. The formation of such an interface bond explains the

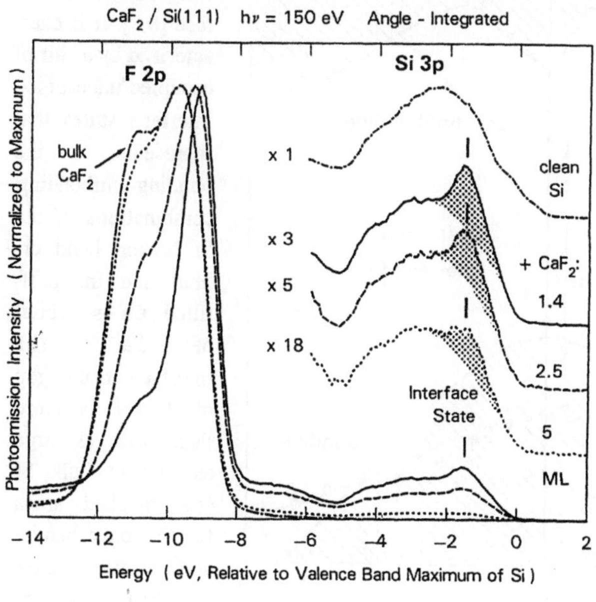

Fig. 3: Angle-integrated valence band spectra of epitaxial CaF_2 overlayers on Si(111). An occupied interface state sits on top of the Si valence band. The F 2p valence band of CaF_2 is narrower in the monolayer [4,5] due to reduced coordination.

unusual 1+ oxidation state for Ca. If Ca remained in the 2+ oxidation state with its rare gas configuration it would not be able to tie up the Si dangling bond.

This type of interface bond is consistent with a structural model obtained from ion scattering (Fig. 2 [22]), where a F layer is missing at the interface compared to the $F^-Ca^{2+}F^-$ sandwich structure of $CaF_2(111)$. In order to decide about covalent versus ionic bonding we have to quantify the charge transfer. It has been an elusive quantity since it depends on the size of the volume given to various atoms. The best empirical way to define charge transfer may actually be via the core level shifts that are induced by charge transfer. Although there are additional contributions to the core level shifts, such as screening and Madelung energy, one finds a strong correlation of the Si 2p core level shift with electronegativity difference for many adsorbates, including Ca and F (see Ref. 24 for calibration experiments with Ca and F overlayers on Si).

If one takes the magnitude of the F-induced shift of -1.1eV as a measure of fully-ionic adsorption, the 0.4 eV shift at the CaF_2/Si(111) interface corresponds to a charge transfer of less than half of an electron. That is an upper limit, since the actual charge transfer for the Si-F bond is only half of an electron according to Pauling. Therefore, one can argue that Si is nearly covalent at the interface. Ca, on the other hand, exhibits a much larger core level shift of 2.1 eV relative to bulk CaF_2, where Ca is in the 2+ oxidation state. Such a core level shift is consistent with a conversion form Ca^{2+} to Ca^{1+} at the interface. The extra electron at the interface Ca^{1+} (relative to bulk Ca^{2+}) comes from the F^- ion of the missing F layer at the interface (see Fig. 2 and [11]). Thus we may say that Si is

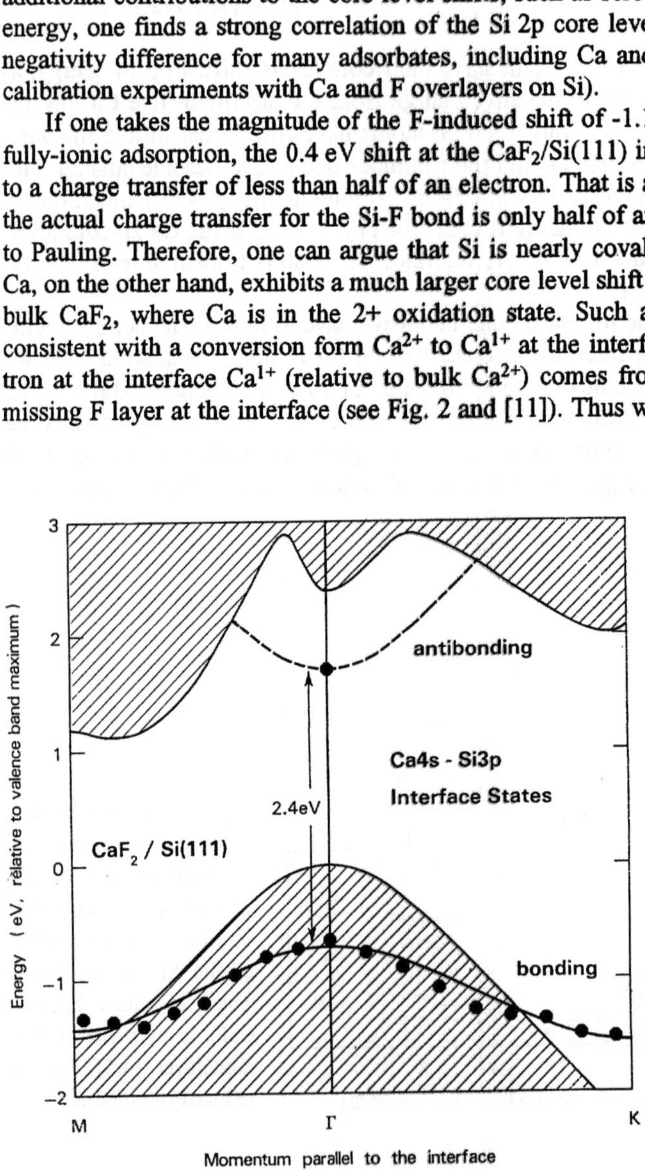

Fig. 4: Electronic structure of the CaF_2/Si(111) interface [8,9]. It is characterized by a pair of occupied/unoccupied interface states that represents the bonding-antibonding combinations of the Si broken bond orbital and the half-filled Ca 4s orbital of Ca^{1+}. The interface band gap of 2.4 eV is more than twice as large as that of bulk Si (see hatched areas for the bulk bands) and five times smaller than that of CaF_2.

bonded covalently and Ca ionically.

The interface bond orbitals predicted by this model can be seen with photoemission and other techniques. Figure 3 [5] shows that there is an occupied interface state centered at about 1.5 eV below the valence band maximum of Si (which is close to the Fermi level at this interface). As expected for an interface state, this feature becomes strongly attenuated by the CaF_2 overlayer (up to a factor of 18 for the thickest film), but remains roughly constant in intensity relative to the Si substrate. To get the complete information about this state we have measured its two-dimensional $E(k_{||})$ band dispersion with angle-resolved photoemission (dots in Fig. 4 [8]). We have also determined the energy gap of 2.4 eV between the occupied, bonding and the unoccupied antibonding interface state using optical second harmonic generation [7]. This technique is particularly useful to obtain the electronic structure of an interface since the effect is dipole-forbidden for inversion-symmetric bulk media, such as Si and CaF_2. The interface gap is more than twice as large as that of Si and five times smaller than that of CaF_2, giving a good example of how solids can be manipulated at interfaces. The main reason behind this large change in the electronic structure is the unusual 1+ oxidation state of Ca, which was forced onto it by the interface geometry. Similar unstable oxidation states become stabilized at other Si insulator interfaces, e.g. the Si^{1+}, Si^{2+}, and Si^{3+} oxidation states at the SiO_2/Si [25].

3. Band Dispersion of Insulators and Self-Energy

High quality epitaxial films of insulators make it feasible to obtain the energy-versus-momentum band dispersions E(**k**) for insulators with angle-resolved photoemission, an experiment that is very difficult for bulk crystals due to charging. In addition, one can determine the difference between the three-dimensional bands in the bulk and the two-dimensional bands of a monolayer. Large gap insulators, such as LiF with its record 14.1 eV gap [26,27], are particularly interesting for determining the influence of the self-energy on the band structure. The effects are expected to become strong when the band width is small compared to the energy gap between bands. The large band gap of LiF also causes the conduction band minimum (CBM) to lie well above the vacuum level, thereby producing the largest negative electron affinity E_A of ordinary solids (Fig. 5 [26,27]).

The band dispersion of the narrow valence band of LiF can be obtained from the angular photoemission patterns shown in Fig. 6 [28]. The result is shown in Fig. 7, with the angle-integrated valence band spectrum from Fig. 5 turned on its side for comparison. To focus on one, fundamental parameter of the band structure we consider the valence band width (3.5eV from Γ_{15} to $X_{4'}$).. It has been taken as crucial measure of many body/self-energy effects for a variety of other systems, e.g. high temperature superconductors and simple metals. The width measured in LiF is 17% larger than in ground state calculations, indicating a significant role of self-energy effects in determining the band width. This agrees qualitatively with the 10% band width increase calculated for LiCl in state-of-the

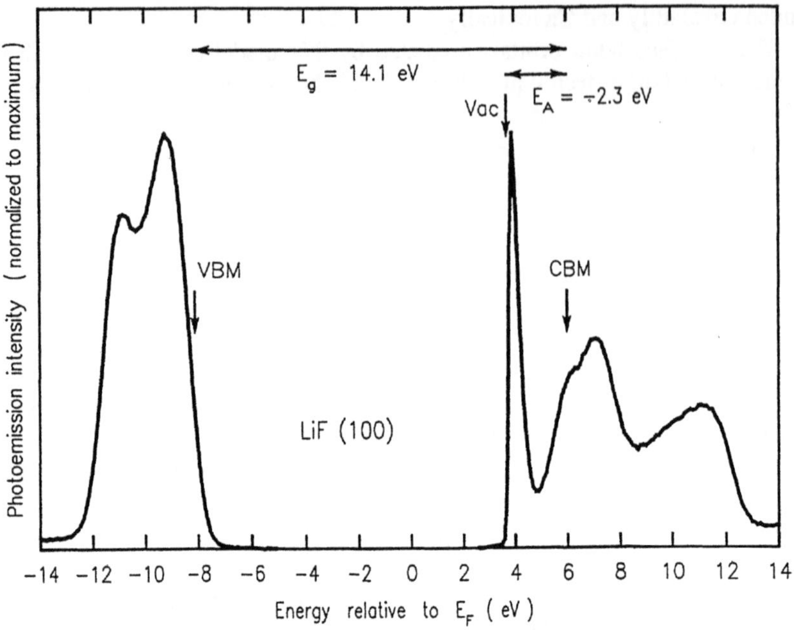

Fig. 5: Overview spectrum of the valence and conduction bands of LiF(100), grown epitaxially on Ge(100). The former is measured by angle-integrated photoemission (hν = 125 eV, left), the latter by secondary electron spectroscopy (right) [26,27]. The vacuum level (Vac) lies below the conduction band minimum (CBM), giving rise to a record negative electron affinity E_A = -2.3 eV [26,27]. The emission peak at the vacuum level corresponds to a surface state, probably induced by the image potential.

art quasiparticle calculations [29]. Interesting enough, the sign of the self-energy effect is opposite to that seen in free-electron-like simple metals, which represent the other end of the localization scale.

The band dispersion of the CaF_2 valence band is more complicated than in LiF since there are two F atoms per unit cell and, consequently, twice as many bands. A determination of bulk band dispersions analogous to LiF is still in progress [30]. However, we can directly see the influence of film thickness on the band width from the angle-integrated photoemission data in Fig. 3 [4,5]. The band width shrinks for thin films due to the reduced interaction perpendicular to the film and becomes much narrower for the monolayer.

4. Insulator Surfaces

At a first glance one would not expect pronounced surface states for alkali halide surfaces, due to the inert, rare gas configurations of the ions. However, there are

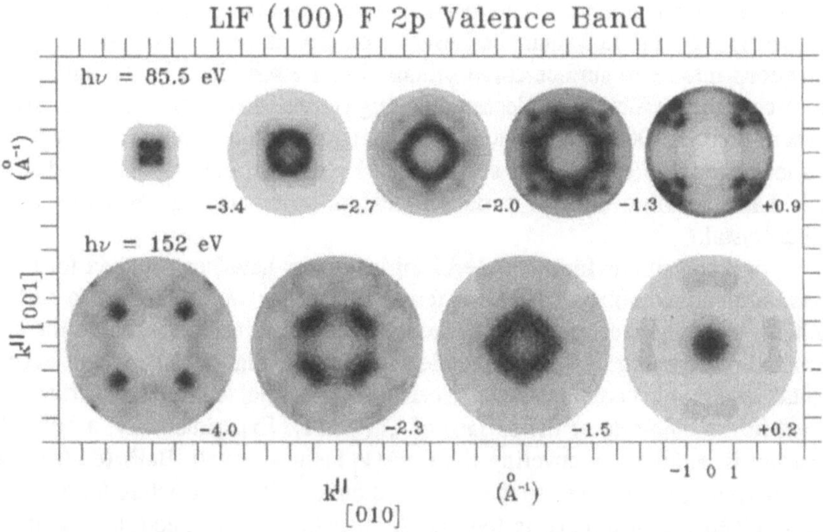

Fig. 6: Angular distribution of photoelectrons [28] from epitaxial LiF(100) at different energies across the valence band (in eV relative to the valence band maximum VBM). The two photon energies hv correspond to different cuts in k-space. Dark areas represent high emission intensity. The two pictures at +0.9 and -4.0 eV are taken in the tails of the spectrum beyond the band edges.

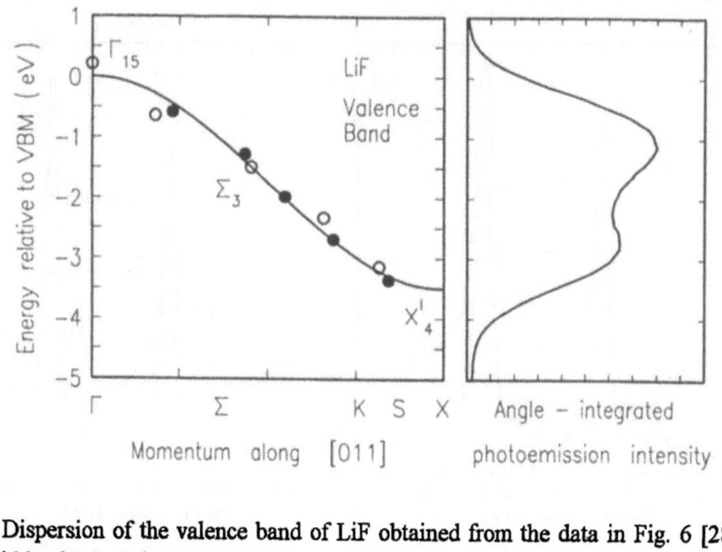

Fig. 7: Dispersion of the valence band of LiF obtained from the data in Fig. 6 [28]. The band width of 3.5 eV is 17% larger than in ground state calculations, showing the importance of self-energy effects in these localized bands.

several peculiar features in the electronic structure that give rise to surface effects. First we will show that the lower symmetry at the surface allows optical transitions into extra surface core excitons that are forbidden in the bulk. Then we discuss the possibility of electrons getting trapped in image potential surface states at the LiF(100) surface due to the negative electron affinity. Finally it will be shown how the surface can be modified by photon and electron irradiation, such that a metallic layer of surface F-centers terminates an insulating alkali halide crystal.

Optical transitions from core levels of insulators have been studied for many years, and surface effects have been found [31]. They were assigned to surface excitons, since the electron hole interaction is large in insulators due to the lack of metallic screening. A good example is shown in Fig. 8 [10], where experimental results are combined with a calculation of the multiplet structures expected in the bulk and at the surface of $CaF_2(111)$. Experimentally, a bulk-like spectrum is obtained by covering the $CaF_2(111)$ surface with BaF_2 (dotted line, left panel). It compares well to the calculated bulk multiplet structure (dotted line, right). When the symmetry is lowered from cubic O_h to three-fold C_{3v} at the surface, two things happen: Extra transitions appear that were forbidden in the bulk (labelled surface in Fig. 8), and the absorption becomes anisotropic. The largest surface feature should only be excited by the electric field component perpendicular to the surface. The experiment exhibits all these effects. The clean $CaF_2(111)$ surface has an extra feature at the expected position, and its intensity grows when the surface sensitivity is enhanced by detecting (positive) ions, which essentially come from the outermost layer. The expected polarization dependence

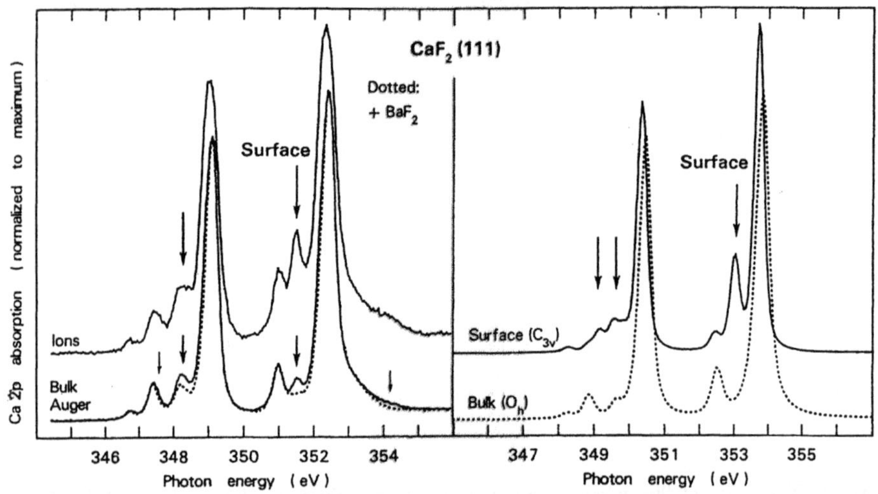

Fig. 8: Transitions from the Ca 2p core level to empty Ca 3d states [10] at an epitaxial $CaF_2(111)$ surface (left: experiment, right: calculation). The transitions marked by arrows are due to the lowering of the symmetry from cubic O_h to three-fold C_{3v} at the surface. Detecting ions emphasizes surface transitions while a coating of BaF_2 suppresses them.

is also observed (not shown).

Clear evidence for a surface state in the band gap comes from the sharp spike seen for LiF(100) near the vacuum level (Vac) in Fig. 5 [26,27]. It lies well below the conduction band minimum (CBM) and is separated from it by a deep valley. It also is found to be very sensitive to surface quality, e.g. radiation damage. The best explanation for this state is that of an image potential state, where an electron is trapped at the surface by its image potential (reduced in a dielectric by a factor of $(\varepsilon - 1)/(\varepsilon + 1)$). Since LiF has no bulk states near the vacuum level these electrons can remain at the surface forever and charge it up to form a two-dimensional electron gas, similar to the one observed on liquid helium [32]. The lowest state is bound slightly below the vacuum level and cannot emit electrons, but a bound electron may be lifted by phonons into the continuum of the image potential just above the vacuum level, where we see intense secondary electron emission. If this scenario is correct, LiF(100) could become a prime substrate for two-dimensional electron gas physics, since it has the capability of holding a much higher density of electrons than liquid He, due to its higher dielectric constant. An order of magnitude higher densities are required to reach the expected quantum Wigner crystal, i.e. a triangular lattice of electrons dominated by Fermi statistics, and not Boltzmann statistics. The latter has already been observed on liquid He [32]. Other, more exotic phases such as fractional statistics quantum liquids are also possible.

The surface of ionic crystals can be modified by irradiation with ultraviolet radiation or electrons. As in the bulk, F-centers are formed by exciting the halogen ion, transferring an electron to the alkali, and desorbing a neutral halogen. For the $CaF_2(111)$ surface it has been found [7] that the complete F surface layer can be desorbed in this fashion, producing the equivalent of an ordered array of surface F-centers (Fig. 9, top). The electronic structure of the surface changes dramatically since we now have an extra unpaired electron at each alkali surface atom, which gives rise to a half-filled, metallic band (Fig. 9, bottom). This band exhibits a dispersion characteristic of the expected 1x1 surface periodicity (not shown). Its bottom at 0.5 eV below the Fermi level at $k_{||} = 0$ is seen in the uppermost spectrum of Fig. 9. Symmetric extrapolation above E_f gives a total band width of 1 eV. This band width represents the lateral interaction between the surface F-centers.

5. Overlayers on Insulators

The task of growing a flat overlayer on an insulator surface is complicated by a thermodynamic trend towards island formation. Essentially, insulators like their surfaces exposed since they have the lowest surface energies of all materials (typical numbers are 300-500 erg/cm^2). Metals or semiconductors have much higher surface energies due to half-filled, broken band like orbitals (typically 1000 erg/cm^2 for simple metals, 3000 erg/cm^2 for transition metals, and 1200 erg/cm^2 for semiconductors, such as Ge and Si). Therefore one has to activate an insulator surface to make sure that the overlayer spreads out. If that can be achieved there is a wide open field of interesting structures, such as three-dimensional devices, enhanced magnetism in monolayers, and highly anisotropic, layered solids.

There have been many attempts to grow device quality Si and Ge on CaF_2 by molecular beam epitaxy [1,2,33]. The wetting problem has been partially overcome by growing a flat amorphous layer at room temperature, where Si is im-

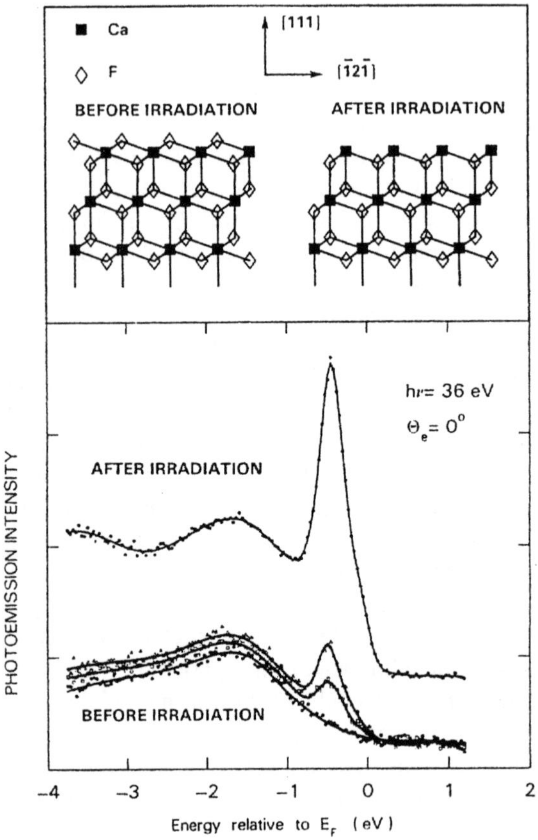

Fig. 9: Transformation of the $CaF_2(111)$ surface from an insulator to a metal by irradiation with photons of 36 eV energy [7]. Valence spectra are shown for saturation, and for two intermediate radiation doses. The removal of a F layer at the surface leaves an extra metallic electron with each Ca surface atom, creating an intense surface state band at the Fermi level. The structure at -1.5 eV is the interface state at the Si(111) substrate (Fig. 3) shining through the CaF_2 overlayer.

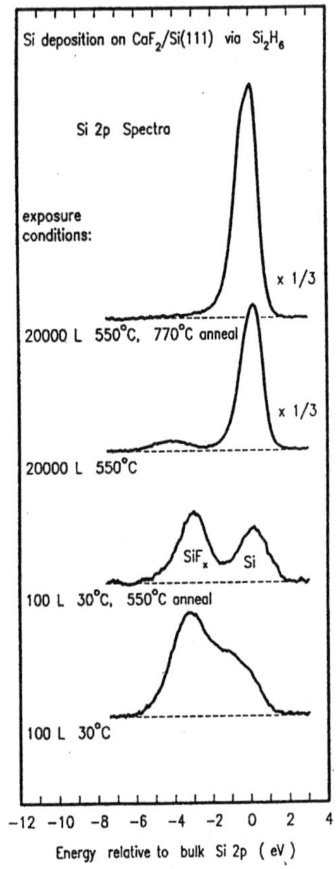

Fig. 10: Growth of Si on $CaF_2(111)$ by adsorption of Si_2H_6 onto a surface activated by irradiation [34]. The initial layer exhibits some Si-F bonding, determined by the Si 2p core level shift. At elevated temperature a continuous Si film can be grown (top), but its epitaxial quality has yet to be characterized.

mobile, and recrystallizing it. A more elegant method uses the activation of the $CaF_2(111)$ surface by irradiation discussed in the previous section. At a dose comparable to that required for establishing the metallic surface layer the growth morphology is found to have an optimum.

We have made an attempt to bias the surface energy even more towards a flat morphology by not only activating the CaF_2 substrate, but also passivating the growing overlayers [34]. This is achieved by a chemical vapor deposition (CVD) method, which uses Si-containing molecules saturated with H or Cl (Si_2H_6 and Si_2Cl_6) instead of pure Si. We find (Fig. 10) that these molecules adsorb only at the activated $CaF_2(111)$ surface at room temperature, forming SiF_x species in the process. That may open up possibilities for patterned deposition, using X-rays passing through a mask as in X-ray lithography, but avoiding all the photoresist and etching operations. In order to deposit more than a monolayer of Si the temperature needs to be raised. Working with Si_2H_6 we obtain a continuous Si film (Fig. 10, top), as evidenced by the disappearance of substrate core level features (not shown). Si_2Cl_6 does not produce Si growth beyond a monolayer, even at elevated temperature, probably since Cl has a tendency towards etching Si. These studies are still in their early phases, and there is a lot more work to be done in characterizing the structure of the growing film and minimizing the defect density. Another promising avenue is the "digital" growth of overlayers. Thereby one adsorbs only a single layer in self-limiting fashion, and activates it in a second step for further deposition.

Acknowledgement

This work was supported in part by the Office of Naval Research. It was conducted at the National Synchrotron Light Source, Brookhaven National Laboratory, which is supported by the Department of Energy (Division of Materials Sciences and Division of Chemical Sciences of Basic Energy Sciences) under Contract No DE-AC02-76CH0016.

References

[1] R.F.C. Farrow, P.W.Sullivan, G.M. Williams, G.R. Jones, D.C. Cameron: J. Vac. Sci. Technol. **19**, 415 (1981)

[2] L.J. Schowalter, R.W. Fathauer: CRC Crit. Rev. Solid State Mater. Sci. **15**, 367 (1989)

[3] U. Barjenbruch, S. Fölsch, M. Henzler: Surf. Sci. **211/212**, 749 (1989)

[4] U.O. Karlsson, F.J. Himpsel, J.F. Morar, D. Rieger, J.A. Yarmoff: J. Vac. Sci. Technol. B **4**, 1117 (1986)

[5] D. Rieger, F.J. Himpsel, U.O. Karlsson, F.R. McFeely, J.F. Morar, J.A. Yarmoff: Phys. Rev. **B34**, 7295 (1986)

[6] F.J. Himpsel, U.O. Karlsson, F.R. McFeely, J.F. Morar, D. Rieger, A. Taleb-Ibrahimi, J.A. Yarmoff: Mat. Sci. and Eng. **B19** (1988)

[7] U.O. Karlsson, F.J. Himpsel, J.F. Morar, F.R. McFeely, D. Rieger, J.A. Yarmoff: Phys. Rev. Lett. **57**, 1247 (1986)
[8] A.B. McLean, F.J. Himpsel: Phys. Rev. **B39**, 1457 (1989)
[9] T.F. Heinz, F.J. Himpsel, E. Palange, E. Burstein: Phys. Rev. Lett. **63**, 644 (1989)
[10] F.J. Himpsel, U.O. Karlsson, A.B. McLean, L.J. Terminello, F.M.F. de Groot, M. Abbate, J.C. Fuggle, B.T. Thole, G.A. Sawatzky: Phys. Rev. **B43**, 6899 (1991)
[11] M.A. Olmstead, R.I.G. Uhrberg, R.D. Bringans, R.Z. Bachrach: Phys. Rev. **B35**, 7526 (1987)
[12] S. Bouzidi, F. Coletti, J.M. Debever, J.L. Longueville, P.A. Thiry: Appl. Surf. Sci. **56-58**, 821 (1992)
[13] Ph. Avouris, R. Wolkow: Appl. Phys. Lett. **55**, 1074 (1989)
[14] K. Nath, A.B. Anderson: Phys. Rev. **B38**, 8264 (1988)
[15] S. Satpathy, R.M. Martin: Phys. Rev. **B39**, 8494 (1989)
[16] H. Fujitani, S. Asano: Phys. Rev. **B40**, 8357 (1989)
[17] H. Fujitani, S. Asano: Surf. Sci. **268**, 265 (1992)
[18] J.M. Salehpour, S. Satpathy, G.P. Das: Phys. Rev. **B44**, 8880 (1991)
[19] S. Ossicini, C. Arcangeli, O. Bisi: Phys. Rev. **B43**, 9823 (1991)
[20] C. Arcangeli, S. Ossicini, O. Bisi: Surf. Sci. **269/270**, 743 (1992)
[21] J.L. Longueville, P.A. Thiry, R. Caudano: J. Vac. Sci. Technol. A **9**, 1670 (1991).
[22] R.M. Tromp, M.C. Reuter, Phys: Rev. Lett. **61**, 1756 (1988)
[23] B.V. King, M. Katayama, M. Aono: Vacuum **41**, 938 (1990)
[24] F.J. Himpsel, B.S. Meyerson, F.R. Mc Feely, J.F. Morar, A. Taleb-Ibrahimi, J.A. Yarmoff: Proceedings of the Enrico Fermi School on Photoemission and Absorption Spectroscopy of Solids and Interfaces with Synchrotron Radiation, eds. M. Campagna and R. Rosei, North Holland, Amsterdam 1990, p. 203
[25] F.J. Himpsel, F.R. Mc Feely, A. Taleb-Ibrahimi, J.A. Yarmoff, G. Hollinger: Phys. Rev. **B38**, 6084 (1988)
[26] D.A. Lapiano-Smith, E.A. Eklund, F.J. Himpsel, L.J. Terminello: Appl. Phys. Lett. **59**, 2174 (1991)
[27] The position of the valence band maximum (VBM) in Fig. 5 is 0.4eV lower than in this reference, taking its more accurate determination in Ref. 28 (Fig. 7) into account. This also lowers the conduction band minimum (CBM) and changes the electron affinity from -2.7 eV to -2.3 eV.
[28] F.J. Himpsel, L.J. Terminello, D.A. Lapiano-Smith, E.A. Eklund, J.J. Barton: Phys. Rev. Lett. **68**, 3611 (1992)
[29] M.S. Hybertsen, S.G. Louie: Phys. Rev. **B34**, 5390 (1986)
[30] L.J. Terminello, D.A. Lapiano-Smith, F.J. Himpsel: to be published
[31] M. Skibowski, G. Sprüssel, V. Saile: Appl. Optics **19**, 3978 (1980)
[32] C.C. Grimes, G. Adams: Phys. Rev. Lett. **42**, 795 (1979)
[33] S. Kanemaru, H. Ishiwara, S. Furukawa: J. Appl. Phys. **63**, 1060 (1988)
[34] L.J. Terminello, F.J. Himpsel, F.R. McFeely, A.B. McLean: J. Vac. Sci. Technol. A **8**, 3038 (1990)

Adsorption of Water Vapor on NaCl(100) and KCl(100) without and with Defects

M. Henzler, A. Stock, and M. Böl

Institut für Festkörperphysik, Universität Hannover, Appelstr. 2, D-30167 Hannover 1, Germany

Abstract. Epitaxial thin films of NaCl and KCl (prepared on Ge(100)) have been used for studies of adsorption of water vapor with photoemission and electron energy loss spectroscopy. Due to the large band gaps amounts of submonolayers to several monolayers are easily detected in UHV up to 10^{-4} mbar in the temperature range of 140 to 300 K. Perfect surfaces show a well ordered 2D-phase with a superstructure on NaCl. Defects are produced as steps due to incomplete annealing or as F-centers with electron bombardment. Water is desorbed from steps reversibly at room temperature and is dissociated irreversibly at F-centers.

1. Introduction

Both alkali halides as substrates and water as adsorbate are rarely studied in surface physics [1-4], although quite important in the every day life. The powerful electron diffraction and spectroscopy techniques could not be used for bulk alkali halides, because these charge up for low electron energies and dissociate for energies where charging is avoided by secondary emission. The discovery of producing nearly perfect and very thin epitaxial films on Ge(100) substrates by proper deposition and annealing opened the investigation of NaCl and KCl with LEED, UPS, and XPS [5,6]. The charging up is avoided by tunneling to the semiconducting substrate, providing always a well defined and measurable Fermi level at the surface.

On those surfaces the adsorption of water in equilibrium with the vapor could be measured with the help of photoemission of the molecular orbitals of the molecule [6-8]. A quite different behaviour is observed for surfaces with defects due to steps [6] or surface F-centers [9,10]. Here those results are summarized including results which are so far only available in theses.

2. Experimental

All experiments have been performed in a UHV system with films prepared in situ by deposition of NaCl on a well cleaned and annealed Ge(100)2x1 surface at about 150 K and annealing of the film at about 400 K. KCl has been deposited on top of NaCl. The thickness (as derived from Auger electron spectroscopy and XPS [5]) was around 3 nm. UPS, XPS and, with an extra electron gun, also EELS were simultaneously available. Additionally, a LEED system and a quadrupole mass spectrometer were in the system. The water vapor pressure was measured with the mass spectrometer.

3. The Alkali Halide Surfaces

The well annealed surfaces have been characterized with LEED, UPS, XPS, and EELS. He I spectra of NaCl (upper curve) and KCl (lower curve) are shown in Fig. 1. The binding energy is given with respect to the Fermi energy (which is directly seen due to emission from the substrate after sufficient magnification of the signal). It should be pointed out that with He I radiation the valence band of NaCl is split (in accordance with literature [11,12]). The peaks for higher binding energies are due to the variations of the density of states in the conduction band (emission of "true secondaries").

When He II is used a broad energy range is free of photoemission and therefore useful for detection of adsorbate states (Fig. 2). The strong emission around -18 eV for KCl is a K 3p core level, with a satellite at -11 eV due to the 48.4 eV photons in the He II spectrum. The XPS and EELS results reproduce well the reported bulk data.

The crystalline order, as determined with a conventional LEED system,

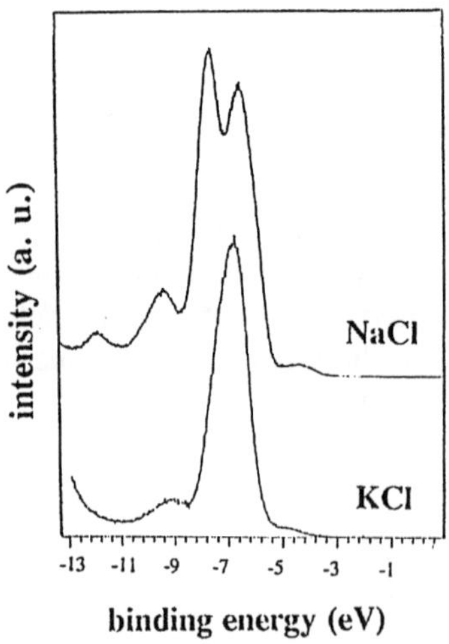

Fig. 1: He I spectra of NaCl on Ge(100) (upper curve) and of KCl on NaCl (lower curve). The binding energy is given with respect to the top of the Ge valence band (not shown in the figure).

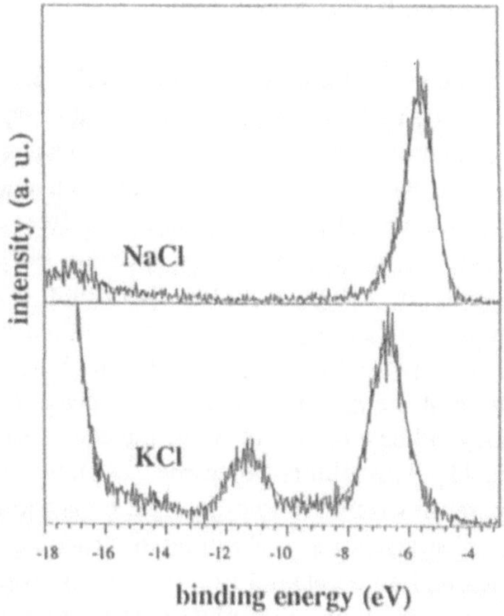

Fig. 2: He II spectra of NaCl on Ge(100) (upper curve) and KCl on NaCl (lower curve).

equals that of the perfect NaCl(100) face [5]. With a high resolution LEED system only minor elastic deformations have been observed due to overgrowth of atomic steps of the substrate [13,14]. With deposition of KCl onto Ge no epitaxial films could be produced. On top of NaCl, however, perfect films of KCl have been grown despite of the large misfit [6]. The film has the same orientation as the NaCl film, it has, however, the lattice constant of the bulk KCl (as derived from the LEED pattern in Fig. 3). This growth mode has been called the "floating" mode [15].

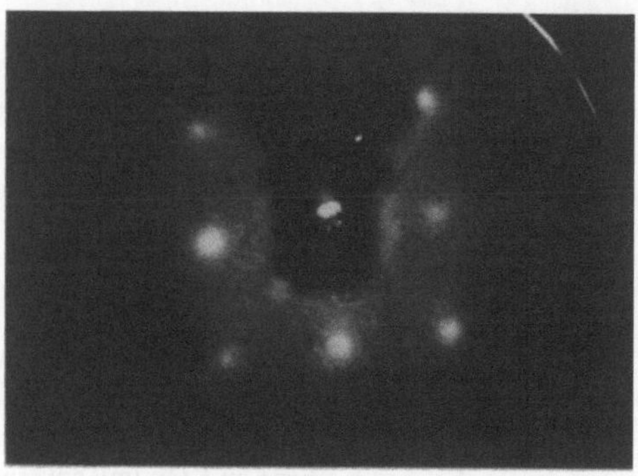

Fig. 3: LEED pattern at 44 eV of a KCl film (approx. 2 monolayers) on top of an epitaxial NaCl film.

4. Adsorption of Water on Defect Free Surfaces

The adsorption has been studied both on NaCl and KCl. It has been shown that water is adsorbed in thermal equilibrium with water vapor, which may be described as reversible physisorption [6,8,9]. Three typical spectra are shown in Fig. 4 for adsorption on KCl at 10^{-8} mbar for different temperatures: At high temperatures (e.g. room temperature) just the emission from the substrate is seen (lower curve). In a special temperature range (around 140 K) three additional peaks are seen, which are assigned to the molecular orbitals of the water molecule (center curve). In this regime a well ordered superstructure c(4x2) is seen with LEED on NaCl [7] and an also well ordered (1x1) structure on KCl [6]. With the XPS analysis of the substrate and oxygen peaks it has been estimated that this water film should be a bilayer of water molecules similar as reported for adsorption on metals [1]. This film is called the 2D-phase of H_2O on NaCl/KCl. At even lower temperatures (upper curve) just the emission from the water film and none from the substrate is seen. No ordered structures are seen in LEED. Here obviously a thick and non-epitaxial film is growing, which is called the 3D-phase. The persistence of the molecular orbitals points to a molecular adsorption without dissociation.

For a quantitative description the strongest molecular orbital ($1b_1$) has been recorded during cooling (rate of 0.2 K/min) at a pressure of 10^{-8} mbar. The result is shown for both NaCl and KCl in Fig. 5. Both curves show the three regimes as described above. Both phase transitions from gas to 2D and from 2D to 3D have a steep transition. The finite slope is given by the cooling rate and increases with decreasing rate [6]. Therefore the phase transition should be a 1st order transition with a broader 2D range for KCl compared to NaCl.

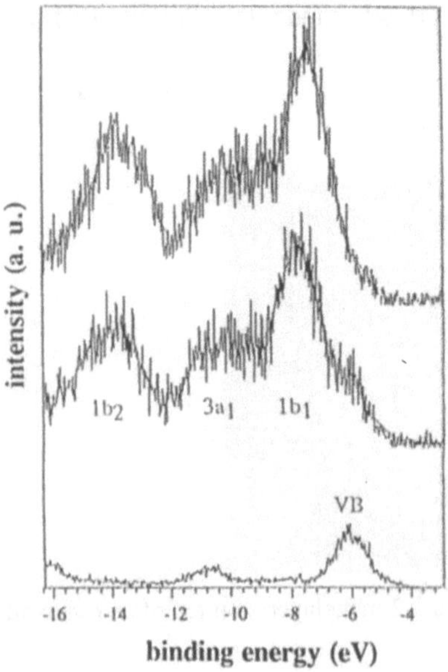

Fig. 4: He II spectra of the water adsorbate phase on KCl: low coverage of 2D-gas ("clean" surface) (lower curve), 2D-phase (center curve), 3D-phase (solid ice) (upper curve).

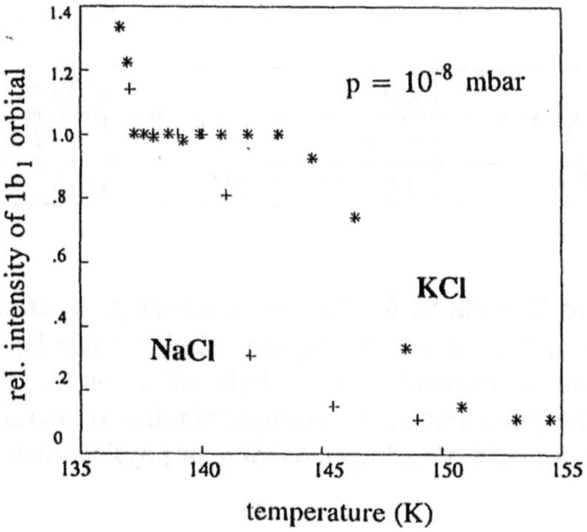

Fig. 5: Adsorption of water on NaCl (+++) and KCl (***) at constant pressure during cooling.

Using the data of Fig. 5 and similar data for pressures from 10^{-5} to 10^{-9} mbar a phase diagram for the three phases of water on the two substrates is derived (Fig. 6). It is seen that the 2D-3D transition is identical for the two substrates (as it should be, since it is given by bulk properties of ice) and different for the gas-

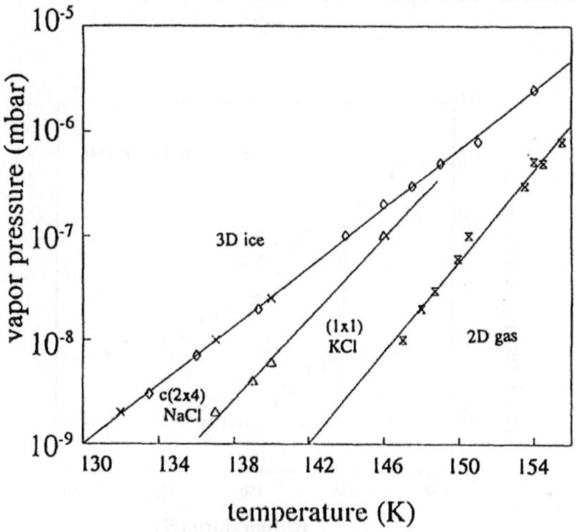

Fig. 6: Phase diagram for water on NaCl and KCl. The symbols for the transitions are: xxx for 2D-3D on NaCl, ◇◇◇ for 2D-3D on KCl, △△△ for gas-2D on NaCl and ⧖⧖⧖ for gas-2D on KCl.

Table 1: Adsorption energies for water on NaCl and KCl (errors estimated from Fig. 6).

	gas => 2D	2D => 3D
H_2O on NaCl	0.77 ± 0.05 eV	0.57 ± 0.05 eV
H_2O on KCl	0.99 ± 0.05 eV	0.57 ± 0.05 eV

2D transition (as it might be for different substrates). Adsorption energies have been derived using the Clausius-Clapeyron equation (table 1). The energy for the 2D-3D transition is identical with the bulk value for ice, the higher binding energy for the 2D-phase indicates a stronger binding to the substrate. Its value is compatible with a model of hydrogen bonding or physisorption.

5. Adsorption on Surfaces with Defects

Two kinds of defects of the clean surfaces have been considered in the experiments. Steps are present at the surface when an incomplete annealing is used after deposition. Quite different defects are produced by an electron beam of sufficient energy.

The influence of atomic steps on the water adsorption is shown with a KCl surface for cooling in a water vapor pressure of $5 \cdot 10^{-9}$ mbar. Figure 7 shows both the adsorption on a well annealed surface (dashed curve) and that on a stepped surface (lower curve with crosses). The highest intensity corresponds to the 2D-phase. A substantial adsorption is seen at temperatures where nearly no

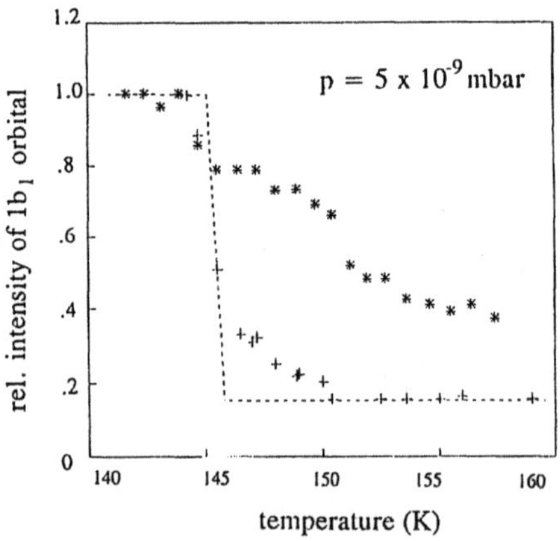

Fig. 7: Adsorption on KCl with steps: *** high step density (no anneal), +++ low step density (weak anneal). The dashed curve shows the expected adsorption for a defect free surface.

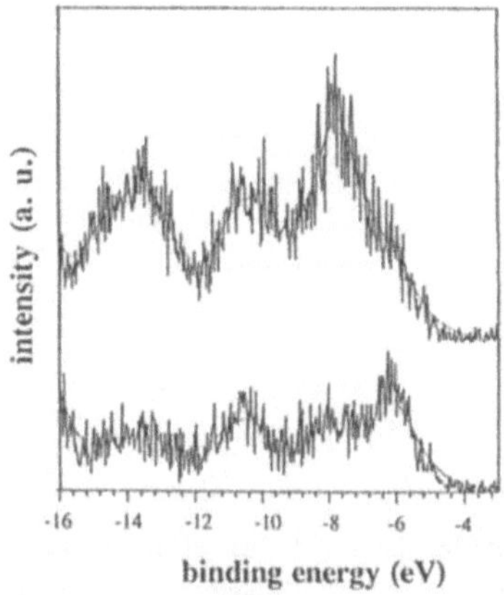

Fig. 8: He II spectra for the 2D-phase (upper curve) and adsorption on steps (lower curve).

adsorption occurs on the step free surface. A UP spectrum in this range compares well with that of the 2D-phase (Fig. 8). It shows the same peaks due to the molecular orbitals and an increased intensity at 6 eV due to the substrate. It is therefore the same type of molecular adsorption as in the 2D-phase. If the slightly different ratios of the intensities of the molecular orbitals are due to the step site, has still to be checked in more detailed experiments. To increase the step density KCl has been deposited onto NaCl at 100 K. The LEED pattern shows a strong broadening of diffraction spots. Since no spot profiles have been recorded, the step density is only roughly estimated to about 10 % step sites out of all surface sites. On that surface the adsorption at temperatures higher than needed for the 2D-phase is about half of the amount in the 2D-phase (upper curve in Fig. 8). This amount is much more than accommodated by the step sites. Obviously each step site is a nucleus for the adsorption of several water molecules. Nevertheless, this adsorption is reversible. Since there is no clear transition from step site to the gas phase, no exact range for this phase can be given in Fig. 6.

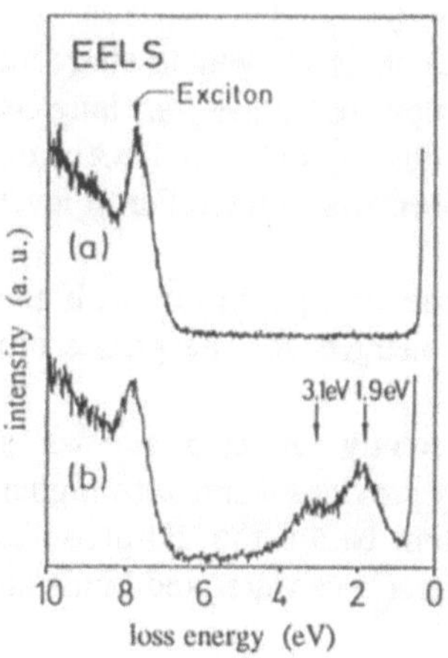

Fig. 9: EEL spectra of a defect free NaCl film (a) and after electron bombardment (b). The two losses in the gap at 1.9 and 3.1 eV are due to color centers (F and F′ resp.) [10].

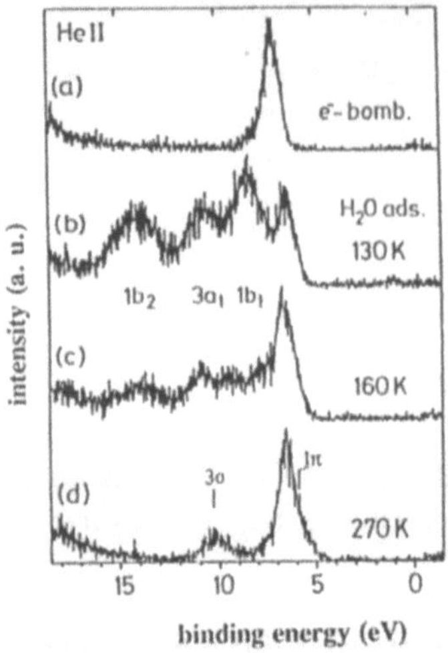

Fig. 10: He II spectra : (a) after electron-bombardment, (b) 2D-phase, (c) water adsorption at defects, (d) adsorption at room temperature (with dissociation).

A quite different defect site may be produced by an electron beam [9,10]. If a freshly prepared NaCl film is bombarded with electrons of 150 eV (1 μA for several minutes), two new losses are seen in the gap region by electron energy loss spectroscopy (EELS) (Fig. 9). These losses agree with optical absorption of F-centers. In UP spectra no additional features are seen (Fig. 10, top curve). If this surface is exposed to water vapor at room temperature, new peaks are seen (lowest curve in Fig. 10). These peaks are not compatible with water molecules, they are well described by OH radicals, which is also supported by the change of the EEL spectra. At lower temperatures water is adsorbed at the OH radicals (curve (c) in Fig. 10), and finally again the 2D-phase is seen (curve (b) in Fig. 10).

6. Conclusion

Single crystalline thin films of NaCl and KCl provide a simple way to study all kinds of surface properties with electron spectroscopy techniques, as long as electrons with energies over 30 - 50 eV and current doses over about 1 μA·s are avoided. Both structural and electronic data are obtained with a fixed Fermi level position provided by the substrate.

Water adsorption yields a well defined 2D-phase in equilibrium with the water vapor. From a phase diagram the activation energies for the phases are derived.

The two types of defects behave quite differently. A step site as a stoichiometric defect site does not dissociate water. It provides a site with higher adsorption energy, which nucleates further adsorption before the 2D-phase is formed. The F-center (a non stoichiometric defect) dissociates water and binds an

OH radical, which is strongly bound and not easily desorbed. The F-center with OH provides sites for further water adsorption the same way as the step edges at temperatures, at which the 2D-phase does not yet form.

In this way epitaxial films of alkali halides open the field of surface studies in the same way as it is used so far for metals and semiconductors.

Acknowledgements

Discussions with H. Pfnür and J. Heidberg are gratefully acknowledged. The investigations have been possible only with the support of the Deutsche Forschungsgemeinschaft, Bad Godesberg, and the Kali & Salz AG, Kassel.

References

[1] P.A. Thiel, T.E. Madey: Surf. Sci. Rep. **7** (1987) 211
[2] J. Estel, H. Hoinkes, H. Karmann, H. Nahr, H. Wilsch: Surf. Sci. **54** (1976) 393
[3] P.B. Barraclough, P.G. Hall: Surf. Sci. **46** (1974) 393
[4] H. Frank: Surf. Sci. **63** (1977) 121
[5] S. Fölsch, U. Barjenbruch, M. Henzler: Thin Sol. Films **172** (1989) 123
[6] A. Stock: PhD thesis, Universität Hannover 1992
[7] S. Fölsch, A. Stock, M. Henzler: Surf. Sci. **264** (1992) 65
[8] A.Stock: diploma thesis, Universität Hannover 1990
[9] S. Fölsch: PhD thesis, Universität Hannover 1991
[10] M. Böl: PhD thesis, Universität Hannover 1993
[11] F.J. Himpsel, W. Steinmann: Phys. Rev. **B 17** (1978) 2537
[12] W. Pong: Phys. Rev. **B** 9 (1974) 2674
[13] C. Schwennicke: diploma thesis, Universität Hannover 1992
[14] H. Pfnür: this volume
[15] M. Henzler: Progr. in Surf. Sci. 1993, in press

Phases and Phase Diagrams of Xenon Adsorbed on Epitaxial NaCl(100) Films and on Ge(100)

H. Pfnür, C. Schwennicke, and J. Schimmelpfennig

Institut für Festkörperphysik, Universität Hannover, Appelstraße 2,
D-30167 Hannover, Germany

Abstract. Thin films of NaCl were grown as epitaxial films on Ge(100). Analyzing LEED profiles we are able to show quantitatively that the NaCl films accommodate monoatomic steps of the Ge substrate by elastic strain which is distributed over many lattice constants, forming an elastic "carpet". Ordered phases and phase diagrams of Xe adsorbed on NaCl(100) and on Ge(100) were investigated in quasi-equilibrium with room temperature gas using high resolution LEED (SPA-LEED). First order phase transitions for Xe/NaCl are observed between a disordered lattice gas and a distorted hexagonal phase and between this phase and the hexagonal multi-layer bulk phase. The distorted hexagonal phase is commensurate with the substrate in only one direction. The hexagonal phase only forms small crystals (average grain size 80 Å), but is still oriented along [110] with one of its axes. On Ge(100) no additional superstructure spots are induced by Xe before the bulk phase grows, but characteristic jumps in intensity of integral order as well as of half order beams are observed as a function of temperature and pressure.

1. Introduction

The interplay of adsorbate-adsorbate interactions and adsorbate-substrate interactions has been extensively studied on substrates with hexagonal symmetry for physisorbed adsorbates like noble gases and weakly bound molecules using diffraction techniques like neutron scattering [1], helium scattering [2] and electron diffraction [3] as well as uptake methods to study anomalies of the specific heat at 2-dimensional phase transitions [4]. For rare gases both adsorbate and substrate potentials have the same symmetry, but the mismatch between the lattice constants of the substrate and a (floating) adsorbate leads to the formation of domain walls with either hexagonal or striped symmetry, or to the formation of floating 2-dimensional solid phases [5].

The competition between these adsorbate-adsorbate interactions and the adsorbate-substrate interactions with different symmetry should lead to new ordering phenomena, and seems to be interesting, therefore. Attempts in this direction have been made before, e.g. in investigations of Xe on metals like

Springer Series in Surface Sciences, Vol. 33 Eds.: E. Umbach · H.-J. Freund
Adsorption on Ordered Surfaces of Ionic Solids and Thin Films

Cu(100) [6] and on Pd(100) [7], and also on the ionic surface of MgO(100) [8], but only hexagonal structures were found on the metallic surfaces. In this paper we will focus our interest on the ordering of an adsorbed rare gas, Xe, on surfaces with fourfold and twofold symmetry. As substrate we have chosen the NaCl(100) and the Ge(100) surfaces. As is well known from detailed structural investigations [9], the Ge(100) surface forms a mixture of (2x1), p(2x2) and c(4x2) reconstructions, whereas the NaCl(100) surface remains unreconstructed [10].

Ionic crystals with a wide band gap can be studied with low energy electrons only by using thin films so that charging effects can be avoided to a large extent. Because of the small mismatch in lattice constants of less than 0.5% Ge(100) is a favourable substrate for epitaxial growth of NaCl [10]. The use of such films has not only technical advantages but also makes them accessible to high resolution structural investigations so that details of the processes of growth and annealing of defects can be studied [11,12]. In the context of the adsorption studies presented here the quality and perfection of the substrate is of importance. In particular, the steps of different height have to be accommodated due to the different crystal structures of Ge and NaCl. We therefore first demonstrate how this can be achieved, using a model which fits the structural data quantitatively. Xenon phases and phase transitions are then described and shortly discussed.

2. Experimental

Experiments are described in detail elsewhere [13]. In brief, they were performed in a standard UHV system (base pressure $6 \cdot 10^{-9}$ Pa) equipped with a high resolution LEED (SPA-LEED) instrument, an Auger electron spectrometer and a quadrupole mass spectrometer. Profile analysis of the diffracted beams was carried out with the SPA-LEED instrument with an effective transfer width larger than 1100 Å.

The sample was mounted to a cryostat, which was cooled with l-He, using a molybdenum frame which was attached by small Ti screws to a Mo sheet. Temperature was measured on the Mo frame by a Chromel-Alumel thermocouple. The crystal assembly was heated from the back side by radiation from a filament (used for the measurements) or by electron bombardment. Temperature was regulated by a computerized temperature control to ± 0.1 K and could be ramped linearly up and down. Temperature was calibrated using the onset of bulk Xe formation during condensation at a Xe pressure of 10^{-4} Pa on the NaCl sample. The error in absolute temperature was estimated as ±1 K.

Our Ge crystals showed, after repeated sputtering/annealing cycles, the well known (2x1) reconstruction at room temperature and both c(4x2)- and p(2x2) reconstructed regions at lower temperatures [14]. The step density was determined by LEED profile analysis. Predominantly steps of monoatomic height d = 1.4 Å were found in agreement with ref. [15]. They had a mean separation of about 250 Å.

NaCl was evaporated onto the substrate at a substrate temperature of 200 K and the deposition rate was monitored by a quartz microbalance. The calibration procedure is described elsewhere [10]. The film thickness used in all our measurements described below was 17 Å, i.e. three double layers.

Xenon was admitted to the vacuum chamber through a variable leak valve. Adsorption experiments were performed in quasi equilibrium with a 3D room temperature gas. Xe pressure was measured using an ion gauge.

3. Results and Discussion

3.1 Morphology of Thin NaCl Films

In order to determine the morphology of the NaCl films, step distributions and terrace widths we analyzed spot profiles of the (00) beam of evaporated and annealed NaCl films at energies between 42 eV and 170 eV. The profiles produced in this way are broadened and show satellites but are isotropic. A one-dimensional cut through the two dimensional spot profile can be separated into a central peak and 2 satellite peaks (see Fig. 1). The central peak contains 3 different contributions which can be uniquely identified: The central spike is well reproduced by a Gaussian contribution with a FWHM identical to the instrumental width. Further, a Lorentzian contribution Lr_1 with a full width at half maximum (FWHM) of 1.5% BZ was found to be characteristic of terraces separated by steps of 1.4 Å height, and vanishes at in-phase scattering conditions [16]. Assuming a geometric distribution of terrace lengths an average terrace width of 160 Å is obtained. The occurrence of a second Lorentzian function Lr_2 with a FWHM of 4.5% BZ is coupled to the existence of the satellites. The satellite peaks, on the other hand, can be well described by a sum of two Lorentzians with maxima at $\pm\, \Delta K$ and half widths which increase as a function of kinetic energy. A model assuming that the NaCl film covers the Ge surface like an elastic carpet, i.e. by gradual accommodation of these steps over several lattice constants, can quantitatively account for these satellites.

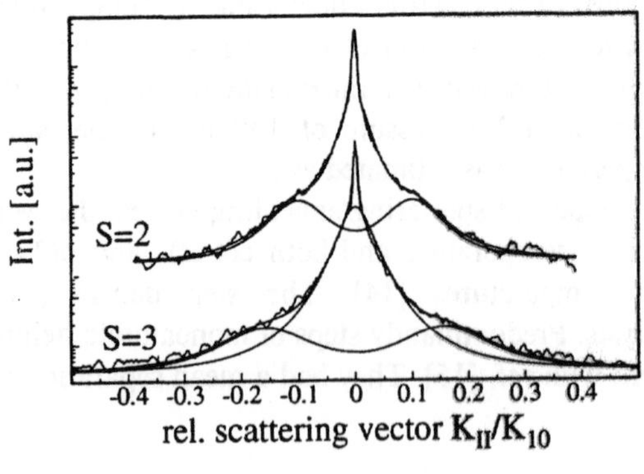

Fig. 1: Typical profiles of the (00) spot of NaCl at in phase conditions ($S = K_Z d/2\pi$ = integer) with respect to the smallest step height of the Ge(100) surface. Lines are fits to the experimental data as described in the text.

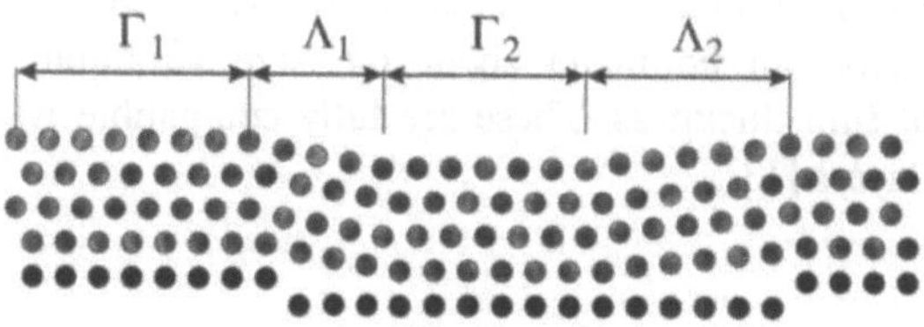

Fig. 2: "Carpet" model of the epitaxial NaCl film which is elastically distorted at monoatomic steps of the Ge(100) surface. Lateral widths of flat areas are given by Γ_i, lateral widths of inclined areas by Λ_i.

This includes the shape of the central peak described by the sum of the Gaussian and the broader Lorentzian function $L_{\Gamma 2}$ which remains unchanged for different in-phase conditions, i.e. integer values of $S = K_Z\, d/2\pi$. A schematic drawing of this model is shown in Fig. 2.

Our films also turned out not to be flat. On the average, the topmost layer consists of terraces on 4 different levels separated by mono-atomic steps. The behaviour of position and half width of the satellite spots as a function of phase S is qualitatively identical to that of surfaces containing facets with an orientation different from the main orientation. We carried out quantitative tests using a one-dimensional description, which is sufficient in this case because the satellite spots are incoherent to each other (see below). These tests have been performed by first using an analytic model which restricts us to the in-phase scattering condition between different Ge terraces [17], and to only 2 terrace heights, but is able to describe both the broadening of the central spike and the behaviour of the satellite peaks. This model is fully corroborated by our simulations shown as lines in Fig.1 where these restrictions were dropped.

As an additional simplification for the analytic model we assumed that the inclined regions are flat and the scatterers have identical form factors *f*. Since the scatterers in the inclined regions do not occupy lattice sites the diffracted electron intensity can be calculated by an incoherent addition of the diffracted intensities from a surface with holes of length Λ_i and terraces with lengths Γ_i and from surfaces with inclined faces of length Λ_i separated by Γ_i. Therefore, a broad contribution is also expected for the central peak, the inverse halfwidth of which should correspond to that of both satellites and is a measure of the size of the distorted regions. Assuming a geometrical distribution of (flat) terrace lengths and a Gaussian distribution of the lateral width of the inclined areas, we can best fit our experimental data with an average length of the inclined areas of $\overline{\Lambda} = 20$ lattice constants and an average variation σ of the lengths Λ_i of $\sigma = 5$ lattice constants.

The validity of the analytical calculation was checked by simulating spot profiles in one-dimension. Similar to the analytical calculations we assumed a geometrical distribution of the terrace lengths and a Gaussian distribution of the length of inclined regions. Spot profiles of surfaces with 1024 scatterers were calculated which roughly corresponds to the transfer widths of our instrument, and averages of more than 20 runs were taken. This simulation yields identical results. Corresponding tilt angles of these areas are less than 1° and local changes of the lattice constants are around 2.2%. Treating the NaCl layers as elastically deformable predictions can be made about the size variation of distorted areas as a function of film thickness. These are fully compatible with our results using thicker NaCl layers [13].

3.2 Phases and Phase Diagram of Xe Adsorbed on NaCl

Two different Xe-induced superstructures were observed on NaCl. Decreasing the temperature below 60 K at a constant xenon pressure above $3 \cdot 10^{-5}$ Pa the left LEED pattern shown in Fig. 3a appears first. Characteristic are pairs of superstructure spots which are, for the lowest orders of diffraction, positioned exactly in the middle between NaCl substrate spots. From the intensity variation as a function of electron energy we tentatively consider the spots at the positions $(1/2, 1-x)$ with $x \ll 1$ and at equivalent positions as first order spots of the superstructure. These and equivalent spots always appear equally bright and with high intensity. From symmetry considerations the first order substrate spots also must belong to this superstructure. This superstructure has to exist in 2 domains, because the square symmetry is broken. A real space representation of one domain of this Xe superstructure is shown in Fig. 3b. Since the (01) spots from substrate and adsorbate coincide, the y-components of $\vec{a}_2^{ads}$ and $\vec{a}_2^{subs}$ must be equal (see Fig. 3b). Thus the distance of two adjacent xenon rows is identical to the lattice constant a_0 of the NaCl(100) surface.

On the other hand, the distance of two adjacent xenon atoms within a row can be determined from the distance ΔK of two innermost adjacent multiple scattering spots using

$$d_{Xe-Xe} = \frac{a_0}{1-(a_0 \Delta K / 4\pi)} \tag{1}$$

The distance of the multiple scattering spots depends on pressure and temperature, as discussed later, but varies only slightly below 50 K. At these temperatures we determined the distance d between two adjacent Xe atoms within a row as 4.37 Å, in excellent agreement with the well known bulk lattice constant in this temperature range [18]. Hence, the distance between adjacent Xe atoms is determined in one direction by the Xe bulk lattice constant and does not lock into the NaCl substrate periodicity along this direction, whereas in the other direction the Xe layer is *expanded* by 5.5% so that it fits on the NaCl substrate and is commensurate in this direction. This results in a distorted hexagonal structure of the Xe overlayer with an angle between primitive Xe vectors of 118.7° instead of 120° in the bulk.

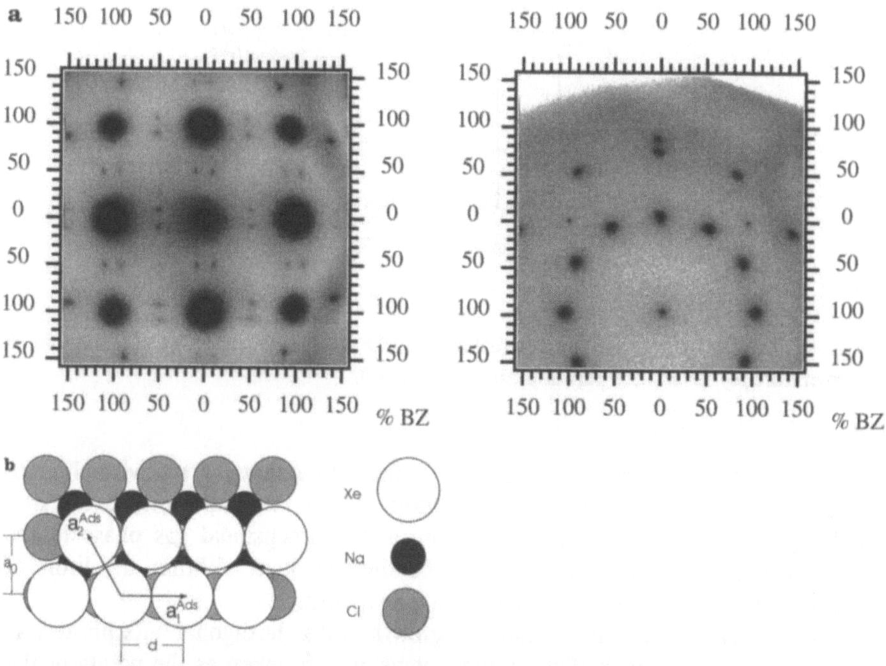

Fig. 3: a) LEED patterns of Xe-induced superstructures on NaCl(100)
- 2D phase at $p = 1 \cdot 10^{-5}$ Pa, T = 58 K and E = 82 eV (left pattern)
- 3D hexagonal bulk structure at $p = 1 \cdot 10^{-4}$ Pa and at the same temperature and beam energy (right pattern)

b) Real space representation of the structure model of the Xe monolayer adsorbed on NaCl

After further decreasing the temperature a second LEED pattern (see the right pattern of Fig. 3a) appears with twelve-fold symmetry consisting of two orthogonal hexagonal domains. One crystallographic direction of these Xe layers is still oriented along the [011] direction of the substrate. From the splitting between (10)-NaCl- and (10)-xenon spots we determined the lattice constant of xenon at a temperature of 58 K as 4.37 Å, which again agrees with the value of the bulk lattice constant [18]. As the formation of this phase is coupled with strong damping of the NaCl diffraction spots, and their gradual disappearance as a function of time once the critical pressure is exceeded, it is most likely that this bulk phase is only formed in the second Xe layer. The orientation of thicker Xe layers are the same as for the second layer. From the half widths of the hexagonal spots we get a mean lateral size of the 3D xenon crystallites of about 80 Å.

In order to determine the phase diagram integrated intensities of superstructure spots as a function of temperature were determined at partial pressures between $3 \cdot 10^{-6}$ Pa and $1 \cdot 10^{-4}$ Pa. Temperature was lowered in steps of 0.1 K and profiles of both the (10) substrate spot and the $(01)_A$ superstructure spot were recorded at constant temperature. In Fig. 4 integrated intensities of the

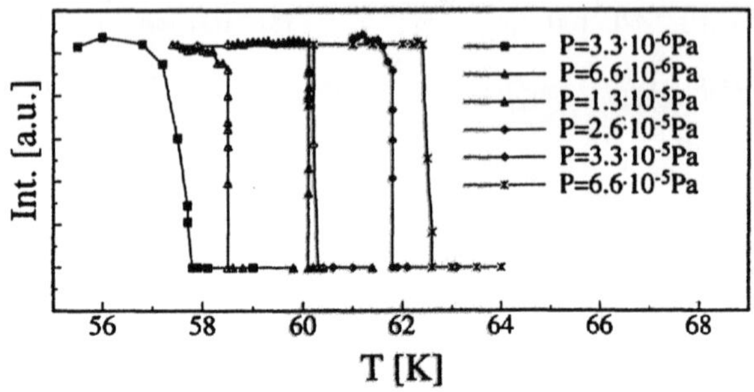

Fig. 4: Xe/NaCl-Integrated superstructure LEED spot intensities as a function of temperature for several Xe partial pressures (isobars).

$Xe(01)_A$ spot are plotted as a function of temperature showing stepwise changes of intensities at characteristic temperatures. These are characteristic for first order phase transitions, in this case from a two-dimensional gas phase to the ordered distorted hexagonal phase of the first Xe layer. Within the limit of experimental resolution no hysteresis could be observed.

The isobars were fully reversible if growth of the hexagonal bulk phase was avoided. The points of half maximum intensity were taken as the points of the phase transitions. These were plotted in Fig. 5. A hysteresis cannot be avoided for the second phase transition even at cooling rates of 0.1 K/min. Therefore, the uncertainty in the determination of the transition points 2D-3D is much higher than that of the transition points gas-2D.

Isosteric heats of adsorption were calculated for the phase boundaries using the Clausius-Clapeyron equation. We obtain 17.3 kJ/mole for the transition from 2d gas to 2d solid, and a value of 10.3 kJ/mole for the 2d/3d phase transition.

An unambiguous identification of a phase being incommensurate is possible from the variation of the lattice constant with temperature, if the thermal expansion of the substrate is negligible in a small temperature range. The Xe-Xe spacing d_{Xe-Xe} was determined along the incommensurate direction with an uncertainty of about 0.005 Å using equation (1) by measuring the distance ΔK

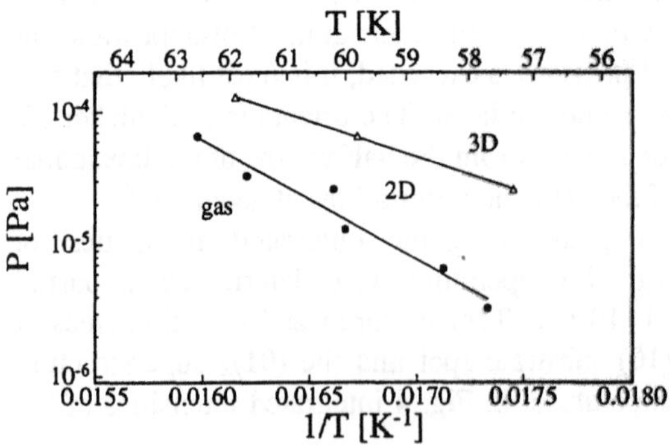

Fig. 5: Phase diagram of Xe adsorbed on NaCl

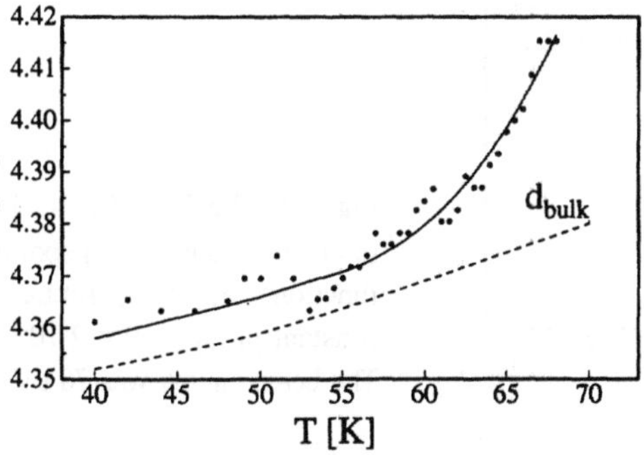

Fig. 6: Xe-Xe spacing on NaCl along the incommensurate direction as a function of temperature (see Fig. 3b).

between adjacent multiple scattering spots. Figure 6 shows the variation of d_{Xe-Xe} as a function of temperature for the hexagonal distorted phase. Below 55 K the measured xenon-xenon spacing was found to be approximately equal to the bulk lattice constant. Above 55 K a drastical increase of d_{Xe-Xe} with increasing temperature was found. Note that the expansion of the overlayer is uniaxial since the xenon monolayer is in registry to the substrate along one direction. The variation of the lattice constant as a function of pressure was found to be small if the measurements were not carried out very close to the phase boundaries. The scatter in the data above 55 K of Fig. 6 reflects these variations at constant temperature, which is in the limit of the experimental accuracy of ± 0.005 Å.

Our results for Xe/NaCl show that a well ordered Xe monolayer can be formed. Due to the interaction with the fourfold substrate potential a distorted hexagonal layer is formed. A very similar behaviour was observed for Ar/MgO(100) [8] where the mismatch of lattice constants between bulk Ar and MgO is about as large as for Xe/NaCl(100). An even larger uniaxial change of lattice constants as a function of gas pressure than in our case was found there. The isosteric heats of adsorption obtained in our measurements for the phase transition to a 2-dimensional solid are about 5% smaller than those obtained by Ewing et al. [19] with photoelectron spectroscopy, if we assume that they observed the same phase transition, and 10% smaller than the results from thermal programmed desorption (TPD) [20]. This might be due to methodological differences like non-equilibrium effects, but also due to the fact that we do not exactly work at constant coverage. The value obtained here for the 2d-3d transition, on the other hand, is considerably smaller than the condensation enthalpy of a Xe solid where a value of 15.5 kJ/mole was found [21]. The discrepancy clearly stems from the non-equilibrium effects mentioned.

3.3 Xe Adsorption on Ge(100)

Xenon adsorption on the Ge(100) surface was studied at temperatures between 40 and 80 K and xenon pressures between 10^{-6} and $2 \cdot 10^{-4}$ Pa. In the whole pressure and temperature range no additional superstructure spots due to xenon adsorption could be observed before 3D condensation started. With Xe adsorbed, all

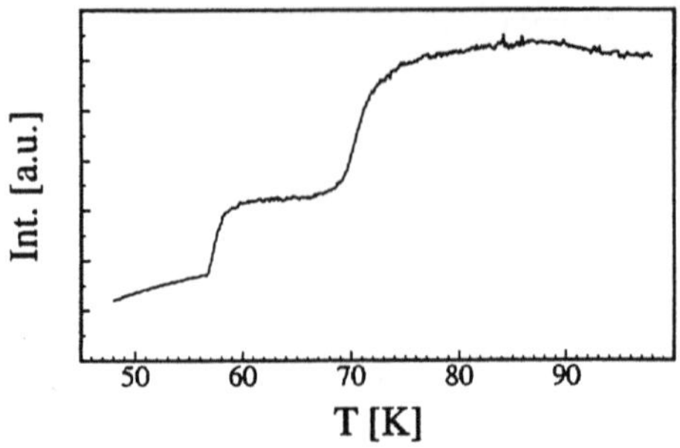

Fig. 7: Xe/Ge(100) - Peak intensity of the (1/2,0) spot as a function of temperature at constant pressure $P = 2\cdot10^{-5}$ Pa. The beam energy was 70 eV.

reconstructions of the clean Ge surface could still be observed. Peak intensities of the (10) and (1/2,0) spot were recorded as a function of temperature for constant pressure. Figure 7 shows typical data for the (1/2,0) beam for a pressure of $2\cdot10^{-5}$ Pa. The temperature was linearly ramped down with a rate of 0.05 K/s. Below 77 K we observed two jumps of the intensity and a plateau between them. Stepwise changes in intensity are also observed for other diffracted beams. Changes in the I-V curve of the (10) beam already at the first jump of intensity indicate that the Xe layer is ordered after this jump. In contrast, a disordered layer (not adsorbed on defined sites) would only lead to damping of substrate spots and not change the I-V curves. Both steps occurred before bulk xenon started to grow. This phase can easily be discriminated, if it forms the well known hexagonal LEED pattern described above. This is not the case for small cooling rates (0.05 K/s). Under this condition only an amorphous layer is formed at temperatures below 56 K. The hexagonal pattern was only observed if the sample was cooled fast from 77 K, or if adsorption was started on the clean sample at temperatures below 50 K. Only the first jump in spot intensity is reversible, whereas a hysteresis was observed for the second one.

From a set of data as shown in Fig. 7 (measurements during adsorption) the isosteric heats of adsorption were determined from plots of ln P versus 1/T as shown in Fig. 8. We obtain values of 22.5 ± 0.8 kJ/mole for the left, and 16.9 ± 0.8 kJ/mole for the right curve. The onset of 3-dimensional growth as an amorphous layer could not be determined since the decrease of the spot intensity of the substrate spots was too slow during the bulk condensation, and no well defined threshold could be detected.

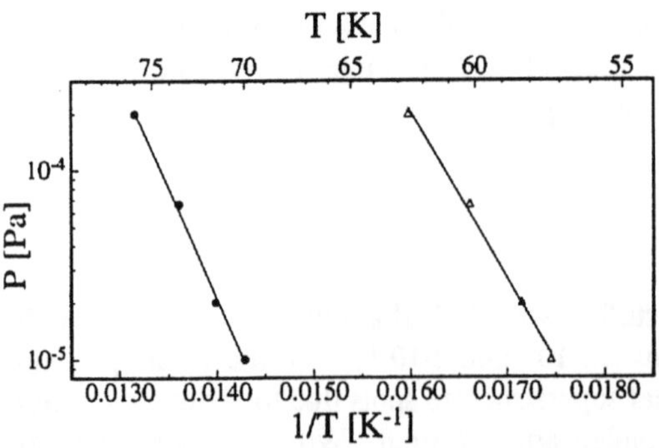

Fig. 8: Plots of ln p versus 1/T for the first and second jump of the spot intensity (see Fig. 7) which can uniquely identified in the measured pressure and temperature range.

If the phases observed are ordered, and I-V curves indicate that they are, they must also form (2x1) structures on the surface with occupation of nearest neighbour (NN) substrate sites. As Xe is about 10% bigger than the Ge NN-distance the Xe layers have to be compressed by this amount in one direction. A very similar behaviour was found during Xe adsorption on the Ge(111) surface by Packard and Webb [22] where again only jumps in intensity of substrate spots were observed during Xe adsorption. These authors carried out model calculations and found that this compression is energetically feasible on Ge(111). This finding can be transferred to Ge(100) if the assumption of a similar corrugation of the substrate potential surface in Ge(100) as on Ge(111) can be made. This kind of calculation still has to be carried out.

In TPD spectra [20], on the other hand, 3 desorption peaks were observed within desorption from the first monolayer. Whereas the desorption energies are not far from our isosteric heats for the two peaks at lower temperatures the desorption energy of the first peak is higher and strongly dependent on coverage. This is consistent with our findings if one assumes that this first desorption peak is characteristic of a decompressed disordered layer, and the other two correspond to desorption out of ordered layers. This correspondence still has to be shown explicitly.

Concluding, high quality NaCl single crystal films can be produced on Ge(100), which still contain some elastic strain due to the different types of lattices. This does not seem to influence the phase transitions of Xe observed on these films. The fourfold corrugation of the NaCl surface is responsible for the formation of a distorted hexagonal phase in the first Xe monolayer. The stronger corrugation on Ge(100) seems to allow even a one-dimensional compression of Xe by 10%.

Acknowledgement

Useful discussions with M. Henzler are gratefully acknowledged. This work has been supported by the Deutsche Forschungsgemeinschaft through project Pf 238/1.

References

[1] H. Freimuth, H. Wiechert, H.P. Schildberg, H.J. Lauter: Phys. Rev. **B42**, 587 (1990) and references cited therein

[2] K. Kern, G. Comsa in: Chemistry and Physics of Solid Surfaces VII (Springer Series in Surface Science Vol. 10), eds. R. Vanselow, R. Howe, Springer, Berlin 1988, p.65

[3] J. Cui, S.C. Fain: Phys. Rev. **B39**, 8628 (1989)

[4] R. Marx: Phys. Rep. **125**, 1 (1985)

[5] M. den Nijs in: Phase Transitions and Critical Phenomena, Vol. 12, eds. C. Domb, J.L. Lebowitz:, Academic Press, New York 1988

[6] A. Glachant, U. Bardi: Surf. Sci. **87, 187 (1979)**
M.A. Chesters, J. Pritchard: Surf. Sci. **28**, 460 (1971)
[7] P.W. Palmberg: Surf. Sci. **25, 598 (1971)**
[8] T. Meichel, J. Suzanne, C. Girard, C. Girardet: Phys. Rev. **B38**, 3781 (1988)
[9] R. Rossmann, H.L. Meyerheim, V. Jahns, J. Wever, W. Moritz, D. Wolf, D. Dornisch, H. Schulz: Surf. Sci. **279, 199 (1992)**
[10] S. Fölsch, U. Barjenbruch, M. Henzler: Thin Solid Films **172**, 123 (1989)
[11] M. Henzler, M. Horn v. Hoegen, U. Köhler: Festkörperprobleme/Advances in Solid State Physics **32**, 333 (1992)
[12] J.A. Venables, G.D.T. Spiller, M. Hanbücken: Rep. Progr. Phys. **47**, 399 (1984)
[13] C. Schwennicke, J. Schimmelpfennig, H. Pfnür: to be published
[14] S. D. Kevan: Phys. Rev. **B32, 2344 (1985)**
[15] J.A. Kubby, J.E. Griffith, R.S. Becker, J. S. Vickers: Phys. Rev. **B36**, 6079 (1987)
[16] C.S. Lent, P.I. Cohen: Surf. Sci. **139, 121 (1984)**
[17] J. Wollschläger, J. Falta, M. Henzler: Appl. Physics **A50**, 57 (1990)
[18] G. Pollack: Rev. Mod. Phys. **36, 748 (1964)**
[19] J.P. Hardy, G.E. Ewing, R. Stables, C.J.S.M. Simpson: Surf. Sci. **159**, L474 (1985)
[20] A. Klekamp, E. Umbach: Surf. Sci. (in press)
[21] C.W. Leming, G.L. Pollack: Phys. Rev. **B2**, 3323 (1970)
[22] W.E. Packard, M.B. Webb: Surf. Sci. **195**, 371 (1988)

Kinetic and Electronic Properties of Physisorbates on Epitaxial NaCl(100) Layers: SF_6, Xe, and CO_2

*A. Klekamp, R. Reißner, and E. Umbach**

Physikalisches Institut, Universität Stuttgart, Pfaffenwaldring 57,
D-70569 Stuttgart 80, Germany

* new address: Experimentelle Physik II, Universität Würzburg, Am Hubland, D-97074 Würzburg, Germany.

Abstract. Ultrathin epitaxial NaCl(100) films on Ge(100) are ideally suited for adsorption studies on insulators using conventional surface techniques. By means of TPD, XPS, UPS, and XAES we have investigated the kinetic and electronic properties of different adsorbates like SF_6, Xe, and CO_2. Only physisorption is observed and hence low substrate temperatures (< 90 K) are required. Layer growth mode, desorption kinetics, and adsorbate orientation (for CO_2) are dependent on the adsorbate and on the preparation parameters. The thickness of the NaCl layer strongly influences both the probability for radiation-induced dissociation of SF_6 and the relative intensity of two different decay channels arising from different Auger initial states one of which is due to charge transfer screening.

1. Introduction

Adsorption studies on well-defined surfaces of insulators have recently attracted considerable attention (see, e.g., the articles of this book). There are certainly several reasons for the activity in this expanding field of surface science some of which are evident. For example, most 'real' surfaces are oxides, or consist of other insulating materials, or are at least covered by a thin film of insulating compounds. Many chemical reactions, e.g. in heterogeneous catalysis, require oxide surfaces. Insulating substrates or interface layers play an important role in (micro-) electronics, and their detailed understanding and control is of increasing importance as the size of the active components decreases.

Although considerable interest in insulator surfaces obviously exists basic research in surface science nearly exclusively concentrated on metal or semiconductor surfaces in the past. This has changed very recently, mainly for two reasons. The properties of clean and adsorbate covered insulator surfaces are in many respects very different from those of metals; thus an exciting new field has opened. And, ultrathin and nearly perfectly ordered films of oxides or halides

Adsorption on Ordered Surfaces of Ionic Solids and Thin Films

can now easily and reproducibly be prepared. Thus insulating substrates are now available that allow both, the study of well-defined surfaces as well as the application of all surface techniques including those that are less suited for insulator crystals because of charging or thermal conductivity problems.

In addition to these reasons there was another reason that made us start the investigation of insulator surfaces. Organic materials have a considerable potential for being used in electronics. This new research area called 'molecular electronics' presently suffers from the fact that the order of nearly all molecular systems is hardly controllable on microscopic or mesoscopic scales. In particular, it is unclear which role the interfaces and the order in the molecular layer(s) play for, e.g., the transport properties. Therefore it is of prime importance to prepare highly ordered layers of suitable organic molecules on well-defined insulator surfaces. Our first attempts using perylene-tetracarboxylicacid-dianhydride (PTCDA) prepared as thin layers by vacuum-sublimation on an ultrathin NaCl(100) film have been successful so far, but the basis for handling and understanding such systems comes from adsorption studies as described below.

For our experiments we chose the substrate NaCl(100) which can be epitaxially grown on Ge(100) layer-by-layer, starting with a double layer. We followed the preparation method developed by Henzler and coworkers [1] and controlled the layer thickness by XPS, LEED, and a quadrupole mass spectrometer, as described elsewhere [2]. The properties of thin films of alkali halides, in particular NaCl, are discussed in the articles by Himpsel, Pfnür and Henzler in this book, and hence are not mentioned here.

In this short article we review our adsorption studies of SF_6, Xe, and CO_2 on NaCl(100) using TPD, XPS, XAES, and polarization-dependent UPS as characterization methods. Experimental details are given in references [2-8] and hence are not discussed here. We concentrate on the comparison of three adsorbates (sections 2.1 - 2.3), and on the thickness-dependence of NaCl layers with respect to charge transfer processes affecting dissociation (section 3) and screening of the Auger initial state (section 4).

2. Adsorption Studies

The choice of the adsorbates was governed by various aspects. We wanted to compare different rare gases (Xe and 'SF_6'), rare gases and (reactive) molecules, and a ball-like (SF_6) with a linear (CO_2) molecule. Moreover, we liked to study the dissociation behaviour upon irradiation and the charge transfer screening process for which SF_6 was a proper choice (see below).

The first surprise was that all three adsorbates are only very weakly adsorbed on the NaCl(100) surface despite the considerable local electric field on an ionic substrate and the reactivity of Na and Cl ions. The electronic data clearly show pure physisorption on a well-prepared NaCl surface since all core and valence levels are observed at very similar binding energies and with very similar intensities as in the gas phase [2,5,7]. In particular, the peak separations are identical demonstrating the absence of bonding or chemical shifts. If the reference levels are taken into account a very small shift between the binding energies of gas phase and adsorbate remains which is due to polarization screening. Moreover, the physisorption bond to the substrate is much weaker compared to that to metal surfaces, as derived from the kinetic (TPD) data, and these are very different for the three adsorbates. This is discussed in the following subchapters.

2.1 Adsorption of SF_6

The most significant information about the SF_6 adsorption comes from the TPD data. Figure 1a shows a set of TPD curves obtained after different exposures. The coverages are given as fractions of a monolayer (ML) and have been calibrated using XPS and TPD by comparison with SF_6 on Ni(111) [3] and Ge(100) [4] where monolayer coverages can be clearly distinguished by the appearance of a well separated TPD peak at higher temperatures. Interestingly, on NaCl mono- and multilayers cannot be distinguished by separate TPD peaks; the interaction with the surface is apparently much weaker than on Ge or metals.

At first glance the shape of the TPD peaks does not change as a function of coverage and resembles that of typical zero order desorption. Such a coverage-independent desorption is expected (and in fact observed) for evaporation of

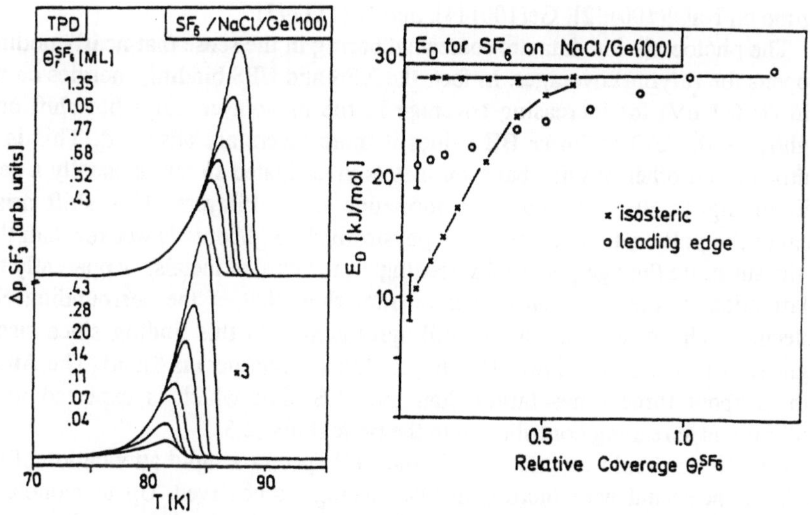

Fig. 1a, left: TPD spectra of SF_6 after adsorption on clean NaCl(100) at 40 K. The SF_6 coverages given in the figure are referred to a saturated monolayer ($\theta_r(SF_6)$=1.0). Fig. 1b, right: Activation energies for desorption, E_D, as a function of relative coverage, θ_r. Two different evaluation procedures have been used, leading edge (open circles) and isosteric method (crosses). Continuous lines are guide lines to the eye.

multilayers but not for desorption of the monolayer. A similar behaviour has by now been found for many physisorbates on atomically flat and densely-packed surfaces. It arises from the coexistence of two or three phases (one 2-D solid and one or two 2-D gas phases) from which desorption takes place. These phases may be in equilibrium with each other, or not. They lead to zero-order like TPD curves, to two types of straight lines in the isosteric Arrhenius plots ($\ln(-d\theta/dt)$ vs. $1/T$ for constant θ), and hence to two different values for the activation energy of desorption, E_D, at least at low coverages [2,3].

The $E_D(\theta)$ data for SF_6 on NaCl(100) are shown as crosses in Fig. 1b and are compared with E_D values derived from a leading edge analysis. (The latter is obviously not applicable in the case of two phases.) It is remarkable that the activation energy of the 2-D gas phase increases with coverage from 10 kJ/mol at very low coverages to 28 kJ/mol, the value of the 2-D condensed phase which is reached around $\theta \sim 0.6$. Note that the behaviour observed here is very similar to that found for SF_6 [3] or rare gases [9] on Ni(111), and that it is nearly perfectly described by theoretical calculations by Kreuzer and Payne [10] which are based on the two/three phase model mentioned above. Thus we can conclude that indeed the coexistence of at least two phases at the desorption temperature determines the desorption kinetics of SF_6 on NaCl(100). This is only possible if the lateral potential corrugation is weak and if few steps or defects are present.

Some other observations should also be mentioned. The adsorption kinetic is apparently independent of the surface. We found a linear uptake curve, i.e. a constant sticking coefficient of $s = 0.6$, for SF_6 in the mono- and multilayer regime on NaCl(100) [2], Ge(100) [4], and Ni(111) [3].

The photoemission data are somewhat boring in the sense that nearly nothing happens for varying coverages. In fact, the XPS and UPS binding energies do not shift ($\ll$ 0.1 eV) for increasing coverage in the monolayer range but shift only slightly (~ 0.1 eV) to *lower* BE values if more layers are adsorbed. This is in contrast to all other physisorbates on metals investigated so far as usually a peak shift to higher BE is found upon formation of multilayers. This shift arises because the polarization screening by physisorbed neighbours is weaker than that by the substrate ('image potential screening' in the case of metals). Apparently the polarization of the NaCl surface is smaller than that of the surrounding SF_6 molecules. The Auger data are in full agreement with this finding since also a slight shift (~ 0.4 eV) to lower BE (higher kinetic energies) is found. The Auger shift is about three times larger than the PES shift which is expected for a predominant screening contribution to the peak shifts [2,3].

Finally we should mention that in the HeII spectra a rapid attenuation of the NaCl valence band with increasing SF_6 coverage is observed. Up to monolayer coverage the Cl 2p-derived peaks nearly completely disappear indicating a complete wetting of the surface (Frank-van der Merwe or Stranski-Krastanov growth) and a preferential adsorption of the SF_6 molecules ontop of the Cl ions.

2.2 Adsorption of Xe

The adsorption and desorption behaviour of Xe on NaCl(100) is rather different from that of SF_6. This is clearly seen in the TPD series depicted in Fig. 2a. Again a single desorption peak is observed, and mono- and multilayers cannot be distinguished. However, in contrast to SF_6 the shape of this peak initially resembles that of first order desorption kinetics up to $\theta \sim 0.6$ but then changes to zero-order-like desorption. Neither a separate monolayer peak (as for Xe on metal surfaces) nor a clear two-phase coexistence can be concluded for this system. This is also reflected by the isosteric Arrhenius plots, which show a single straight line for each coverage, and by the coverage dependence of the activation energy, E_D, displayed in Fig. 2b. Just one activation energy is derived for each coverage, and the data points from isosteric and leading edge analyses coincide.

This different kinetic behaviour can be understood if the results from the XPS, UPS, and XAES experiments are consulted [2]. In all spectra a continuous shift of 0.3-0.4 eV to lower BE values is observed. In contrast to SF_6 this shift starts already at the lowest coverages measured and continues up to $\theta \sim 3$. The fact that XAES and PES data show the same shift and that in parallel the work function increases by 0.28 eV indicates that mainly a change of the local potential occurs upon increasing the coverage, and that screening effects do hardly contribute. The latter observation is a hint that non-wetting behaviour

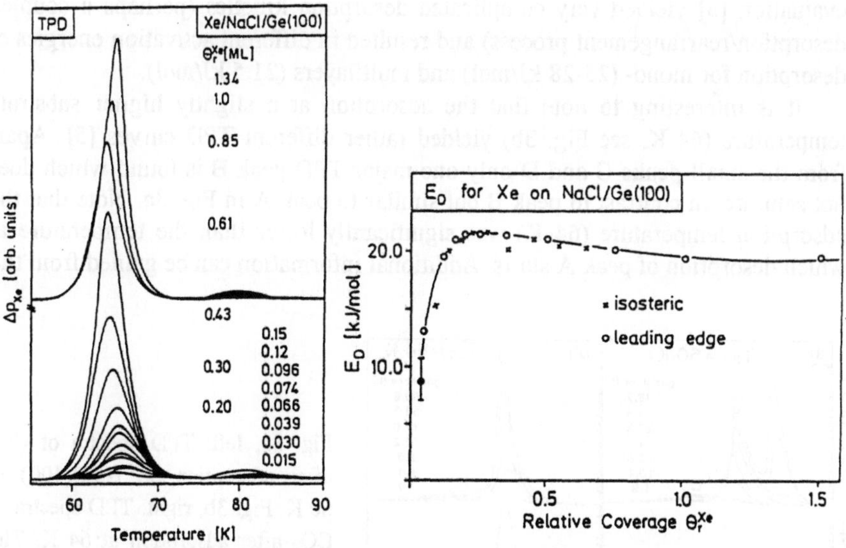

Fig. 2a, left: TPD spectra of Xe after adsorption on clean NaCl(100) at 40 K. 2b, right: Activation energies for desorption, E_D, as a function of relative coverage, θ_r. The Xe coverages are given in units of a nominal "monolayer" calibrated by comparison with a true monolayer of Xe/Ni(111) using XPS and TPD [2,9].

occurs, i.e. that three-dimensional clusters are formed even in the submonolayer range (Volmer-Weber growth). This finding is corroborated by the observation that the substrate derived VB structures in the HeII spectra are much less attenuated by adsorbed Xe than expected in the case of complete wetting. In fact, significant Cl 2p intensity is found even for $\theta \sim 4$ proving the growth of 3-D islands. The Xe adsorption behaviour on NaCl(100) is very similar to that on a Na-covered Ni(111) substrate [11] but very different to that on clean Ni(111) [9] or Ge(100) [4] and to that on the same NaCl surface but following a different preparation method (equilibrium conditions [12]).

2.3 Adsorption of CO_2

The adsorption behaviour of CO_2 on NaCl(100) [5] is very different from that of the other two adsorbates discussed above. This can, for instance, be clearly seen if we compare the TPD data displayed in Fig. 3. In Fig. 3a, which shows the desorption of layers prepared at or below 60 K, four different TPD peaks are observed. While the two very small peaks labelled C (desorption from the sample holder) and D (desorption from defects and steps) [5] are less important for this comparison the occurrence of two peaks which are attributed to the mono- (B) and multilayer (A), respectively, is very interesting. It indicates that the monolayer is more strongly bound, or exhibits different desorption kinetics than the multilayers. In fact a detailed isosteric, isothermic, and leading edge evaluation [5] yielded very complicated desorption kinetics (perhaps a coupled desorption/rearrangement process) and resulted in different activation energies of desorption for mono- (25-28 kJ/mol) and multilayers (21.5 kJ/mol).

It is interesting to note that the adsorption at a slightly higher substrate temperature (64 K, see Fig. 3b) yielded rather different TPD curves [5]. Apart from the small peaks C and D only one major TPD peak B is found which does not saturate, in contrast to peak B but similar to peak A in Fig. 3a. Note that the adsorption temperature (64 K) was significantly lower than the temperature at which desorption of peak A starts. Additional information can be gained from the

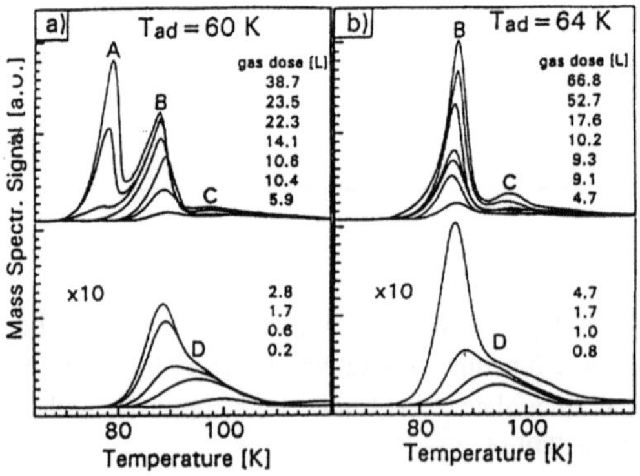

Fig. 3a, left: TPD spectra of CO_2 after adsorption on NaCl(100) at 60 K. Fig. 3b, right: TPD spectra of CO_2 after adsorption at 64 K. The relative quantities are given as exposures (in Langmuir; 15 L correspond approximately to a monolayer).

quantitative evaluation [5], which yielded slightly smaller (by 2-3 kJ/mol) activation energies of desorption for the 64 K layer, and from the sticking coefficients. These have a constant value for submono-, mono-, and multilayers (θ = 0-7 ML) in the case of T_{ad} = 60 K; for T_{ad} = 64 K $s(\theta)$ is slightly smaller in the monolayer regime but significantly smaller for $\theta > 1$ ML (by a factor 5).

At present we have no further information about the origin of this marked difference. We assume that CO_2 grows in different morphologies in the two temperature regimes but we have no geometrical information about the layers prepared at T_{ad} = 64 K. After low temperature preparation we took polarization- and angle-dependent UP spectra and angle-dependent XAE spectra [5]. They all showed marked angular variations for $\theta \leq 0.2$ but very little variation for $\theta = 1$. For $\theta \leq 0.2$ the angular dependences are consistent with a preferentially flat lying CO_2, in agreement with the FTIR data of Heidberg et al. [13] and Ewing et al. [14] for CO_2 on NaCl single crystals. There is some evidence from the UPS and XAES data [5] that in our case either the orientation is not as parallel to the surface as in the single crystal studies or that the layer is less well ordered.

Moreover, for a tilt angle around 25°-35° with respect to the surface which was derived in the FTIR studies for the monolayer regime we would have expected a stronger angular dependence than that observed. Hence we must conclude that our preparation at temperatures $T_{ad} \leq 60$ K resulted in a different adsorbate structure than the preparation used by Heidberg et al. and Ewing et al. (adsorption-desorption equilibrium at elevated temperatures and CO_2 pressures). We may speculate that our layers prepared at T_{ad} = 64 K were similar to the latter but a proof is still missing.

3. Substrate-Dependent UV-Induced Dissociation of SF_6

In a previous publication [6] we presented UPS and XPS data which clearly demonstrated the radiation-induced dissociation behaviour of SF_6 adsorbed in mono- and multilayers on Ni(111). While core holes may always lead to fragmentation (atomic S and F as well as molecular SF_4 fragments have been identified), independent of the distance of SF_6 with respect to the metal surface, this is different for HeI or HeII excitation. In this case dissociation only occurs for multilayer SF_6 whereas the excitation is apparently transferred into the substrate ('quenched') for SF_6 molecules that are close enough to the metal before the fragments can sufficiently move apart. The reason for the different dissociation behaviour of core and valence holes, respectively, is probably that double (or multiple) valence holes which result from the rapid Auger decay of a core hole remain long enough localized (self-trapped) on the adsorbate whereas a single hole in valence orbitals can rapidly tunnel to the metal valence band before dissociation occurs [6].

The dissociation probability of SF_6 in multilayers is drastically enhanced for HeI as compared to HeII or X-ray irradiation (a factor 7 as compared to HeII). The reason for this is a resonant transition from the bonding $5t_{1u}$+ $1t_{2u}$ orbitals to the antibonding $2t_{2g}$ orbital that occurs for hν ~ 21 eV and is apparently very effective in breaking intramolecular bonds. The consequence of this resonant excitation process is that the HeI spectra of multilayers rapidly degrade on the time scale of a few minutes. In Fig. 4 a typical UP spectrum of undissociated SF_6 (2nd from bottom; identical for SF_6 on Ni) is compared with a degraded multilayer spectrum (bottom). One can clearly see that the structure of the various valence peaks which are labeled in the figure is 'washed out' after irradiation indicating dissociation. Indeed an XPS analysis of an irradiated multilayer yielded peaks from S and F atoms, and from SF_4 fragments [6].

The interesting point in the present context is the question whether the one-hole excitation can also be transferred to a semiconductor or insulator substrate, and how thick an insulator film must be to suppress a possible quenching process. Figure 4 gives the answer [7]. It shows spectra obtained with HeII radiation (which also contains HeI with ~3 times higher intensity) taken from SF_6 monolayers adsorbed on clean Ge(100) and on 2, 5, and 9 layers of NaCl on Ge(100). It is clearly seen that the valence structures are not, or hardly, affected for a SF_6 monolayer on bare Ge or on the NaCl double layer. In fact, there was no change even for much longer irradiation times than those needed for recording a spectrum.

The spectrum taken from SF_6 adsorbed on 9 NaCl layers, however, showed drastic changes even after short irradiation periods. Although in this case slight charging also occurs the major contribution to the 'blurring' of structure arises from dissociation. This can be concluded from the similarity of the topmost and bottom spectra of Fig. 4, from the continuous and observable degradation with time, and from an XPS analysis which revealed the usual fragments [7]. For 5 NaCl layers degradation was also observed but on a considerably longer time

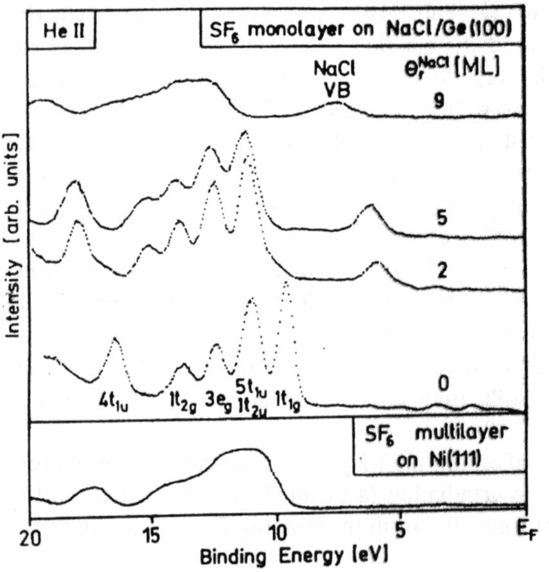

Fig. 4: UP spectra (HeII) taken with unmonochromatized radiation (HeI/HeII ratio ~ 3:1) from SF_6 monolayers on Ge(100), on epitaxial NaCl(100) films of different thickness, and from SF_6 multilayers on Ni as indicated in the figure. All spectra were taken from a freshly prepared SF_6 monolayer with the exception of the multilayer spectrum at the bottom which was recorded after 100 min of UV irradiation.

scale. Thus we can summarize that single-hole dissociation processes which occur on a time scale of $\geq 10^{-13}$ s can be quenched by adsorption on metals, semiconductors and very thin insulator films, for example with a thickness of less than ~15 Å.

4. Substrate-Dependent Charge Transfer Screening of the Auger Initial State

It is well known that core holes can be very effectively screened by charge transfer from the vicinity to an ionized atom provided that the coupling between them is good enough and that unoccupied orbitals are available. In this way binding energy shifts of several eV have been found, for instance for small adsorbates. SF_6 is a molecule with three unoccupied orbitals close to threshold with at least one of them lying below the ionization threshold. If the coupling would be good enough we would hence expect a significant shift of the photoemission peaks of SF_6 adsorbates with respect to the gas phase values because of possible charge transfer screening. However, if we take the different reference levels into account, the comparison shows that there are only very small shifts (0.2-1.2 eV) for all kinds of substrates (Ni, O/Ni, Na/Ni, Ge, NaCl) which are due to polarization screening [2-4,15]. Thus we must conclude that on the time scale of the photoemission process (~10^{-17} s) a charge transfer does not occur, probably because of the weak coupling and the (F-induced) Coulomb wall around the molecular center.

The question is what happens after photoemission. If we could wait long enough there would of course be charge transfer some time. But we can only wait about 10^{-15} s until a core hole decays via an Auger transition. Does charge transfer occur before? Figure 5 gives an answer [15]. It shows a series of Auger spectra of the main peak ($KL_{2,3}L_{2,3}$) of the F1s Auger fine structure for different substrates. Apart from the topmost spectrum which shows an SF_6 bilayer all spectra were taken from monolayers. The structures can be divided into two groups labeled A and B in Fig. 5. Structure A which is dominating in the gas phase (bottom) and in most adsorbate spectra represents the $KL_{2,3}L_{2,3}$ Auger channel of a poorly screened molecule as its energy is rather close to that of the isolated molecule. It is very interesting that the spectrum taken from SF_6 adsorbed on about 5 layers of NaCl (2nd from bottom) is identical to the (broadened) spectrum from the gas phase.

Structure B appears for all other substrates. It is the more intense (with respect to structure A) the stronger the bonding to the substrate is. In particular for a Na precovered surface structure B is dominating, in agreement with a likely hybridization of Na 3s and SF_6 orbitals [8]. Thus we must conclude that structure B is due to an Auger initial state which has been prepared by charge transfer screening from the substrate into the SF_6 molecule. This conclusion is directly proved by very recent threshold experiments using synchrotron radiation: Direct resonant excitation into the first unoccupied SF_6 orbital yielded Auger spectra which are dominated by structure B. Structure B is hence attributed either to a participant decay or to a spectator-shifted decay of the main channel (or to both) the second interpretation being the more likely [15].

In conclusion we emphasize that we observe charge transfer screening between photoionization and Auger decay, and that its probability depends on the coupling to the substrate and on the thickness of an insulating substrate layer. Note the difference between the 2nd and 3rd spectrum from the bottom in Fig. 5. Charge transfer screening is apparently absent for 5 NaCl layers but is still observed, though as weak feature in the Auger spectrum, for a NaCl double layer [15]. Thus the SF_6/NaCl experiments play a key role in understanding charge transfer processes and give new insight into their time scale.

5. Summary

In conclusion we have briefly discussed three different physisorbate systems on NaCl(100) surfaces, SF_6, Xe, and CO_2. They are all very weakly adsorbed but show surprising differences in their adsorption behaviour ranging from complete monolayer wetting for SF_6 (Frank-van der Merwe or Stranski-Krastanov growth) to non-wetting behaviour for Xe (3-D island or Volmer-Weber growth) and in their desorption kinetics ranging from relatively simple desorption kinetics for

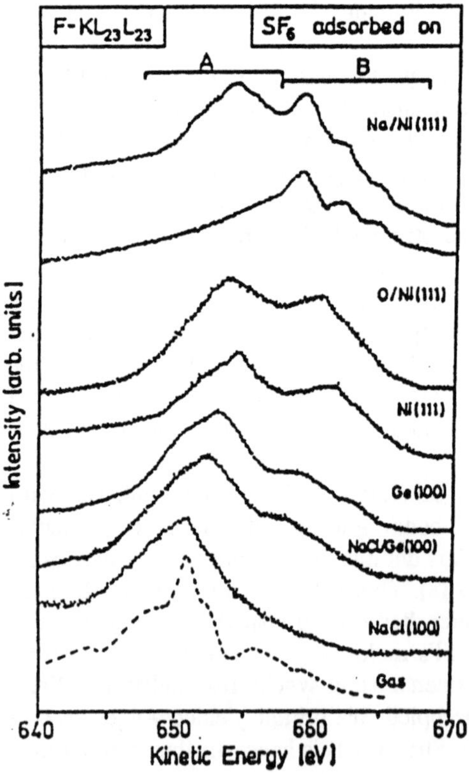

Fig. 5: XAES spectra of the main decay channel of an F1s core hole ($KL_{2,3}L_{2,3}$) for SF_6 on different substrates as indicated in the figure. Besides the topmost (bilayer) and the bottom spectrum (gas phase) all other data are from monolayers. 'NaCl/Ge(100)' means a double layer of NaCl on Ge(100), while NaCl(100) indicates ≥ 5 layers. Two different regions of structures are labeled A and B in the spectrum.

Xe over two-phase desorption for SF_6 to complex desorption behaviour for CO_2 probably involving a desorption/ rearrangement process.

The comparison of dissociation processes following resonant excitation by HeI radiation for SF_6 monolayers on different NaCl(100) layers revealed a thickness-dependent dissociation probability. In other words, the tunneling (quenching) process of charge depends on the thickness of the insulating layer and is suppressed by ≥ 15 Å NaCl on the time scale of the motion of the fragments ($\geq 10^{-13}$ s).

A similar comparison for SF_6 on different substrates has been made by monitoring charge transfer screening between the XPS final and the Auger initial state. Here the time scale is shorter ($\geq 10^{-15}$ s) and hence the NaCl thickness for which charge transfer is suppressed is smaller, between 6 and 15 Å.

Acknowledgement

We are greatly indebted to Prof. Dietrich Menzel for the loan of equipment. This work was supported by the Deutsche Forschungsgemeinschaft (project Um 6/3-2 and 6/3-3) and by the Ministerium für Wissenschaft und Kunst of Baden-Württemberg. One of us (E.U.) gratefully acknowledges support by the Fonds der Chemischen Industrie.

References

[1] S. Fölsch, U. Barjenbruch, M. Henzler: Thin Solid Films **172**, 123 (1989); U. Barjenbruch, S. Fölsch, M. Henzler: Surf. Sci. **211/212**, 749 (1989)

[2] A. Klekamp, E. Umbach: Surf. Sci. **284**, 291 (1993)

[3] A. Klekamp, E. Umbach: Surf. Sci. **249**, 75 (1991)

[4] A. Klekamp, E. Umbach: submitted to Surf. Sci.

[5] R. Schwämmle: Diploma Thesis, Universität Stuttgart, 1992
R. Reißner, A. Klekamp, E. Umbach: in preparation

[6] A. Klekamp, E. Umbach: Chem. Phys. Lett. **171**, 233 (1990)

[7] A. Klekamp, E. Umbach: J. Phys.: Condens. Matter, **5**, L67 (1993)

[8] A. Klekamp, E. Umbach: Surf. Sci. **271**, 555 (1992)

[9] M.J. Breitschafter: Doctoral thesis, TU München 1985

[10] H.J. Kreuzer, S.H. Payne: Surf. Sci. **198**, 235 (1988); **200**, L433 (1988); **205**, 153 (1988); **222**, 404 (1989)

[11] W. Weiss, E. Umbach: Surf. Sci. **249**, L333 (1991)

[12] H. Pfnür, C. Schwennicke, J. Schimmelpfennig: article in this book and references therein

[13] J. Heidberg, L. Cabigon, E. Kampshoff, M. Kandel, R. Kühnemuth, D. Meine, B. Redlich, O. Schönekäs, M. Suhren, H. Weiss, D. Wetter: article in this book and references therein

[14] G.E. Ewing: article in this book and references therein

[15] A. Klekamp, E. Umbach: Phys. Rev. B, in preparation

Adsorption of Gases on Ionic Single Crystals: NaCl(100) and MgO(100)

J. Heidberg, L. Cabigon, E. Kampshoff, M. Kandel, R. Kühnemuth, D. Meine, B. Redlich, O. Schönekäs, M. Suhren, H. Weiss, and D. Wetter*

Institut für Physikalische Chemie und Elektrochemie der Universität Hannover
Callinstraße 3-3A, D-30167 Hannover 1, Germany
*Chemistry Department, Central Mindanao University, Musuan, Bukidnon, Phillipines

Abstract. Collective vibrations and ordering of adsorbed gases on NaCl(100) and MgO(100) surfaces, prepared by cleaving single crystals in situ, were observed by polarization infrared spectroscopy under high resolution. Furthermore, the system CO_2/MgO(100) was studied by profile analysis of low energy electron diffraction spots. The vibrational lines of CO and CO_2 on NaCl(100) measured, are the sharpest known for adsorbates. Systematic isotopic mixture experiments with complete line shape simulations reveal Davydov splittings due to long-range correlation fields, yielding the structure of the adsorbed layers in concert with results of electron and helium atom beam diffraction experiments. Orientational phase transitions, which have attracted little attention in the experimental literature, were detected. Dynamic processes, including resonant desorption by selective vibrational excitation of the adsorbate, were observed. The simplicity of these dynamic systems allows testing of theoretical models: We proposed interaction potentials and an extension of the classical model for dipole-dipole coupling including correlation field multiplets and arbitrary tilt angles, not limited to normal orientation of the adsorbate molecules.

1. Introduction

Adsorption has been studied intensively at metallic and to lesser degree at semiconductor surfaces. In our work we have concentrated on ionic insulator surfaces. Exploratory work on molecular vibrations and potentials at ionic surfaces was performed by Folman's and Heidberg's groups [1,2]. The study of physisorption at ionic insulator surfaces can yield unique insight into weakly interacting chemical systems.

Weak surface forces on ionic insulator crystals can be sufficient to generate long-range order in the molecular overlayer, detectable only on extended single crystal surfaces which are as perfect as possible. Such long-range ordering effects observed are correlation fields, leading to Davydov splittings, and phase transitions.

Springer Series in Surface Sciences, Vol. 33 Eds.: E. Umbach · H.-J. Freund
Adsorption on Ordered Surfaces of Ionic Solids and Thin Films

In our investigations, we concentrated on the adsorption of gases CO, CO_2, H_2, CH_4, H_2O on NaCl(100), an inert and nearly ideal surface, and on MgO(100), which is much more reactive. H_2O on NaCl(100) and MgO(100) shows polarized, very broad vibrational bands indicating saturation of all OH-groups by H-bonding [3]. On MgO(100) sharp absorption features are observed as well, attributed to dissociative adsorption of H_2O. In the following, a few selected results concerning the adsorption of CO and CO_2 are presented.

2. Experimental

Polarized high-resolution FTIR-spectra were recorded on a Bruker IFS 113 v spectrometer equipped with transfer optics according to Heidberg, Hoge and Warskulat [4]. NaCl(100) and MgO(100) single crystals were cleaved in situ at a base pressure $< 2 \cdot 10^{-10}$ mbar. Low energy electron diffraction was performed on both freshly UHV-cleaved, clean MgO(100) single crystals and on the adsorbate CO_2/MgO(100) using a Leybold SPA-LEED instrument (spot profile analysis of LEED after Henzler [5]). The emission current of the electron gun was minimized to reduce electron beam damage. The simultaneous use of FTIR and SPA-LEED was not possible, but the experimental conditions (T, p(CO_2)) were kept comparable.

3. Results and discussion

3.1 CO Adsorbed on NaCl(100)

Spectra of CO at natural abundance adsorbed on NaCl(100) at monolayer coverage were observed in the temperature range between 55 and 6 K. Details on the monolayer preparation and experimental conditions are given in [6-8]. Typical spectra of the monolayer CO/NaCl(100) as a function of temperature are shown in Fig. 1.

At 40 K one single absorption at about 2155 cm^{-1} is detected with p-polarized radiation only. Upon lowering the temperature a narrowing and a slight lineshift are found. At temperatures below 35 K a second absorption redshifted by about 6 cm^{-1} is observed, which is stronger at s- than at p-polarization. Intensity increases with decreasing temperature and as for the high-frequency band a narrowing and a slight lineshift are detected. From the ratio of integrated s- and p-polarized absorptions A_s/A_p a perpendicular orientation of the transition dipole of the high frequency band and a parallel orientation of the transition dipole of the low frequency band with respect to the surface plane were determined.

The single absorption at high temperatures at about 2155 cm^{-1} has been the subject of many theoretical and experimental investigations (see e.g. [6-11] and

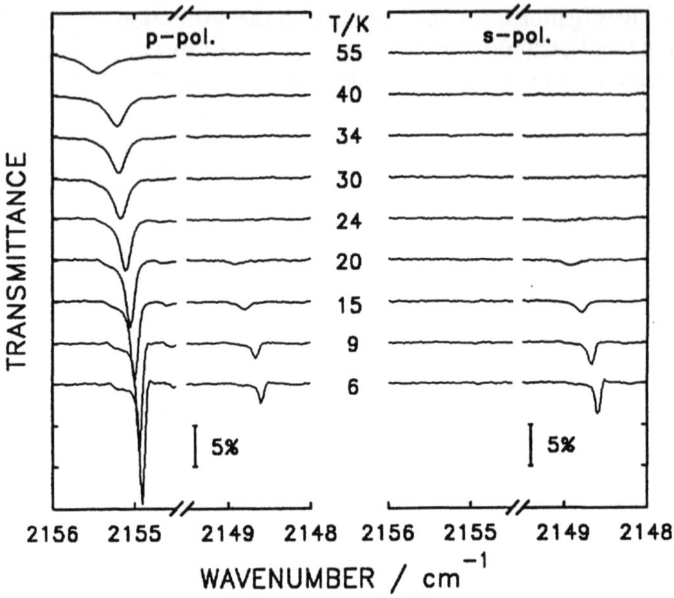

Fig.1: Transmittance spectra of the monolayer natural CO/NaCl(100) for different temperatures. Angle of incidence 50 ± 1°, resolution 0.065 cm^{-1}.

references therein). It was attributed to CO molecules oriented perpendicular to the surface, which form a p(1x1) structure with one molecule per Na^+Cl^- ion pair.

In order to explain the appearance of an additional absorption at low temperatures in the adsorbate CO/NaCl(100) different models have been discussed [7]. Isotopic mixture experiments clearly indicate that the spectral features have to be attributed to a correlation field splitting, due to a (2x1) structure with two molecules on energetically equivalent but translationally inequivalent adsorption sites per unit cell. A second layer adsorption and adsorption on defect sites were excluded. Thus the spectral features are due to a 2D continuous phase transition of the monolayer structure in the temperature range around 30 K. From the polarization dependency of the absorptions the angle ϑ between the molecular axis and the NaCl(100) plane and the intermolecular angle φ between the projections of the molecular axes of translationally inequivalent molecules onto the surface were determined. The two molecules per unit cell are oriented antiparallel, i.e. $\varphi = 180°$. From spectra simulation using equations of classical electrodynamics a tilt angle ϑ between the molecular axis and the surface of 65° at 6 K was deduced [6]. The calculated peak wavenumbers agree with measured values within 1 cm^{-1}, the peak intensities within 10%. The temperature dependency of the lineshapes can not be simulated through dephasing according to the exchange model. The orientational phase transition and its influence on the absorption spectra have to be taken into account. Detailed work considering this aspect is currently in progress.

Dynamic processes including vibrational dephasing and vibrational predesorption were observed: Dephasing of the internal modes induced by thermal excitation of an external mode was studied by measuring the temperature dependency of the internal line shape. The mechanism, however, is not unraveled and no theory is available for multiplet systems as yet. Vibrational predesorption was stimulated in the system CO/NaCl(film)/NaCl(100) by excitation of the internal CO fundamental using tunable frequency doubled pulsed CO_2-laser radiation. Desorption was monitored with time-of-flight techniques [11]. Figure 2 illustrates the results of an experiment using a 1:1 mixture of $^{12}C^{16}O$ and $^{13}C^{16}O$ in the adsorbate monolayer. The laser transmission, linear infrared transmission and desorption yield spectra are shown. It is evident that the desorption is effected through a resonant process, i.e. through a vibrational predesorption state. Excitation of one isotope leads to desorption of both, which can be due to a diffusive v-v transfer in the phase relaxation of the collective excitation and/or energy relaxation. The translational temperature of the desorbing CO-molecules was found to be small, e.g. 30 K for a surface temperature of 20 K.

3.2 CO_2 Adsorbed on NaCl(100)

Typical spectra of the ν_2-vibration of the monolayer CO_2 on NaCl(100) are shown in Fig. 3. The monolayer was prepared at 77 K with an equilibrium CO_2 pressure of $p = 10^{-8}$ mbar. The band at 663.2 cm^{-1} shows a FWHM of 0.04 cm^{-1},

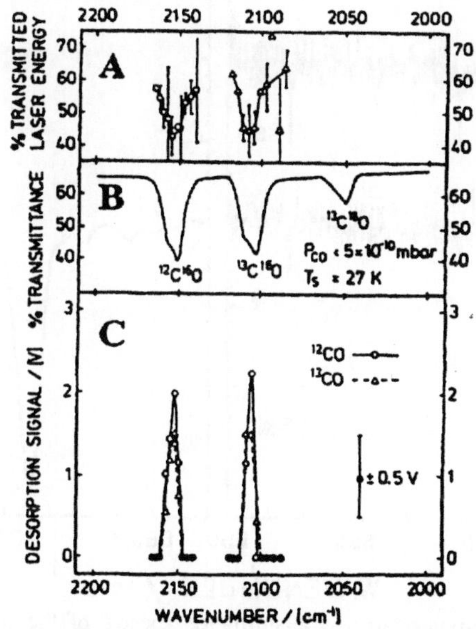

Fig. 2: CO_2-laser transmission spectrum (A), linear infrared spectrum (B) and laser desorption yield spectrum (C) of a 1:1 mixture $^{12}C^{16}O$:$^{13}C^{16}O$ adsorbed of NaCl(film)/NaCl(100). Surface temperature 27 K, p(CO) < $5\cdot10^{-10}$ mbar.

the sharpest of adsorbates known. The quartet is caused by a site symmetry splitting due to the removal of the two-fold degeneracy of the ν_2 gas-phase absorption and an additional correlation field splitting caused by the in-phase and out-of-phase vibration of two energetically equivalent molecules in a (2x1) adsorbate unit cell. This was shown by complete spectra simulation of the ν_3 stretching vibration taking into account results from potential energy calculations and from low energy electron or helium atom beam diffraction experiments [12,13]. The molecules are tilted by 34 ± 5° with respect to the surface, their projections onto the surface enclose an angle of 80 ± 5° (cf. [14]). The monolayer structure changes only slightly between 5 and 80 K. In the spectral range of the ν_2 bending vibration the tilt angle of the molecules has to be reduced to 26° to achieve reasonable agreement in peak position between theory and experiment. By introducing interactions of higher order than dipole-dipole coupling we hope to remove this discrepancy.

The thermodynamics was investigated at submonolayer and monolayer coverage under systematic variation of equilibrium pressures and temperatures. A 2D first order phase transition was observed on air cleaved as well as on UHV-cleaved crystals. On the latter, adsorption isotherms show a stepwise increase of coverage at the transition pressure $p = p_t$ from $\Theta \approx 0.01$ to $\Theta \approx 0.95$, the step height being larger than on surfaces prepared in air (Fig. 4). Due to the reduced defect density a 2D-lattice gas at $p \leq p_t$ could be detected. From the polarization dependency of its singlet ν_3-absorption an orientation parallel to the surface was

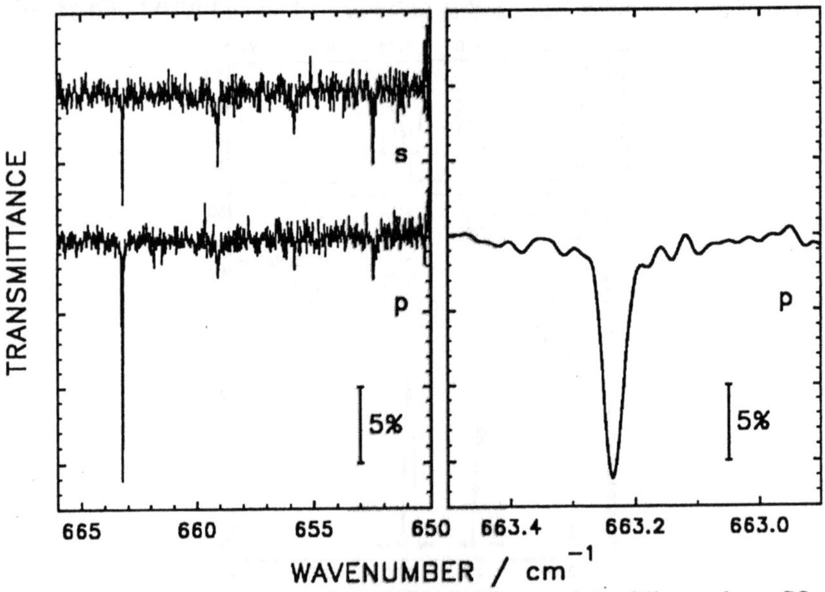

Fig. 3: Left: p- and s-polarized ν_2-transmittance spectra of the full monolayer CO_2 on NaCl(100) at 5 K. Resolution 0.029 cm^{-1} (no apodisation).
Right: p-polarized transmittance spectrum of the ν_2 component at 663.22 cm^{-1}. Enlarged part of the left figure.

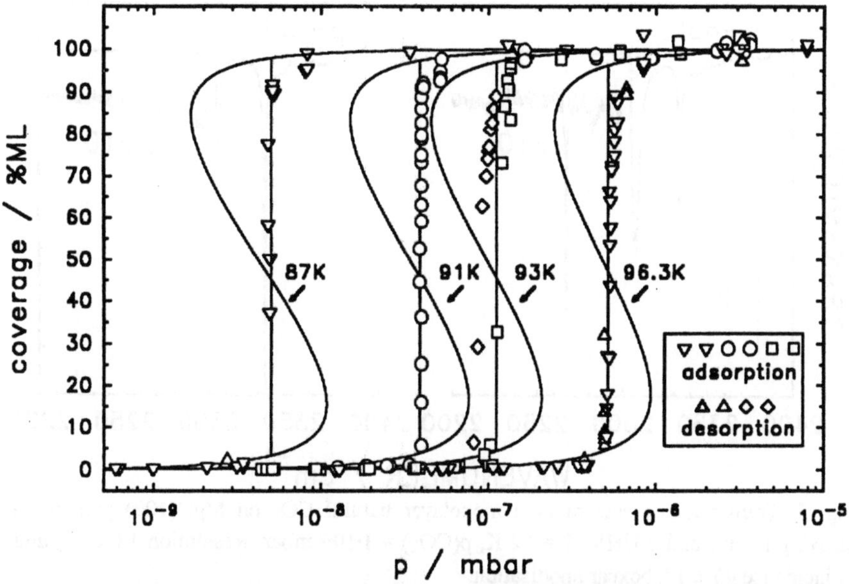

Fig. 4: Adsorption and desorption isotherms for CO_2 on NaCl(100), prepared by cleaving a single crystal in situ under UHV. Hollow symbols denote s-, full symbols p-polarization, the lines represent Fowler-Guggenheim isotherms fitted to the data.

derived, in concert with theoretical calculations. At $p = p_t$ islands, having constant density and the monolayer structure, grow in thermodynamic equilibrium with the 2D-lattice gas phase. A closer investigation of the 2D phase transition at $p < 1 \cdot 10^{-10}$ mbar yielded the enthalpy of 2D - sublimation $\Delta H_V^{2D} = (3.0 \pm 0.1)$ kJ/mol. The difference of the isosteric heats of adsorption of both phases of $q_{st}^s = (35.6 \pm 1.3)$ kJ/mol (2D-solid) and $q_{st}^s = (32.4 \pm 1.1)$ kJ/mol (2D-lattice gas) confirms this result.

3.3 CO_2 Adsorbed on MgO(100)

FTIR-measurements. CO_2 adsorption on MgO(100) surfaces prepared by cleaving in situ under UHV at 82 K results in sharp infrared absorptions at 2334, 2308, 2306 and 2267 cm^{-1} (Fig. 5) [15,16]. All bands are strongly polarization dependent. In particular the band at 2308 cm^{-1} is detectable only with s-polarization, obscuring the 2306 cm^{-1} signal, which is visible only as a weak shoulder. The absorptions can not be assigned to three energetically different adsorption species, because the multiplet structure does not exist in the spectrum of the 1% of $^{13}C^{16}O_2$ present in the natural adsorbate. In the region from 2300 to 2200 cm^{-1} only the single, sharp line at 2267 cm^{-1} is observed. We infer, that the splitting of the ν_3-absorption is due to a correlation field present in a highly

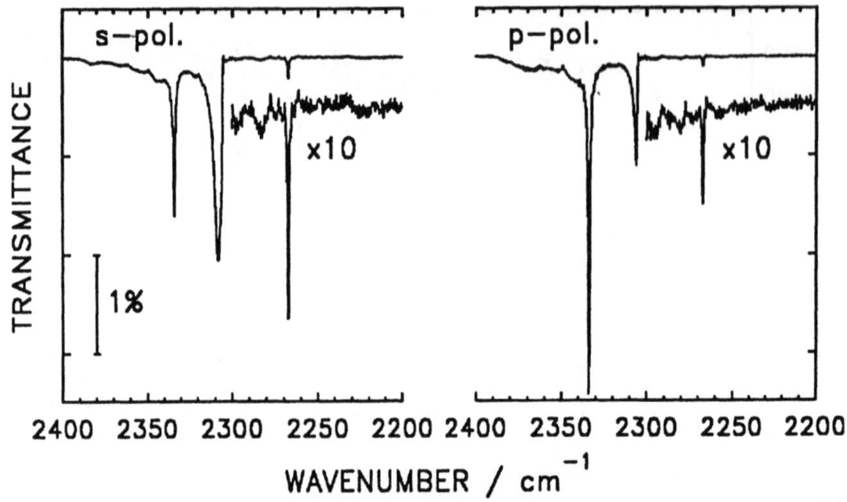

Fig. 5: Transmittance spectra of a monolayer natural CO_2 on MgO(100) prepared by cleaving in situ under UHV. T = 82 K, p(CO_2) = $1 \cdot 10^{-8}$ mbar. Resolution 1.0 cm^{-1}, angle of incidence 45 ± 1°, boxcar apodisation.

ordered adsorbate. Hence, there must be at least two CO_2-molecules per adsorbate unit cell, similar to the system CO_2/NaCl(100).

The polarization dependency of the bands, measured at an angle of incidence of 0°, shows, that in the adsorbate the fourfold symmetry of the ideal MgO(100) surface does no longer exist. A preferential orientation of adsorbate domains with twofold symmetry of typically 80% is inferred. For the different domains the projections of the transition dipole moments of the absorptions onto the surface are oriented along the ⟨001⟩ or ⟨010⟩ azimuthal crystal directions, where ⟨100⟩ denotes the surface normal, s - polarization is parallel to ⟨001⟩ and ⟨010⟩ coincides with the cleavage direction. We assume that the predominance of one sort of adsorbate domain over the other is effected by a corresponding preferential orientation of atomic steps on the surface, induced by the cleavage process. The steps then act as nucleation centers for the adsorbate domains. An asymmetry in the distribution of atomic steps on UHV-cleaved MgO(100) surfaces has indeed been observed in metal cluster decoration studies [17,18]. Taking into account the measured domain distribution, from the A_s/A_p value of the singulet at 2267 cm^{-1} the tilt angle of the adsorbate molecules with respect to the surface was deduced to be 27 ± 10°. The transition dipole moment of the band at 2334 cm^{-1} was found to be tilted by 54 ± 8° with respect to the surface and for the predominant domain its projection onto the surface is oriented parallel to the ⟨001⟩ direction.

SPA-LEED-measurements. Figure 6A shows the SPA-LEED-data obtained from a MgO(100) single crystal surface 1 h after cleavage in situ under UHV.

The well known clear MgO(100)(1 x 1) diffraction pattern is observed [19]. The spots all show a diamond shape, the points of which are oriented towards the ⟨001⟩ and ⟨010⟩ azimuthal crystal directions. The sharpness of the spots decreases as the diffraction order increases, which reflects a distortion of the pattern presumably caused by charging of the insulating crystal surface. A detailed analysis of the (00)-spot was done, which was focussed to a minimal diameter. With the assumption, that the surface charging does not affect the spot shape in this case, the mean undistorted terrace width of the MgO(100) surface was determined to be 47 and 17 nm for the ⟨001⟩ and ⟨010⟩ directions, respectively. This can be explained in terms of atomic steps on the MgO(100) surface oriented along ⟨001⟩ and ⟨010⟩: The cleavage of the crystal, which is performed in the ⟨001⟩ direction, induces a predominance of ⟨001⟩-oriented steps. The SPA-LEED-measurements thus support the interpretation of the FTIR spectra presented above.

Figure 6B shows the SPA-LEED-pattern obtained upon adsorption of CO_2 on UHV-cleaved MgO(100) at 82 K. The pattern is somewhat distorted at the edges of the picture due to fringe field effects at the SPA-LEED octupole deflector, but clearly superstructure spots corresponding to a rectangular $(\sqrt{2} \times 2\sqrt{2})$ R 45° structure can be observed. The spots related to domains with the long side of the rectangle parallel to the ⟨010⟩ direction were found to be much stronger than those of domains oriented along ⟨001⟩, again in agreement with the FTIR experiments. Missing adsorbate (0k)-spots of odd diffraction order indicate the presence of a glide plane in the adsorbate unit cell parallel to the short side of the rectangle. This adsorbate structure was also observed in a very recent LEED study [20]. The symmetry of the structure implies that the adsorbate unit cell contains two CO_2 molecules, i.e. one CO_2 molecule per two MgO formula units. Using this information, the infrared absorption at 2334 cm^{-1} can be assigned to

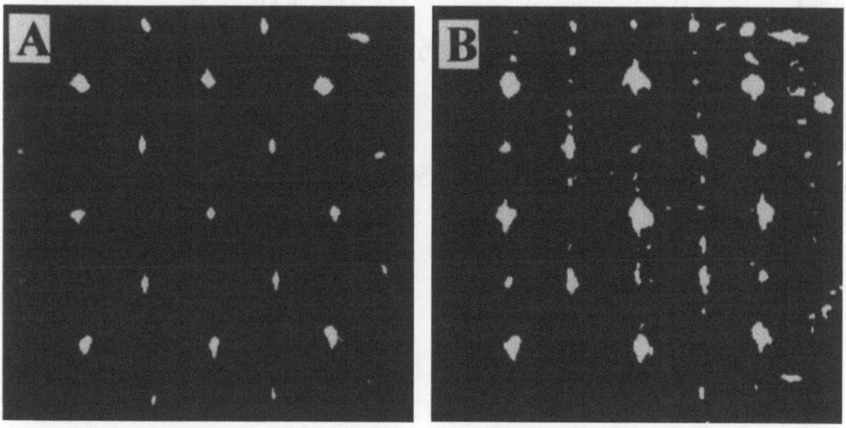

Fig. 6: SPA-LEED pattern of (A) clean, UHV-cleaved MgO(100) surface, filament current 2.30 A; (B) adsorbate CO_2/MgO(100), filament current 2.25 A, $p(CO_2) = 1.3 \cdot 10^{-8}$ mbar. Electron energy 214.8 eV, T = 82 K

an in-phase vibration of the two molecules in the adsorbate. From the two measured tilt angles mentioned above, simple geometrical considerations then allow to estimate the angle enclosed by the projections of the two molecules onto the surface to be 137^{-23}_{+36}°. The molecules are thus arranged in a herringbone structure.

3.4 CO Adsorbed on MgO(100)

IR-spectra of CO adsorbed on UHV-cleaved MgO(100) surfaces are presented in Fig. 7. The monolayer was prepared at a CO-pressure of $5.3 \cdot 10^{-9}$ mbar and a temperature of 40 K. Spectra were recorded at different temperatures upon warming the sample to 60 K. In all spectra a broad weak background between 2165 cm^{-1} and 2120 cm^{-1} is observed, which is attributed to adsorption on surface defects. In the temperature range between 40 and 50 K three sharp absorptions are detected at 2151.3, 2137 and 2132 cm^{-1}. The first two of them are found with p-polarization only, which indicates transition dipoles oriented perpendicular to the surface. From the A_s/A_p ratio of the third band a tilt angle of the transition dipole of approximately 20° with respect to the surface is inferred. Upon increasing the temperature above 55 K a single, broader absorption at 2150.5 cm^{-1} is detected with p-polarized radiation only. This is consistent with observations of the adsorption of CO on sintered polycrystalline MgO [21]. The spectra can be interpreted in terms of a structural 2D phase transition occurring around 50 K, in concert with a recent LEED investigation [22], where a commensurate (2x3) adsorbate structure was found between 40 and 48 K. Above this temperature range a phase transition to an incommensurate structure was observed.

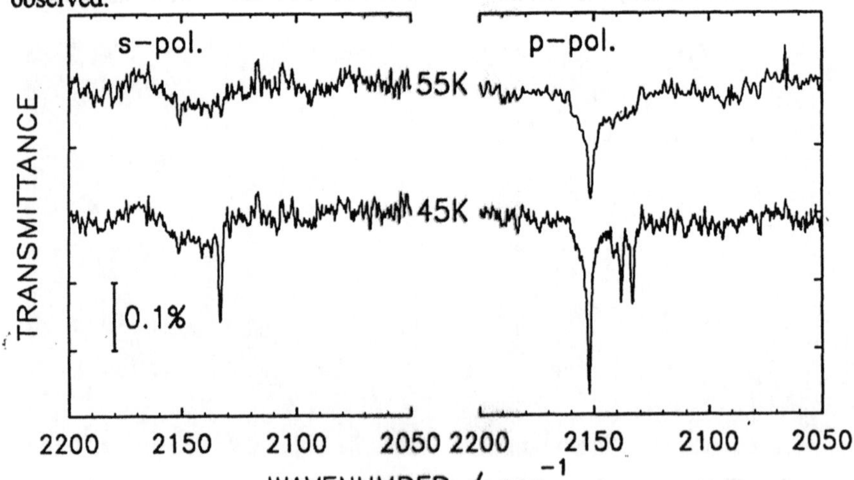

Fig. 7: Transmittance spectra of a monolayer CO adsorbed on UHV-cleaved MgO(100) at 45 and 55 K. Resolution 0.55 cm^{-1}, angle of incidence 45°.

4. Outlook

The main difference between physisorbed and gaseous molecules is the high degree of alignment caused by the substrate. Due to the absence of surface selection rules, polarized FTIR-spectroscopy of adsorbates on ionic insulator surfaces is able to reveal a variety of details of this alignment, e.g. azimuthal and polar molecular angles, 2D orientational phase transitions. In addition, the modelling of these systems by potential energy calculations can be performed rather reliably [23-25]. Both these aspects do not apply to adsorption on metals and semiconductors.

Vibrational energy flow is likely to have a significant and in some cases dominant influence on surface chemistry. A detailed knowledge of the time scales and paths of vibrational energy flow allows to explore and design conditions, where reactions can follow non-thermal pathways, which have a strong effect on chemical reactivity. The competition between dissociation and desorption of an adsorbed molecule is an example. Most theories of the rates of thermal reactions are based on the assumption that vibrational energy flow is fast compared to the reaction rate. It is an important objective to prove these assumptions.

Technological interest in the adsorption on salt systems has arisen from the disposal of highly radioactive waste from nuclear power plants in permian rock salt domes. NaCl aerosol may play a role in the sea spray chemistry. The catalytic activity of doped MgO is important in the oxidative coupling of methane.

Acknowledgements

We thank Prof. C. Girardet, Prof. A.J. Stone and Prof. J. Suzanne for preprints and discussions. Support by the Deutsche Forschungsgemeinschaft, the Land Niedersachsen and the Fonds der Chemischen Industrie is gratefully acknowledged.

References

[1] Y. Kozirovski, M. Folman: J. Chem. Phys. **41**, 1509 (1964)

[2] J. Heidberg, S. Zehme: Ber. Bunsenges. Phys. Chem. **72**, 1070 (1968)

[3] J. Heidberg, W. Häser: J. Electron Spectrosc. Relat. Phenom. **54/55**, 971 (1990); J. Heidberg, D. Wetter: unpublished results

[4] J. Heidberg, E. Kampshoff, O. Schönekäs, H. Stein, H. Weiss: Ber. Bunsenges. Phys. Chem. **94**, 112 (1990)

[5] U. Scheithauer, G. Meyer, M. Henzler: Surf. Sci. **178**, 441 (1986)

[6] J. Heidberg, E. Kampshoff, M. Suhren: J. Chem. Phys. **95**, 9408 (1991)

[7] J. Heidberg, E. Kampshoff, R. Kühnemuth, M. Suhren, H. Weiss: Surf. Sci. **269/270**, 128 (1992)

[8] J. Heidberg, E. Kampshoff, R. Kühnemuth, O. Schönekäs, M. Suhren: J. Electron Spectrosc. Relat. Phenom. **54/55**, 945 (1990)
[9] D. Schmicker, J.P. Toennies, R. Vollmer, H. Weiss: J. Chem. Phys. **95**, 9412 (1991)
[10] G.E. Ewing: Acc. Chem. Res. **25**, 292 (1992)
[11] J. Heidberg, K.-W. Stahmer, H. Stein, H. Weiss: J. Electron. Spectrosc. Relat. Phenom. **45**, 87 (1987)
[12] J. Schimmelpfennig, S. Fölsch, M. Henzler: Surf. Sci. **250**, 198 (1991)
[13] G. Lange, D. Schmicker, J.P. Toennies, R. Vollmer, H. Weiss: in preparation
[14] E. Umbach, H. Pfnür: this conference
[15] J. Heidberg, D. Meine: Ber. Bunsenges. Phys. Chem. **97**, 211 (1993)
[16] J. Heidberg, D. Meine: Surf. Sci. Lett. **279**, L175 (1992)
[17] C. Duriez, C. Chapon, C.R. Henry, J.M. Rickard: Surf. Sci. **230**, 123 (1990)
[18] J.L. Robins, T.N. Rodin, R.L. Gerlach: J. Appl. Phys. **37**, 3893 (1966)
[19] V.E. Henrich: Surf. Sci. **57**, 385 (1976)
[20] J. Suzanne, V. Panella, D. Fery, M. Sidoumou: submitted to Surf. Sci.
[21] D. Scarano, G. Spoto, S. Bordiga, S. Coluccia, A. Zecchina: J. Chem. Soc. Faraday Transact. **88**, 291 (1991)
[22] P. Audibert, M. Sidoumou, J. Suzanne: Surf. Sci. Lett. **273**, 467 (1992)
[23] J. Heidberg, E. Kampshoff, R. Kühnemuth, O. Schönekäs: Surf. Sci. **251/252**, 314 (1991)
[24] A. Lakhlifi, C. Girardet: Surf. Sci. **241**, 400 (1991)
[25] S. Picaud, P.N.M. Hoang, C. Girardet, A. Meredith, A.J. Stone: submitted to Surf. Sci.

Excitons and High Resolution Infrared Spectroscopy of Adlayers on Ionic Surfaces

G. E. Ewing

Department of Chemistry, Indiana University, Bloomington, IN 47405, USA

Abstract. Infrared spectroscopy of ordered molecular adlayers on ionic substrates often reveals structured absorption bands with unusually sharp features. This rich spectroscopy can be interpreted by exciton theory. Using the example of CO_2 on NaCl(100), we show how the molecular arrangement within the adlayer, temperature dependent geometry changes, and domain size can be revealed by application of exciton theory. Interpretation of the spectroscopy of other systems by exciton theory is also briefly discussed.

1. Introduction

The interpretation of infrared spectra of adsorbed molecules as coupled oscillators began nearly thirty years ago [1]. Through the years the coupled oscillator model has been developed to account for increasingly sophisticated spectroscopic measurements of bandshapes, bandshifts, photometry, and adlayer structures on a variety of surfaces. Recent infrared spectroscopy of ordered molecular adlayers on ionic surfaces reveals unusually sharp features. Multiple features, often observed, can be consequences of different adsorption sites, degeneracy breaking, correlation splitting or domain size. This rich spectroscopy deserves a thorough interpretation. The quantum mechanical treatment of coupled oscillators - the exciton model - can provide this interpretation.

The goal is to understand the structure of molecular adlayers on ionic surfaces. Here the experimental interrogation probe is infrared spectroscopy. The theoretical analysis must acknowledge that the adsorbed molecules are influenced by their neighbors in the adlayer and by the substrate. Hammaker et al. [1] first suggested that molecules within an adsorbed layer are coupled to each other through their transition dipoles. In a classical treatment, they developed a model to account for the infrared spectroscopy of CO and its isotopes chemisorbed to silica-supported platinum. Coupling of transition dipoles within an adlayer and to the substrate to account for spectroscopic data has been treated by Mahan and Lucas [2], Persson and Ryberg [3], Chen and Schaich [4] and many others as a recent review by Hollins [5] shows. In the language of quantum mechanics this coupling takes the form of exciton theory [6-9].

In our own work [10,11], and independently by Chen and Schaich [12] and

Springer Series in Surface Sciences, Vol. 33 Eds.: E. Umbach · H.-J. Freund
Adsorption on Ordered Surfaces of Ionic Solids and Thin Films

by Heidberg and his group [13-15], we have invoked transition dipole coupling to account for details of the infrared spectroscopy of CO_2 on NaCl(100). The spectroscopy of this system, as does the spectroscopy of other adlayers on well-defined ionic crystal surfaces [16-20], reveals sharp features with bandwidths orders of magnitude narrower than those originally observed for CO on supported Pt. Adlayer CO_2 on NaCl(100) then becomes a useful system to test the abilities of exciton theory to unravel the structures of monolayer and thin film adsorbates.

In our presentation we will first look at some spectroscopic laboratory experiments with CO_2 on NaCl(100). The exciton model will next be briefly reviewed. We will then use the model to interpret the laboratory measurements of CO_2 adlayers on NaCl(100). Exciton theory is suitable for interpretation of spectra of physisorbed adlayers on ionic surfaces since molecular properties are only slightly perturbed by the substrate. Predicted exciton behavior of random, island and fractal adlayers is also reviewed.

2. Some Data: Laboratory Experiments with CO_2 on NaCl(100)

We begin by looking at data from our laboratory. In Fig. 1 we have monolayer spectra of CO_2 on the (100) faces of single crystals of NaCl at 5 K and 86 K. The

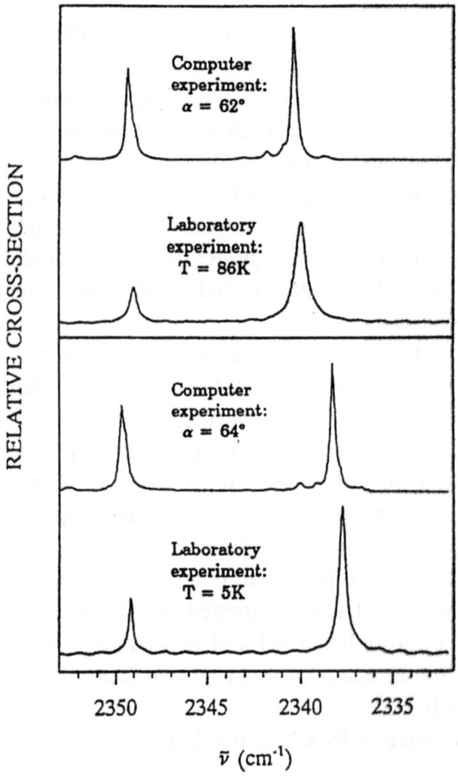

Fig. 1: Monolayer CO_2 on NaCl(100). Computer experiments of a square planar array of 900 transition dipoles are labeled by the molecular tilt angles α = 64° or α = 62°. Laboratory experiments for monolayer CO_2 on NaCl(100) are at temperatures 5 K and 86 K. From Berg et al. [11].

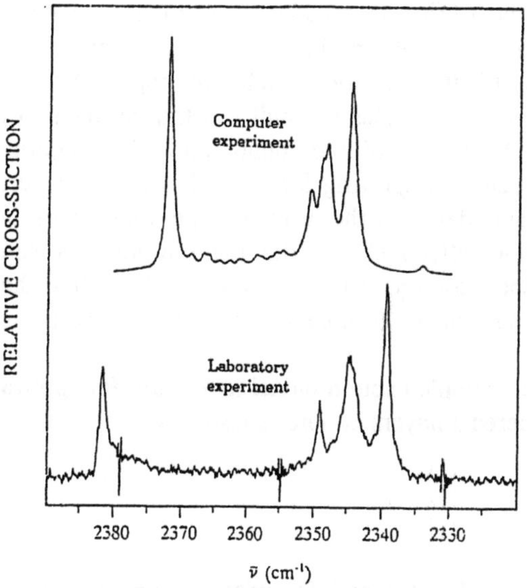

Fig. 2: Multilayer CO_2 on NaCl(100). The computer experiment with four planes of transition dipoles, for a total of 1800 transition dipoles, arranged to model the structure of bulk crystalline CO_2. The laboratory experiment with ~4 layers of CO_2 on NaCl(100) at 65 K from Berg and Ewing [10].

experimental means for preparing the monolayer and recording the spectra are described elsewhere [10,17]. The prominent doublet at 2349.1 cm^{-1} and 2337.5 cm^{-1} at 5 K is associated with the ν_3 mode of the asymmetric stretching vibration of the isolated (gas phase) $^{12}CO_2$ molecule. This mode is non-degenerate, yet two features appear in the adsorbed phase. The low frequency component, that we [10] have called the a_u mode, shifts upward by 2.5 cm^{-1} to 2340.0 cm^{-1} as the temperature is raised to 86 K. The high frequency component, assigned to the b_u mode, shifts downward slightly to 2349.0 cm^{-1}. The bandwidths of these doublet features depend on sample preparation and temperature. For the data shown at 5 K, the a_u mode has a bandwidth of 0.5 cm^{-1} while the b_u mode is 0.3 cm^{-1}. By contrast, the ν_3 vibration for $^{13}CO_2$ in natural abundance appears as a singlet (not shown on Fig. 1) when in the adsorbed phase and its bandwidth, of 0.09 cm^{-1} at 5 K, is instrument limited. Its frequency is nearly invariant with temperature.

Figure 2 shows multilayer CO_2 on NaCl(100). Here the temperature is 65 K and the experimental conditions have allowed several additional layers of CO_2 to condense on the original monolayer. The original doublet components near 2349 cm^{-1} and 2338 cm^{-1} remain, but new features near 2380 cm^{-1} and 2344 cm^{-1} have appeared. Experiments in our laboratory show that the multilayer spectra change continuously with film thickness. In the limit of a thick film (~1 μm or ~100 layers) the feature near 2344 cm^{-1} dominates the spectroscopic profile [21], the other complicating spectroscopic features are no longer apparent.

The rich spectroscopy in Figs. 1 and 2 can be explained in a number of equivalent ways. In their original interpretation, Berg and Ewing [10] and Heidberg et al. [13] proposed that the monolayer consisted of pairs of CO_2

molecules canted with respect to each other to produce a herringbone pattern. This structure was also consistent with helium diffraction measurements for monolayer CO_2 on NaCl(100) [22]. It is the transition dipole coupling of in-phase and out-of-phase vibrations of the two molecules per unit cell that accounts for the doublet observed in the $^{12}CO_2$ spectra of the monolayer. The coupling interpretation also explains the absence of splitting for the $^{13}CO_2$ feature since, because of the low (~1%) natural abundance of the heavy isotope, these sparsely distributed molecules cannot couple effectively. While these arguments can account for certain details of the spectroscopy of Figs. 1 and 2 with facility, the bandshapes of the monolayer features and the complexity of multilayer spectra is not so easy to interpret.

We offer here the conceptionally simple exciton model to account for spectra of CO_2 on NaCl(100) and other ordered adlayers on ionic substrates.

3. The Exciton Model

The exciton model goes back to Frenkel [6] and Davydov [7] and has been extended to discuss electronic or vibrational effects in crystals by Craig and Walmsley [8], Decius and Hexter [9] and Hexter [23]. In its simplest form, the transition dipoles (vibrational or electronic) of a collection of *n* molecules are coupled together. In the absence of this (weak) coupling, the properties of the condensed phase molecules are essentially those of the isolated (i.e. gas phase) molecules. A patch of *n* molecules is *n*-fold degenerate since capture by a photon could excite any one of the equivalent (but indistinguishable) molecules. It is the coupling among the molecules that can remove this degeneracy. The excited state system is then treated by *n*-fold degenerate perturbation theory and conveniently solved by the linear variation method [24].

A single molecule, call it j, is assumed to have a transition at $E^{(0)} = h\nu_0$, where ν_0 (s^{-1}) is the monomer optical transition frequency. This, for example, might be the vibrational frequency of the ν_3 mode of a gas phase CO_2 molecule. The isolated molecule photon capture occurs through its transition dipole moment (aligned along the molecular axis) whose magnitude is given by

$$\mu_j^{01} = \langle \Phi_j | \mu_j | \Phi'_j \rangle \tag{1}$$

where the prime indicates the excited state (v=1) and its absence identifies the ground state (v=0). We now place *n* molecules on a surface, e.g. the (100) face of NaCl. The ground state wavefunction becomes

$$\Psi_0 = \Phi_1 \Phi_2 \Phi_3 \ldots \Phi_n \tag{2}$$

An excited state wavefunction may be written as

$$\Psi_m = c_{m,1} \Phi'_1 \Phi_2 \Phi_3 \ldots \Phi_n + c_{m,2} \Phi_1 \Phi'_2 \Phi_3 \ldots \Phi_n + c_{m,3} \Phi_1 \Phi_2 \Phi'_3 \ldots \Phi_n + \ldots c_{m,n} \Phi_1 \Phi_2 \Phi_3 \ldots \Phi'_n \tag{3}$$

There are n of these excited state wavefunctions; the coefficients $c_{m,1}$, $c_{m,2}$, ... $c_{m,n}$ of level m = 1, or 2, ... or n may be determined once the coupling among the molecules is specified and the secular determinant solved [24].

In the simplest approximation, we imagine the coupling is only through the transition dipoles. The perturbation Hamiltonian becomes

$$H^{(1)} = (4\pi\varepsilon_0)^{-1}\sum_{i=1}^{n}\sum_{j<i}^{n}\mu_i^{01}\mu_j^{01}\left\{\mathbf{l}_i\mathbf{l}_j - 3(\mathbf{l}_i\mathbf{r}_{ij})(\mathbf{l}_j\mathbf{r}_{ij})\right\}/R_{ij}^3 \tag{4}$$

written in SI units, with $\varepsilon_0 = 8.85 \times 10^{-12}$ J^{-1} C^2 m^{-1} (the vacuum permeability); where R_{ij} is the separation between transition dipoles (molecules) i and j; $\mathbf{l}_i$, $\mathbf{l}_j$ are unit vectors for these transition dipoles; and $\mathbf{r}_{ij}$ is a unit vector from j to i [25]. The diagonal matrix elements are all relative to the ground vibrational state,

$$H_{ii}^{(0)} = h\nu_0 \tag{5}$$

and the off-diagonal matrix elements are

$$H_{ij}^{(1)} = (4\pi\varepsilon_0)^{-1}\left|\mu^{01}\right|^2\left\{\mathbf{l}_i\mathbf{l}_j - 3(\mathbf{l}_i\mathbf{r}_{ij})(\mathbf{l}_j\mathbf{r}_{ij})\right\}/R_{ij}^3 \tag{6}$$

where we have used eq. (1) but dropped the j (or i) subscript on μ^{01} since all the transition dipoles are assumed to be the same. Diagonalization of the $n \times n$ secular determinant gives the n E_m eigenvalues and the corresponding n Ψ_m eigenfunctions of first order degenerate perturbation theory.

The mth optical transition is at energy $E_m = E^{(0)} + E_m^{(1)} = h\nu_m$. The integrated cross-section $\bar{\sigma}_m$ (m molecule^{-1}), which is proportional to transition probability, is

$$\bar{\sigma}_m = \frac{8\pi^3\nu_m}{3n(4\pi\varepsilon_0)hc^2}\left|\left\langle\Psi_m\left|\sum_{j=1}^{n}\mu_j\mathbf{l}_j\right|\Psi_m\right\rangle\right|^2 \tag{7}$$

Use of eqs. (1), (2) and (3) transforms eq. (7) into

$$\bar{\sigma}_m = \frac{8\pi^3\nu_m\left|\mu^{01}\right|^2}{3n(4\pi\varepsilon_0)hc^2}\left|\sum_{j=1}^{n}c_{mj}\mathbf{l}_j\right|^2 \tag{8}$$

We can approximate $\nu_m \approx \nu_o$ from the gas phase. The transition dipole moment is related [26] to the integrated molecular cross-section, $\bar{\sigma}_g$, of the isolated (e.g. gas phase) molecule by

$$\left|\mu^{01}\right|^2 = \frac{3hc^2\overline{\sigma}_g(4\pi\varepsilon_0)}{8\pi^3\nu_0\eta} \tag{9}$$

The index of refraction η is that of the bulk solid adlayer. This term is intended to account for the bulk polarizability of the adlayer which effectively screens the transition dipoles. Alternative and more involved polarizability corrections are also available [2,4,12,26-28].

4. Some Data: Computer Experiments

In order to generate computational spectra we must specify the parameters of the isolated (gas phase) molecules that appear in eq. (9). These are $\tilde{\nu}_m \approx \tilde{\nu}_0 = 2349$ cm^{-1} and the integrated cross section $\overline{\sigma}_g = 1.1 \times 10^{-18}$ m $molecule^{-1}$ for the ν_3 mode [29]. The bulk index of refraction of bulk solid CO_2 is $\eta = 1.4$ [30]. Finally, the molecular positions in the monolayer or multilayer must be specified. The qualitative view of these adlayers is given in Fig. 3(a). In Fig. 3(b) the angles α and β define the orientation of the CO_2 molecular axis (coincident with the transition dipole, μ^{01}, axis). The justification of this adlayer geometry is given elsewhere [10,14,22,31]. Polarized infrared spectroscopy estimates the tilt angle of the monolayer near $\alpha = 66\pm2°$ [10,14]. In the herringbone pattern, neighboring molecules have a relative azimuthal separation of $\Delta\beta = 90°$. Finally the molecules in the adlayer are in registry with the NaCl(100) substrate so nearest neighbor separation is that of the nearest neighbor Na^+ separation of 398 pm [31]. These distances and angles allow the geometric terms in eqs. (4) and (6) to be specified.

Now that the molecular and adlayer parameters have been specified, diagonalization of the secular determinant yields eigenenergies and eigenvectors. The spectrum consists of $\tilde{\nu}_m$ and $\overline{\sigma}_m$ values, which follow directly from coefficients $c_{m,j}$ of eq. (3). For *n* molecules, *n* separate vibrational transitions are calculated over a wide range of cross sections. The resulting quantum mechanical "stick" spectrum can be more easily compared with experiment after accounting for lineshape broadening mechanisms. A Lorenztian bandwidth of 0.3 cm^{-1}, typical for heterogeneous or dephasing values [33,34], has been convoluted onto our calculated spectra.

We have performed a series of calculations on a square two-dimensional array containing $n = 30 \times 30 = 900$ CO_2 molecules. Two theoretical spectra, using tilt angles $\alpha = 62°$ and $64°$ with $\Delta\beta = 90°$, are compared in Fig. 1 to experimental data taken at 86 K and 5 K. The calculation predicts two features, with the lower energy band having slightly greater intensity. The agreement in frequencies is excellent. Furthermore, values of α which compare favorably with experiment are very close to the value determined from polarized infrared measurements: $\alpha = 68\pm5°$ [10], and $\alpha = 64\pm5°$ [14]. It can be seen that smaller

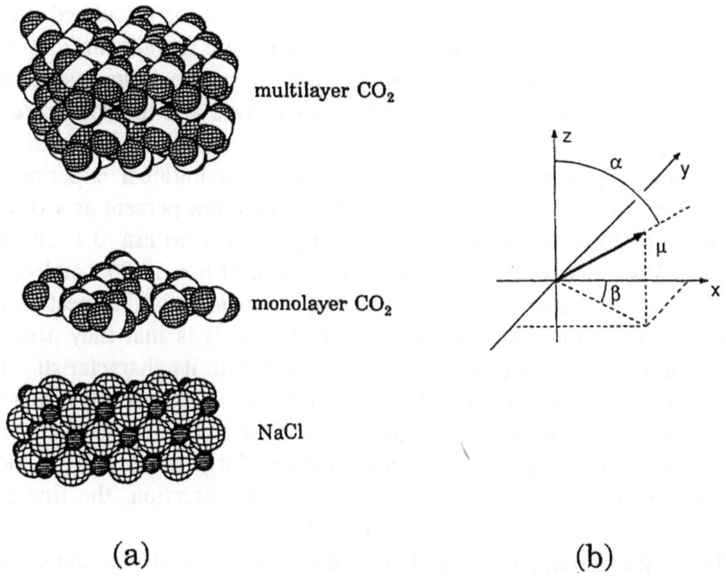

Fig. 3: The structure of monolayer and multilayer CO_2 on NaCl(100). In (a) a segment of substrate NaCl is shown with larger spheres Cl^- and smaller spheres Na^+. Monolayer CO_2 is shown displaced from the (100) face of NaCl for ease in visualization. Multilayer CO_2 is likewise displaced above the monolayer. In (b) the angular coordinates of CO_2 are shown. The transition dipole, μ, through the CO_2 molecular axis is tilted by angle α from the z-axis. The z-axis is orthogonal to the (100) face of NaCl. The x- and y-axes coincide with the unit cell axes in the (100) face of NaCl. The azimuthal angle, β, defines the second CO_2 angular coordinate. From Berg and Ewing [10].

tilt angles (i.e. the molecular axis more nearly normal to the surface) decrease the calculated exciton splitting. Larger tilt angles increase the splitting. Notice that not only is the proposed monolayer structure verified by the exciton model, but a temperature change in the geometry i.e. the molecular tilt change, is identified. Elsewhere [11] we have associated this tilt change at higher temperatures with excitation of an anharmonic frustrated rotational mode of CO_2 within the monolayer. A related interpretation has been provided by Heidberg et al. [15].

An aspect of the model we have not yet considered is the extent to which generated spectra are influenced by the size of the domain. In the examples considered, the monolayer domains consist of n = 900 molecules in a square array. While all the generated spectra are dominated by a doublet, a certain amount of fine structure is present. This is apparent in the calculated spectra for $\alpha = 62°$ or $\alpha = 64°$ in Fig. 1, where weak features are found near 2340 cm^{-1} and 2350 cm^{-1}. Scores of other features are evident in the "stick" spectra before broadening by Lorenztian profiles. This fine structure and the doublet separation is a consequence of the finite nature of the domains. We can test the spectroscopic dependence on domain size by computational experiments. For example, the overall exciton splitting decreases by a few percent as n is increased from 900 to 2000 molecules and convergence (to within 0.1 cm^{-1}) to an extrapolated doublet separation is not achieved until $n \approx 10,000$ molecules. Our interpretation of the low temperature limiting bandwidths of the spectroscopically observed $^{12}C^{16}O_2$ doublet features (0.3 and 0.5 cm^{-1}) is that they arise from a superposition of domains of different sizes - each with its characteristic splitting. These domains are mosaics consisting of patches of CO_2 molecules each with its herringbone pattern aligned along one of the four equivalent axes in the (100) face of NaCl [10,12,14]. In the limit of a single large patch of CO_2 molecules with the herringbone pattern aligned in the same direction, the fine structure would collapse into two sharp resonances [12].

The exciton interpretation of bandwidths also accounts for the exceedingly narrow absorption of the isotopically dilute $^{13}CO_2$ in the monolayer. Here vibrations of $^{13}CO_2$ couple poorly with those of $^{12}CO_2$ because their frequencies are nonresonant and couplings among $^{13}CO_2$ are weak because typical separation distances, R_{ij}, are much larger than those of the majority isotopomer and the corresponding splittings, as dictated by eqs. (4) and (6), are orders of magnitude smaller. The actual linewidth of the $^{13}CO_2$ feature, determined when suitable high resolution spectroscopic techniques are applied, may likely be limited by the heterogeneities of the monolayer or substrate.

As a final test of the exciton model we attempt to reproduce the laboratory measurement of the ν_3 asymmetric stretching region of multilayer CO_2 adsorbed on NaCl(100) at 65 K. The computer experiment, or simulation, was produced from a structure four molecular layers thick, with 450 molecules per layer for a total of 1800 molecules: each molecule represented by a transition dipole. The exact structure chosen for the computation was a slab of bulk crystalline CO_2 [31], a segment of which is reproduced in Fig. 3(a). Our assumption that multilayer CO_2 on NaCl(100) closely resembles the bulk crystal is based on structural similarities between the monolayer and solid [10,14]. The monolayer has CO_2 molecules in a herringbone pattern, with molecular axes tilted 66°±5° from the surface normal. Each layer of the crystal is also a herringbone arrangement, of similar density, but with molecular axes tilted only 45°. In addition, the multilayer experiment shows an absorption feature at 2344 cm^{-1} with a bandwidth of 3.0 cm^{-1}, both of which are characteristic of bulk crystal CO_2 [21].

Once again, the agreement between the computer experiment and the laboratory experiment as shown in Fig. 2 is qualitatively good. The calculated resonances near 2350 cm^{-1} and 2344 cm^{-1} correspond to the doublet in a monolayer. The signals near 2372 cm^{-1} and 2348 cm^{-1} correspond to the new features associated with multilayer coverage. The simulated spectrum can be brought into better agreement with the experiment by treating the index of refraction as an adjustable parameter and making small variations in molecular orientations. The explanatory value of the exciton model, however, is that it accounts for the principal spectroscopic observations with a small set of physically clear assumptions.

5. Other Systems

Another system for which there is a rich collection of laboratory data is CO on NaCl(100) [17,26,33,34,35]. Both classical electrodynamic models [4,35] and the exciton model [34] have been applied to explain the laboratory data. Fine structure near the principal absorption band can be interpreted in terms of the finite domain size of CO patches adsorbed to (100) face terraces bounded by steps on the NaCl surface [36]. This fine structure interpretation is similar to that we have just offered for the CO_2 on NaCl(100) data. The exciton model applied to CO on NaCl(100) also explores the spectroscopy of isotopic mixtures, submonolayer coverages and the effect of surface defects on absorption profiles [34]. The exciton model in this system goes beyond transition dipole coupling of eqs. (4) and (6) to include transition quadrupole and higher order transition multipoles as well as permanent multipoles. These additional terms must be included in order to understand band profiles of finite domains. The spectroscopic consequences of the low temperature phase transition of CO on NaCl(100) [35] have also been explored by exciton theory [37].

In a recent theoretical study, Wales and Ewing [38] have explored the infrared spectroscopy of one, two and fractal dimensional adlayers. Here the effects of dimensionality are explored. Fractal adlayers will for example have an easily identifiable spectroscopic signature. Submonolayer coverages that give rise to either island formation (lattice crystal) or random adsorption (lattice gas) can also be distinguished by their spectra. These two adsorption phases have been identified in laboratory measurements [10,14,18,19,33].

Exciton calculations are mainly sensitive to the properties of the isolated molecules and their adlayer geometry. The exciton model moreover is applicable to physisorption on a variety of ionic substrates since the molecular properties ($\tilde{\nu}_m$, μ^{01}, etc.) of the adlayer are only slightly perturbed. As a consequence a match of the computational spectra with the laboratory spectra allows a detailed interpretation of adlayer structure.

Acknowledgements

The author recognizes his partners in this work: Dr. Otto Berg, Dr. Robert Disselkamp and Dr. David Wales. This work was funded by the National Science Foundation under Grant number CHE91-15444.

References

[1] R. Hammaker, S. Francis, R. Eischens: Spectrochim. Acta **21**, 1295 (1965)
[2] G.D. Mahan, A.A. Lucas: J. Chem. Phys. **68**, 1344 (1978)
[3] B.J. Persson, R. Ryberg: Phys. Rev. **B24**, 6945 (1981)
[4] W. Chen, W.L. Schaich: Surf. Sci. **218**, 580 (1989)
[5] P. Hollins: Surf. Sci. Rep. **16**, 2 (1992)
[6] J. Frenkel: Phys. Rev. **37**, 17 (1931)
[7] A.S. Davydov: Theory of Molecular Excitons, McGraw-Hill, New York 1962
[8] D.P. Craig, S.N. Walmsley: Excitons in Molecular Crystals, Benjamin, New York 1968
[9] J.C. Decuis, R.M. Hexter: Molecular Vibrations in Crystals, McGraw-Hill, New York 1977
[10] O. Berg, G.E. Ewing: Surf. Sci. **220**, 198 (1989)
[11] O. Berg, R. Disselkamp, G.E. Ewing: Surf Sci. **177**, 8 (1992)
[12] W. Chen, W.L. Schaich: Surf. Sci. **220**, L733 (1989)
[13] H. Heidberg, E. Kampshoff, R. Kühnemuth, O. Schönekäs, H. Stein, H. Weiss: Surf. Sci. Lett. **226**, L43 (1990)
[14] J. Heidberg, O. Kampshoff, O. Schönekäs, H. Stein, H. Weiss: Ber. Bunsenges. Phys. Chem. **94**, 112, 118, 127 (1990)
[15] J. Heidberg, E. Kampshoff, R. Kühnemuth, O. Schönekäs: Surf. Sci. **269/270**, 120 (1992)
[16] J. Heidberg, E. Kampshoff, M. Sukren: J. Chem. Phys. **95**, 9401 (1991)
[17] G. E. Ewing: Acct. of Chem. Res. **25**, 292 (1992)
[18] S.K. Dunn, G.E. Ewing: J. Phys. Chem. **96**, 5284 (1992)
[19] L. Quattrocci, G.E. Ewing: J. Chem. Phys. **96**, 4205 (1992)
[20] L. Quattrocci, G.E. Ewing: Chem. Phys. Lett. **197**, 308 (1992)
[21] D.A. Dows, V.J. Shettino: J. Chem. Phys. **58**, 5009 (1973)
[22] G.-Y. Lui, G.N. Robinson, G. Scoles, P.A. Heiney: Surf. Sci. **262**, 409 (1992)
[23] R.M. Hexter: J. Chem. Phys. **33**, 1833 (1960)
[24] L. Pauling, E.B. Wilson: Introduction to Quantum Mechanics, Chap. 7, Dover, New York 1985
[25] A.D. Buckingham: Adv. Chem. Phys. **12**, 107 (1967)
[26] H.H. Richardson, H.-C. Chang, C. Noda, G.E. Ewing: Surf. Sci. **216**, 93 (1989)
[27] P. Jakob, Y.J. Chabal, K. Raghavachari: Chem. Phys. Lett. **187**, 325 (1991)
[28] Y.J. Chabal: Surf. Sci. Rep. **8**, 211 (1988)
[29] S. Penner: Quantitative Molecular Spectroscopy and Gas Emissivities, Addison-Wesley, Reading 1959
[30] B.E. Wood, J.A. Roux: J. Opt. Soc. Am. **72**, 720 (1982)
[31] W.H. Keesom, J.W.L. Kohler: Physica **1**, 655 (1934)
[32] H.H. Landolt, R. Börnstein: Atom-und Molekularphysik 4. Teil, Kristalle, Springer, Berlin 1955
[33] C. Noda, H.H. Richardson, G.E. Ewing: J. Chem. Phys. **92**, 2099 (1990)
[34] R. Disselkamp, H.-C. Chang, G.E. Ewing: Surf. Sci. **240**, 193 (1990)
[35] J. Heidberg, E. Kampshoff, M. Suhren, H. Weiss: Surf. Sci. **269/270**, 128 (1992)

[36] G. Meyer, N.M. Amer: Appl. Phys. Lett. **56**, 2100 (1990)
[37] D. Dai, G. Ewing: to be published
[38] D.J. Wales, G.E. Ewing: J. Chem. Soc. Faraday Trans. **88**, 1359 (1992)

Part II

Clean Surfaces of Metal Oxides: Geometric and Electronic Structure

Characterization of Single-Crystal α-Al_2O_3 (0001) and $(11\bar{2}0)$ Surfaces and Ag/Al_2O_3 Model Catalysts by Atomic Force Microscopy

G. Beitel, K. Markert, J. Wiechers, J. Hrbek, and R.J. Behm*

Abteilung für Oberflächenchemie und Katalyse,
Universität Ulm, D-89069 Ulm, Germany
*Permanent address: Department of Chemistry, Brookhaven National Laboratory, Upton, New York 11973, USA

Abstract: We present an overview on results of an AFM study on structural aspects of Ag/α-Al_2O_3 model catalysts. Aiming at well-defined, atomically smooth substrates, the structural evolution of clean (0001) and $(11\bar{2}0)$ α-Al_2O_3 surfaces upon annealing is characterized. This includes the development of distinct step-terrace topographies and of well-ordered atomic structures. Upon thermal evaporation of Ag on well-defined $(11\bar{2}0)$ substrates at room temperature, three-dimensional Ag clusters are formed, indicative of a Volmer-Weber growth mode. Size, distribution and thermal stability of these clusters are explored. First results on their chemical stability, with respect to annealing in an oxygen atmosphere, are included.

1. Introduction

Studies of surface phenomena on insulating materials have been hampered for a long time by sample charging [1]. Only in recent times these materials became more easily accessible by either using ultrathin oxide films epitaxially grown on metal substrates [2,3], which allow electron tunneling into the substrate and therefore do not charge up, or by employing newly developed techniques not involving the use of charged particles. Among these insulating materials are a large number of oxides, which are highly relevant because of their technological importance. One of them is α-alumina (α-Al_2O_3), which is used in microelectronics (silicon-on-sapphire technology [4]) and in heterogeneous catalysis, where polycrystalline alumina supports covered by dispersed metal particles are important catalysts. Ethylene oxidation for instance, which makes use of Ag/Al_2O_3 catalyst, is a major industrial process [5]. We have recently started to investigate structural aspects of Ag/α-Al_2O_3 model catalysts, consisting of disperse Ag deposited on atomically smooth α-Al_2O_3 substrates. This study aims at the understanding of substrate effects on the size and distribution of the Ag particles and of the thermal and chemical stability of substrate and deposit. These structural aspects are studied by atomic force microscopy (AFM) in an ultrahigh vacuum (UHV) environment. In this article we present an overview of our results. It will include the evolution of characteristic elements in the surface morphology such as steps and facets on (0001) and $(11\bar{2}0)$ surfaces as a function of preparation conditions, as well as first results on the surface atomic structure under these conditions, and on the growth and stability of Ag clusters on these surfaces.

Previous data on structural properties of α-Al_2O_3 came mostly from electron microscopy studies. The most important crystal faces of α-Al_2O_3, the (0001), $(11\bar{2}0)$ and $(10\bar{1}2)$ (c-, a- and r-cut) surfaces, were investigated by reflection electron microscopy (REM) [6-9] and by transmission electron microscopy (TEM) combined with diffraction techniques (reflection high-energy electron diffraction - RHEED- and transmission electron diffraction - TED) [10]. These studies gave insight e.g. into the facetting behavior. Furthermore a (4×1) reconstruction with a periodicity of 1.73 nm was observed on a $(11\bar{2}0)$ surface after annealing for 24 h to 1400 °C in air [7]. On the other side, low-energy electron diffraction (LEED) studies resolved a number of reconstructions on the (0001) surface after annealing in UHV [11-16]. While for annealing temperatures around 1000 °C a bulk-like (1×1) structure was observed, heating to 1400 °C resulted in a complex $(\sqrt{31}\times\sqrt{31})R9°$ reconstruction. In addition, several structures such as a (2×2), a $(\sqrt{3}\times\sqrt{3})R30°$ and a $(3\sqrt{3}\times3\sqrt{3})R30°$ were observed after annealing to intermediate temperatures [12-14,16]. Finally first AFM results have been published, obtained by imaging the (0001) and the $(10\bar{1}2)$ surface in air [17,18]. Flat terraces and surface steps of 0.4 ± 0.1 nm height were reported for the $(10\bar{1}2)$ surface, while no further details on the atomic structure could be resolved on the (0001) surface.

2. Experimental

The AFM experiments were performed in a UHV system equipped with standard facilities for surface preparation and characterization, a Ag evaporator source, a load lock for rapid sample transfer and a fully UHV-compatible AFM. The AFM was a home-built design which relies on optical beam deflection for detection [19]. The maximum scan range of the instrument was 1 µm × 1 µm with a vertical resolution of 0.03 nm. This set-up allowed in situ Ag evaporation and subsequent AFM imaging under clean conditions. Furthermore, capillary forces due to water condensation between tip and surface were avoided, allowing very low load forces in the AFM measurements.

Commercially available (Kelpin) square slabs ($8 \times 8 \times 1$ mm^3) of c- and a-cut α-Al_2O_3 single crystals, chemo-mechanically polished on one side, were used as sample material. As received, the surface roughness of these samples was about 0.5 - 1 nm, with some scratches of 1 - 1.5 nm depth. Prior to AFM imaging in UHV the samples were cleaned with nitric acid, trichloroethane, acetone and water and then annealed in an electrical furnace at temperatures between 900 °C and 1600 °C for 12 h in air. Ag was deposited from a molybdenum basket with a rate of ~ 0.2 monolayers per minute (ML/min). During silver deposition the pressure was below 4×10^{-10} mbar.

For each preparation three independent measurements were made on different locations of the (0001) and $(11\bar{2}0)$ surface. The AFM results presented in this paper can therefore be considered as representative of the respective surface morphologies.

3. Results and Discussion

α-Al_2O_3 crystallizes into the corundum structure (Fig. 1), which ideally is formed by a hexagonally close-packed (hcp) arrangement of the O^{2-} anions. The Al^{3+} cations occupy 2/3 of the octahedral interstitial sites of the O^{2-} lattice. The radii of the O^{2-} anions and the Al^{3+} cations are 0.14 and 0.053 nm, respectively. In this structure the coordination number is 4 for the O^{2-} anions and 6 for the Al^{3+} cations, which is consistent with the stoichiometry of 2 : 3. Linking OAl_4 tetrahedra and AlO_6 octahedra to form a 3D-crystal leads to a distortion of the coordination polyhedra, resulting in a slightly distorted hcp sublattice of the O^{2-} anions and wavy Al^{3+} layers consisting of open hexagons. The hexagonal unit cell consists of six O^{2-} layers, which leads to a unit cell with lattice parameters of $a_0 = 0.475$ nm and $c_0 = 1.299$ nm [20].

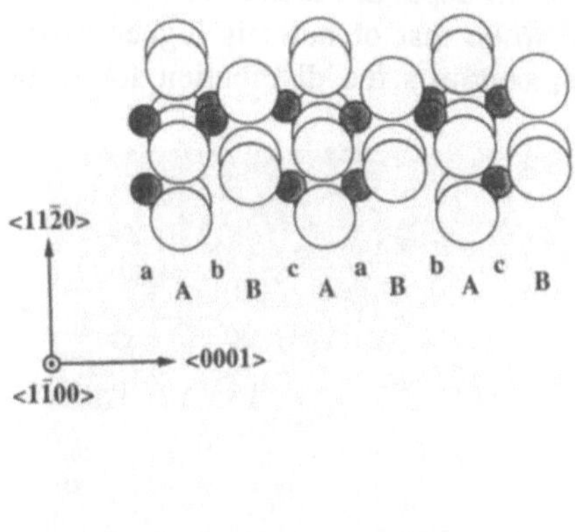

Fig. 1: 3D view of the corundum structure, showing the ABAB packing of the oxygen anions in the hcp arrangement. Al^{3+} cations occupy 2/3 of the octahedral sites in an abcabc packing.

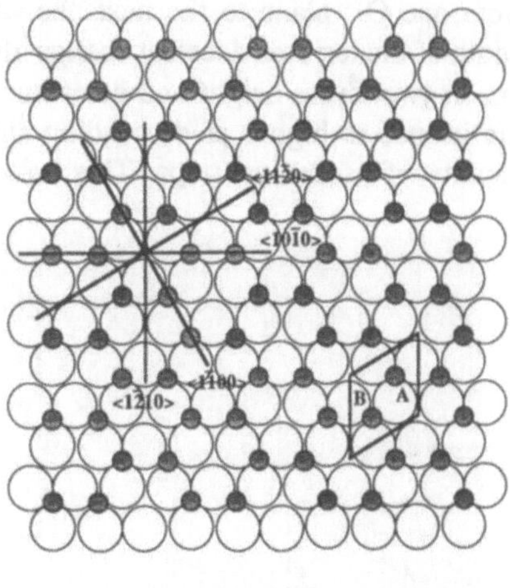

Fig. 2: Model of an Al-terminated and bulk-truncated (0001) surface (top view). The surface unit cell and different threefold hollow sites (A and B) as well as the most commonly observed step directions are indicated.

3.1 α-Al_2O_3 (0001)

Figure 2 shows an idealized top view model of the α-Al_2O_3 (0001) surface, terminated by an Al^{3+} layer with the aluminum cations in bulk positions. (This structure model represents a cut through the bulk structure. Because of inequivalent charges in the topmost layer it would not be possible for the actual surface. A neutral topmost layer could be obtained e.g. by removing half of the Al^{3+} cations.) On this surface the hexagonal arrangement of oxygen trimers leads to two different positions for the aluminium cations: a threefold hollow site on a close-packed oxygen trimer (A) and a similar type site between slightly separated oxygen atoms of three different trimers (B). In this figure the unit cell and the most commonly observed step directions are indicated. Steps in $[1\bar{1}00]$ and $[11\bar{2}0]$ or $[10\bar{1}0]$ and $[1\bar{2}10]$ directions will produce step geometries enclosing angles of 90°, respectively. Note that for compatibility with earlier studies we used the (h, k, -h-k, l) notation not only for the description of lattice planes, but also of lattice directions, contrasting common crystallographic praxis.

After annealing to temperatures between 900 °C and 1100 °C in air, flat terraces and distinct step edges can be resolved, which is illustrated in the 80 nm × 80 nm AFM image in Fig. 3a. The image was recorded on a surface annealed to 1000 °C. We find 20 - 30 nm wide terraces separated by steps of 0.22 ± 0.02 nm height, which corresponds to a surface miscut of ~ 0.5°. This result contrasts earlier AFM findings on a (0001) surface in air where no steps could be detected [18]. Also the width of the terraces is similar after annealing to temperatures within this range. The observed step height corresponds to monoatomic steps, i.e. from one O^{2-} plane to the next one with a theoretical height of 0.22 nm. The width of the terraces does not change after annealing. On these terraces islands of one or two layers height and holes one layer in depth are found. The density of the islands is higher at descending steps, while that of holes is higher in the vicinity of ascending steps. This behavior contrasts the distribution found by

a b

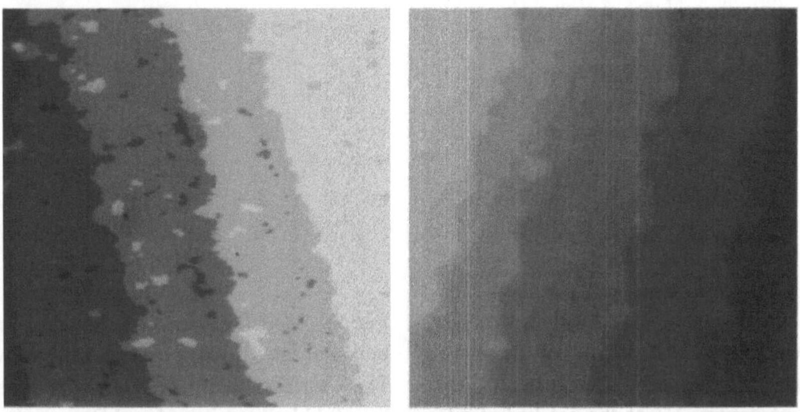

Fig. 3: Large scale AFM images of the (0001) surface annealed at a) 1000 °C (80 nm × 80 nm) and b) 1400 °C (450 nm × 450 nm) for 12 h in air.

Peng et al. [6] for the annealing behavior of etched $(\bar{1}012)$ surfaces, where the accumulation of holes and islands is just opposite. The rough shape of the step edges with no defined step direction is a further characteristic feature of these surfaces. Both the rough step edges and the presence of islands and holes on the terraces indicate that the surface defects caused by the chemo-mechanical polishing process are not removed completely, i.e. these (0001) surfaces are still in a non-equilibrated state. The observation of flat, though defective terraces indicates that at 1000 °C the surface is sufficiently mobile for short range diffusion, but not for longer range mass transport. This agrees well with estimates on the mobility of kinks on the $(\bar{1}012)$ surface, where a diffusion barrier of 5.5 eV for kink migration was determined from REM observation [6]. Using similar values for the diffusion properties in the present case results in a diffusion length of ~ 10^{-1} nm for 12 h annealing at 1000 °C. Even significantly higher mobilities would barely facilitate short range diffusion and not allow any long range transport, consistent with our observations.

The general appearance of the sample is strongly different after annealing at 1400 °C. The many defects observed before are almost completely removed and instead a strongly facetted surface is found, which is illustrated in the 450 nm × 450 nm AFM image in Fig. 3b. The width of these terraces ranges from 100 - 150 nm, separated by steps of 6.5 - 8.0 nm height. The steps are linear over larger distances, forming an overall zigzag geometry. The step directions can be identified as $[1\bar{1}00]$ and $[01\bar{1}0]$, the step edges are most likely formed by $(11\bar{2}0)$ and $(\bar{2}110)$ planes, respectively, which stand vertically on the (0001) terrace plane. These large terraces are subdivided into smaller ones of about 25 nm width by almost straight steps running along $[1\bar{2}10]$, i.e. at an angle of 30° with respect to the major step direction. The height of these steps is measured to be one or two lattice constants in the c-direction, i.e. about 1.3 nm or 2.6 nm. Indexing of the observed step directions is based on surface energies calculated by Tasker [21], where the $(11\bar{2}0)$ plane was found to have the lowest relaxed surface energy among the low-index planes possible. Assuming that this corresponds to the larger facets the other step directions can be indexed as well. Facetting on a scale as observed here, forming steps up to nm height, requires extensive mass transport and hence significant surface mobility. Again, this agrees well with previous REM findings [6,22]. Under conditions of such high surface mobility it is not surprising that defects on the flat terraces can be healed out.

As consistently observed by LEED [11-16], the (0001) surface exhibits a (1×1) bulk-like structure after annealing to temperatures around 1000 °C, with the hexagonal unit-cell parameter $a_0 = 0.475$ nm and an oxygen-oxygen distance of $a_0/\sqrt{3} \cong 0.28$ nm. This is resolved in high resolution AFM images recorded on the terraces of a (0001) surface annealed at 1000 °C, as shown in Fig. 4, which displays a 2 nm × 2 nm scan area. A distorted hexagonal corrugation is seen in this image, with interatomic distances of 0.3 ± 0.02 nm, which agrees well with the structure of the oxygen layer of an α-Al_2O_3 (0001) surface. The corrugation is measured to be 0.02 - 0.03 nm. In addition to the overall hexagonal arrangement a number of defects can be observed, such as a missing surface atom in the lower left corner of the image, and varying interatomic distances, which together result in a locally disordered and distorted hexagonal arrangement. These local distortions should not be confused with the periodic distortions present in a bulk-truncated surface with its two different Al^{3+} adsorption sites. These latter distortions have not been resolved in our data so far.

3.2 α-Al_2O_3 $(11\bar{2}0)$

As a second example we investigated the α-Al_2O_3 $(11\bar{2}0)$ surface. An idealized atomic model is shown in Fig. 5. The most important features of this more complex surface structure are parallel oxygen rows along $[1\bar{1}00]$ with an interatomic distance along the rows of ~ 0.28 nm and a distance of 0.43 nm between neighboring rows. The surface unit cell of the unreconstructed surface, with lattice parameters of 0.82 nm along and 1.3 nm perpendicular to the oxygen rows, is marked by a full line. Equivalent Al^{3+} cations are aligned along the $[\bar{1}101]$ and $[2\bar{2}01]$ directions, rotated by either 58° or 39° in opposite directions with respect to the O^{2-} rows.

This surface was annealed in steps of 100 °C from 1100 °C to 1600 °C for 12 h in air. Its structural evolution during annealing is illustrated in two 900 nm × 900 nm AFM images shown in Fig. 6. From the step-to-terrace ratio it turned out that the crystal surface was cut at an angle of ~ 2° to the $(11\bar{2}0)$ plane. This miscut along $[1\bar{1}02]$ is found in all of the observed AFM images on different positions of the crystal surface. It does not change as a function of the annealing temperature. The morphology of the observed steps, on the other hand, shows a

Fig. 4: High resolution AFM image (2 nm × 2 nm) showing an atomic corrugation consistent with the (1×1) O structure of the (0001) surface, recorded after annealing at 1000 °C.

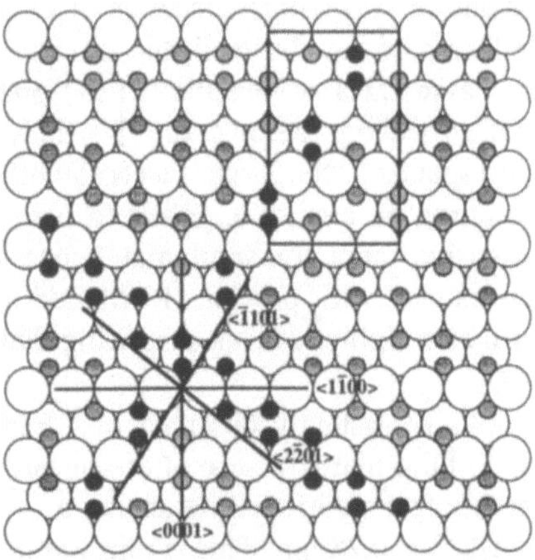

Fig. 5: Model of a bulk-truncated $(11\bar{2}0)$ surface. The surface unit cell and most important step directions are indicated.

strong temperature dependence.

After annealing at 1100 °C (Fig. 6a) a distinct step-terrace topography has developed, but the surface still appears to be comparatively rough. A variety of different step edges can be seen. Straight steps separating larger terraces run mostly along $[\bar{1}101]$, which results in $(1\bar{1}02)$ facets at the step edges. The smaller terraces exhibit a more rugged geometry. These small terraces as well as defects on the large terraces have step edges along $[1\bar{1}00]$, [0001] and high-index directions. The width of the terraces between $(1\bar{1}02)$ facets ranges from 80 to 120 nm and the facet height varies between ~ 0.95 and 2.4 nm. This corresponds to steps of 4 - 10 monolayers if we take the distance of 0.24 nm between two subsequent oxygen layers as a monoatomic step. The smaller terraces on top of the larger ones are in the order of 50 nm wide with varying step heights in the range of a few atomic layers. Although the $(1\bar{1}02)$ steps are most prominent because of their relatively long, straight segments, they clearly represent a minority species as compared to the total amount of steps.

Heating this sample to higher temperatures causes a significant smoothening of step edges and terraces. The fraction of the step edges formed by $(1\bar{1}02)$ facets increases. The same is true for steps along $[1\bar{1}00]$, most likely terminated by (0001) planes. On the other hand, the number of the small, rugged shaped terraces is strongly reduced. The average terrace width between $(1\bar{1}02)$ facet steps remains unchanged as compared to the surface annealed at 1100 °C, but their height now ranges from 8 - 18 monoatomic steps. The increased step height goes along with the removal of smaller steps, reducing the overall step density.

This tendency is illustrated in Fig. 6b, recorded on a surface which has been annealed to 1400 °C. The overall topography of the surface is clearly dominated by $(1\bar{1}02)$ facetted steps, enclosing terraces of 80 - 120 nm width. Both steps and terraces exhibit few defects only. The step height is in the same range as mentioned above, between 8 and 18 atomic layers. Furthermore, small steps along $[1\bar{1}00]$, at an angle of 57.6° with respect to the dominating $(1\bar{1}02)$ steps,

a

b

Fig. 6: Large scale AFM image (900 nm × 900 nm) of the $(11\bar{2}0)$ surface prepared by a) annealing at 1100 °C and b) 1400 °C for 12 h in air.

are resolved on these terraces. These steps along $[1\bar{1}00]$ now mostly extend over the entire width of the large terraces, subdividing these into smaller terraces. The height of these steps is between 2 and 4 atomic layers.

After annealing the crystal to 1600 °C, the highest temperature we investigated, the surface now looks almost perfect. Except for straight $(1\bar{1}02)$ facets only very few defects are found. Also steps along $[1\bar{1}00]$ are almost completely absent. Interestingly, the $(1\bar{1}02)$ facets are found to have a well defined step height of 12 atomic layers now. A few 'defects' exhibit step heights of 24 atomic layers and can be considered as double steps. The physical origin of this preferential step height is unknown so far and can only be speculated on. Most plausibly it is related to a reconstruction of the $(1\bar{1}02)$ facets [13,23] stabilizing these particular step heights. This almost defect-free surface appears to be the equilibrated state of the slightly misoriented α-Al_2O_3 $(11\bar{2}0)$ surface at 1600 °C. It basically exposes two planes, namely $(11\bar{2}0)$ planes on the terraces and $(1\bar{1}02)$ planes of uniform width at the facets. These results, the finding of preferentially formed $(1\bar{1}02)$ facets on this surface, in addition to the $(11\bar{2}0)$ terraces, is in excellent agreement with earlier electron microscopy data and with surface free energy calculations. Based on those calculations the $(1\bar{1}02)$ belongs to one of the most stable surfaces [21]. The authors of one electron microscopy study reported the observation of steps along $[1\bar{1}00]$, $[\bar{1}101]$ and $[2\bar{2}01]$ on annealed, vicinal $(11\bar{2}0)$ surfaces [10]. Hsu et al. [24] found vertical walls of $(1\bar{1}0l)$ type on these samples. These reports agree very well with our observations, supporting our assignment of vertical $(1\bar{1}02)$ walls to the predominant steps along $[\bar{1}101]$, which clearly could not be done on the basis of AFM measurements alone because of the finite tip radius.

The atomic structure of the $(11\bar{2}0)$ surface and possible reconstructions have been investigated both by LEED and electron microscopy, but most probably because of the more complex structure of this plane a consistent picture has not been developed. In an early LEED study Chang reported a 'complex' diffraction pattern from samples annealed to 1400 °C in UHV [13]. In a more recent work Schildbach et al. found a (12×4) reconstruction on surfaces oriented to within 1° after similar treatment [25]. Samples exhibiting a larger misalignment showed a more complex diffraction pattern resembling the observations of Chang. Susnitzky et al. [10] reported weak reflections in electron diffraction patterns of thin $(11\bar{2}0)$ foils annealed to 1400 °C and attributed them to a (1×2) periodicity on the surface. It was speculated that this structure might be induced by the presence of contaminants such as Si or Ca, and not necessarily intrinsic to α-alumina. Hsu et al. [7] on the other hand reported a (4×1) periodicity on a 1400 °C heat-treated $(11\bar{2}0)$ single crystal. They succeeded to directly image this superstructure which shows a periodicity of 1.73 nm in the [0001] direction by REM. No periodic structure was resolved in the $[1\bar{1}00]$ direction, normal to the [0001]. Based on the observed periodicity of 1.73 nm, four times the spacing between neighboring oxygen rows, and the absence of any periodicities in other directions, this surface structure was attributed to a one-dimensional (4×1) reconstruction. A structural model of this reconstruction or a mechanism for its formation have not been proposed so far.

High resolution AFM images taken on the $(11\bar{2}0)$ surface show the development of a periodic structure on the large terraces already after annealing at 1300 °C. At this temperature the terraces are still atomically rough. Nevertheless straight lines along $[1\bar{1}00]$ can be observed with a periodicity of ~ 1.7 nm in the [0001] direction. After 1400 °C annealing this reconstruction is fully developed. The 15 nm × 15 nm AFM scan in Fig. 7 clearly shows the 1.73 nm periodicity as dark lines present on most parts of the terraces and extending to the small $[1\bar{1}00]$ steps on the terraces. Upon more careful inspection one finds that the brighter areas between these lines are actually made up by three protruding rows with two equally spaced, less prominent deeper lines in between. These protruding rows are most plausibly attributed to the close-packed rows of oxygen anions along $[1\bar{1}00]$. This assignment is corroborated by the finding of a second corrugation along the rows with a periodicity of 1.1 nm, which perfectly matches the fourfold interatomic distance between oxygen anions along the rows. These extra periodicities must be associated with the surface reconstruction.

The periodicities of the surface corrugation of 1.1 nm and 1.73 nm, respectively, lead to a (4×4) structure of the reconstruction with a unit cell of 3.28 nm × 5.6 nm. The larger size of the unit cell as compared to the observed corrugation results from including the $(11\bar{2}0)$ projected bulk unit cell. Most likely this structure is identified with the (12×4) reconstruction reported recently [25].

The fact that we observe three equally broad sub-units of 0.57 nm width within the surface unit cell, while in the bulk-truncated $(11\bar{2}0)$ surface the oxygen rows in $[1\bar{1}00]$ direction are separated by 0.43 nm, suggests a complex surface structure involving significant rearrangement from a bulk-truncated surface model. In the moment it can only be speculated on the atomic structure of this phase. A model might be constructed e.g. by removing one of the five oxygen rows within the 1.73 nm period. Further relaxation in the Al^{3+} and O^{2-} positions in the topmost layer(s) are required to account for the observed extra periodicities

Fig. 7: High resolution AFM image (15 nm × 15 nm) of the '(4×1)' reconstructed $(11\bar{2}0)$ surface, recorded on a sample annealed at 1400 °C.

and to change the distance between adjacent oxygen rows from the bulk value to the observed 0.57 nm.

3.3 Ag on α-Al_2O_3 (11$\bar{2}$0)

Ag/Al_2O_3 is well known as an industrial catalyst for selective oxidation of ethylene [26]. As mentioned before, the alumina samples investigated here are intended to serve as well defined, atomically smooth support for Ag clusters in a Ag/Al_2O_3 model catalyst. In first experiments along this line we used α-Al_2O_3 (11$\bar{2}$0) samples annealed to 1400 °C as substrate material. Silver was deposited under UHV conditions on a room temperature substrate, at a rate of ~ 0.2 ML/min, which was then transferred into the AFM. As expected from the large difference in surface free energies Ag grows in a Volmer-Weber growth mode and does not wet the surface. After deposition at room temperature a large number of small clusters is found, which are evenly distributed over the surface. However, with the AFM operated in the repulsive mode and with the applied force of 0.1 nN the islands are easily pushed away from the terraces by the AFM tip, until they lock in at ascending steps. Therefore the sample was annealed for 10 min at 300 °C after deposition. This leads to larger, more stable Ag clusters which are less susceptible to tip induced displacements. They can be imaged under the above conditions, although after repeatedly scanning the same area the cluster distribution changes noticeably. The size and the distribution of Ag clusters following this treatment is shown in two AFM images (400 nm × 400 nm) in Fig. 8. The deposition time in these cases was 5 and 15 minutes, equivalent to coverages of 1 ML and 3 ML, respectively. The even distribution of Ag clusters on the flat terraces, similar to that observed directly after deposition,

a

b

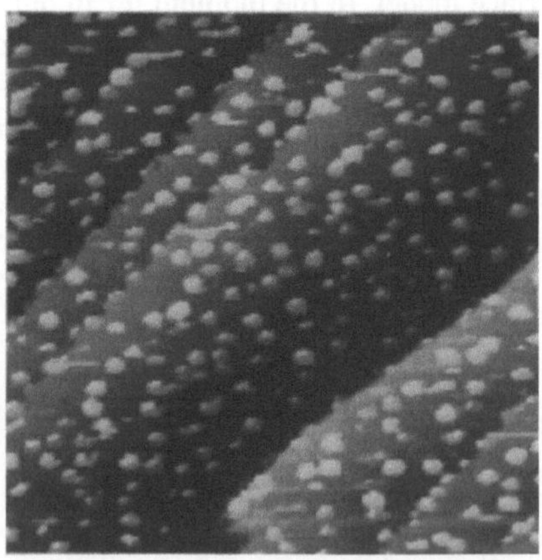

Fig. 8: Large scale AFM image (400 nm × 400 nm) of Ag covered (11$\bar{2}$0) substrates which had been annealed at 1400 °C before deposition. Ag was deposited for a) 5 min and b) 15 min at room temperature. Both samples were annealed in UHV for 10 min at 300 °C after deposition.

indicates homogeneous nucleation on these terraces. In addition, heterogeneous nucleation leads to step decoration. The cluster density of 2×10^{11} cm^{-2} is identical on both surfaces. The clusters have hemispherical shapes. Their average diameter grows from 8 nm at the lower Ag coverage to 12 nm for the higher one.

Finally we began to investigate the chemical stability of the Ag clusters. First results show that after annealing the Ag covered samples in an oxygen atmosphere, at temperatures around 250 °C and pressures between 10^{-6} mbar and 1 bar, the clusters are localized and cannot be moved by the tip anymore. Clearly more experiments have to be done to allow more quantitative conclusions on these effects. These experiments are currently in progress.

4. Conclusions

The structural evolution of clean (0001) and $(11\bar{2}0)$ α-Al_2O_3 surfaces upon annealing, and the growth and stability of deposited Ag clusters as a model system for Ag/Al_2O_3 catalysts were investigated by AFM. The clean surfaces are shown to exhibit regular step-terrace topographies after annealing at temperatures between 1000 °C and 1600 °C. Hence, the surface near regions, which have been amorphized during polishing, are recrystallized even at the lowest annealing temperature.

For the (0001) surface monoatomic rough steps with undefined step geometry are found after annealing at 1000 °C, while after annealing at 1400 °C facetting is observed, with characteristic $(11\bar{2}0)$ and $(\bar{2}110)$ facets of 6.5 to 8 nm height between (0001) terraces. On a 2° misoriented $(11\bar{2}0)$ surface annealing leads to increasingly perfect, facetted surfaces, which are dominated by a periodic sequence of $(1\bar{1}02)$ facets and wide terraces. High resolution AFM images resolve atomic corrugations, which perfectly agree with a (1×1) structure on the (0001) surface and a (4×4) reconstruction on the $(11\bar{2}0)$ surface.

Ag deposited on smooth, 1400 °C annealed $(11\bar{2}0)$ substrates grows in three-dimensional clusters, i.e. in a Volmer-Weber growth mode. Homogeneous nucleation causes a statistic cluster distribution on the flat terraces, in addition, heterogeneous nucleation leads to step decoration. Annealing in oxygen is found to stabilize the Ag clusters against tip induced displacement, which is easily possible in the case of clean Ag clusters.

Acknowledgements

Financial support for this work came from the Deutsche Forschungsgemeinschaft via SFB 338 in the initial phase of the project. GB gratefully acknowledges a fellowship from the Studienstiftung des Deutschen Volkes, Bonn, and JH was supported by the US Department of Energy, Office of Basic Energy Sciences.

References

[1] V.E. Henrich: Rep. Progr. Phys. **48, 1481** (1985)

[2] R.M. Jaeger, H. Kuhlenbeck, H.-J. Freund, M. Wuttig, W. Hoffmann, R. Franchy, H. Ibach: Surf. Sci. **259**, 235 (1991)

[3] see articles of H. Neddermeyer, G.A. Somorjai, H.-J. Freund et al., and of D.W. Goodman in this volume

[4] H. Kahn in: "Handbook of Thin Film Technology", eds. L.I. Maissel, R. Glang: McGraw-Hill, New York 1970, p. 10-1

[5] J.C. Zomerdijk, M.W. Hall: Catal. Rev. Sci. Eng. **23**, 163 (1981)

[6] L.-M. Peng, J.T. Czernuszka: Surf. Sci. **243**, 210 (1991)

[7] T. Hsu, Y. Kim: Surf. Sci. **258**, 119 (1991)

[8] Y. Kim, T. Hsu: Surf. Sci. **258**, 131 (1991)

[9] Y. Kim, T. Hsu: Surf. Sci. **275**, 339 (1992)

[10] D.W. Susnitzky, C.B. Carter: J. Am. Ceram. Soc. **69**, C-217 (1986)

[11] J.M. Charig: Appl. Phys. Lett. **10**, 139 (1967)

[12] C.C. Chang: J. Appl. Phys. **39**, 5570 (1968)

[13] C.C. Chang: J. Vac. Sci. Technol. **8**, 500 (1971)

[14] T.M. French, G.A. Somorjai: J. Phys. Chem. **74**, 2489 (1970)

[15] E. Gillet, B. Ealet: Surf. Sci. **273**, 427 (1992)

[16] M. Gautier, J.P. Duraud: L.P. Van, M.J. Guittet, Surf. Sci. **250**, 71 (1991)

[17] R.C. Barrett, C.F. Quate: J. Vac. Sci. Technol. **A8**, 400 (1990)

[18] M.D. Antonik, R.J. Lad: J. Vac. Sci. Technol. **A10**, 669 (1992)

[19] J. Wiechers: Ph.D. thesis, Universität München (1993)

[20] W.E. Lee, K.P.D. Lagerlof: J. Electron Microsc. Tech. **2**, 247 (1985)

[21] P.W. Tasker: Adv. Ceram. **10**, 176 (1984)

[22] D.W. Susnitzky, C.B. Carter: J. Mater. Res. **6**, 2403 (1991)

[23] M.A. Schildbach, A.V. Hamza: Surf. Sci. **282**, 306 (1993)

[24] T. Hsu, Y. Kim: Ultramicroscopy **32**, 103 (1990)

[25] M.A. Schildbach, A.V. Hamza: Phys. Rev. **B45**, 6197 (1992)

[26] R.A. van Santen, H.P.C.E. Kuipers: Adv. Catal. **35**, 265 (1987)

STM Characterization of Metaloxide Films

Th. Bertrams, A. Brodde, H. Hannemann, C.A. Ventrice Jr., G. Wilhelmi, and H. Neddermeyer

Institut für Experimentalphysik der Ruhr-Universität Bochum,
Postfach 10 21 48, D-44780 Bochum, Germany

Abstract. We describe our efforts to prepare thin ordered metaloxide layers on metal substrates and to characterize them by means of scanning tunnelling microscopy. The primary goal of the study is to obtain oxide layers which are thin enough to allow tunnelling of substrate (tip) electrons yet are sufficiently thick to develop and represent the surface of a bulk oxide. The methods used for thin film oxide growth were controlled oxidation of metal single-crystals (Ni(100) and NiAl(110)) and deposition of Ni in an O_2 atmosphere on Au(111). These techniques result in the growth of ordered layers of NiO(100), Al_2O_3 and NiO(111) or three domains of NiO(100), respectively.

1. Introduction

Thin ordered oxide layers and deposited metal clusters on these oxides are becoming increasingly important as model substrates for adsorption and reaction studies of molecules (see other papers in this volume). Since the structural properties of the oxide surfaces play an essential role for their activity, one is interested in details of their atomic arrangement and the general growth mode of the oxide films. Eventually, it might also be possible to identify the adsorbed species themselves on these model surfaces.

In the present work, we describe our attempts to prepare thin, ordered oxide films by controlled oxidation of metal single-crystals and by deposition techniques and to characterize them by scanning tunnelling microscopy (STM). The application of STM to oxide films is not straightforward because of the existence of a large band gap at the Fermi level E_F. In certain cases it is possible to obtain sufficient conductivity by the presence of atomic defects which are introduced by annealing or by other systematic deviations from the ideal stoichiometry. In some cases the use of a large sample bias voltage allows tunnelling from the oxide valence or conduction band electrons. However, for a thin oxide film the majority of the tunnelling current results from the metal substrate which, for high sample bias, can lead to a very large sample-tip distance which prevents one from obtaining atomically resolved structures with the STM. On the other hand, the lateral modulation of the potential barrier by the thin oxide film should also provide information on the oxide structure. Moreover, the use of STM in its spectroscopical mode rather than atomic force microscopy (AFM, see article by R.J. Behm in this volume) should in principle lead to information on the electronic structure.

Previous STM experiments on metal oxide surfaces (the list may not be complete) have been reported by Brown and You on oxidized Cr [1], by Hanrieder et al. on a Pt covered Ga_2O_3 thin-film supported catalyst [2] and by Rohrer et al. on a reduced $TiO_{2-x}(110)$ surface [3]. The STM investigation of oxidized metal powders by Kwetkus and Sattler should also be mentioned in this context [4] as well as a recent study of Fe_3O_4 [5] (the latter system is conducting at room temperature, however). Surprisingly, in Ref. [1] atomically resolved images are described, which have been acquired in air and could be explained by the presence of an ordered Cr_2O_3 layer on the surface of the Cr crystallites. No special preparation techniques have been found necessary to obtain these results. It is clear that for further adsorption experiments, a cleaning procedure of the surface under UHV conditions is necessary. It would be an interesting question, whether on Cr covered with a natural oxide layer similar STM results may be obtained in UHV.

The present report provides an overview (i) on our recent experiments on the initial oxidation stages of Ni(100) where by heat treatment after oxidation a thin NiO(100) film may be formed, (ii) on the growth of NiO(111) and three domains of Ni(100) by deposition of Ni in an O_2 atmosphere on Au(111) and (iii) of the controlled oxidation of NiAl(110) where a thin ordered layer of Al_2O_3 may develop on the surface. In addition to the measurements of the oxidized surfaces, we have also studied the subsequent growth of a metal adsorbate (Ag) on the oxide/metal substrate systems. The deposition of Ag not only allows a further characterization of the oxide layers but also represents an attempt to model a metal/oxide catalyst.

2. Experiment

The instrumentation used for the present experiments consists of a large UHV chamber containing the STM and facilities for sample preparation (sputtering, annealing, gas exposure, metal deposition by effusion cells) and characterization (Auger electron spectroscopy (AES), ion scattering spectroscopy (ISS) and low-energy electron diffraction (LEED)) [6]. The base pressure of the system is in the low 10^{-11} mbar range and during operation of the Ni and Ag effusion cells better than a few 10^{-10} mbar. In particular, the base pressure during Ni deposition is by two orders of magnitude better than it was in a previous LEED/photoemission experiment where we investigated the possibility of growing epitaxial NiO films by evaporation of Ni in an O_2 atmosphere onto substrates as Ag(100) and Au(111). Those substrates approximately match the lattice parameter of NiO. In this experiment we succeeded in the preparation of thin ordered films of NiO(100) and NiO(111) [7]. Since in the present work we were unable to reproducibly prepare NiO(111) (see below) by this method, we also started with an investigation of the influence of the residual gas atmosphere on the growth mode of NiO. For example, we studied the effect of simultaneous adsorption of H_2O (the partial pressure was 1...4·10^{-8} mbar) in addition to an O_2 partial pressure of 10^{-6} mbar.

The metal substrates (Ni(100), Au(111) and NiAl(110)) are treated in situ by the usual sputter/anneal cycles, until LEED showed the expected pattern of the clean surfaces and in AES contamination could no more be detected. In STM the atomic corrugation of the metal substrates could be obtained after adequate tip formation. The metal deposition (Ni and Ag) was performed with a rate of approximately 0.3 monolayers (ML) per minute as measured with quartz crystal oscillators mounted on the shutters of the effusion cells.

In the STM measurements constant current topographies (CCT's) and local I/U characteristics have been recorded for sample bias voltages between -10 and +10 V and a tunnelling current in the range of 1...10 nA. The presence of O_2 during the oxidation or adsorption experiments produced oxide particles (also on the tip), which often affected the spatial resolution by instabilities of the tunnelling current or sometimes by insufficient sharpness of the tip. In particular, the measurement of local I/U characteristics turned out to be difficult and are not shown in the present work. To demonstrate the electronic effects in the images, we show and discuss CCT's obtained in a wide range of sample-bias voltages U where the contribution of the oxide electronic structure may be extracted at least qualitatively.

3. Results: NiO(100) and NiO(111)

The growth of epitaxial NiO(100) on Ni(100) by O_2 exposure and subsequent annealing has already been observed by Saiki et al. [8] and later studied in more details by Wang et al. [9]. By combining spot analysis LEED (SPALEED) with STM results, it has been shown that the NiO(100) layers normally grow in form of NiO(100) crystallites, which are tilted by approximately 8° against Ni(100) [10]. From the large width of the NiO diffraction spots it could be concluded that the NiO(100) crsytallites grow in various azimuthal orientations. The large lattice mismatch between NiO and Ni is the main reason for the observed misorientation. The actual growth pattern is determined by the step orientation and distribution on the Ni(100) single-crystal.

In Fig. 1 we illustrate the initial stage of NiO formation, i.e., the nucleation on Ni(100) by using STM. The clean substrate surface has been exposed at room temperature (RT) to 10 Langmuir (L, 1 L = 10^{-6} Torr s) of O_2. The main difference from the clean surface is the appearance of a large number of small bright spots (corresponding to protrusions) with a lateral size of a few tens Å or less and a height in the order of a few Å which correspond to the NiO nuclei. Atomically resolved images of surfaces prepared in the same way additionally exhibit ordered adsorbate structures in the p(2x2) symmetry. The LEED pattern at this stage is of the c(2x2) type which can be explained by a specific domain structure of the

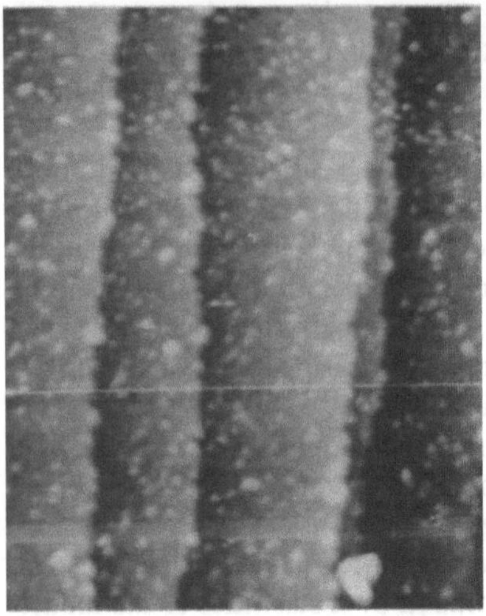

Fig. 1: CCT from Ni(100) which has been exposed to 10 L of O_2 at RT. U = 1 V (sample positive), I = 0.5 nA and area 650x750 $Å^2$.

p(2x2) adsorbate [11]. The formation of both an ordered O adsorption structure and of NiO nuclei is characteristic for O_2 exposure at RT. A closer examination of the spatial distribution of NiO nuclei shows that the largest nuclei are found at the step edges (see also triangular feature on the right bottom edge of the CCT) which led us to conclude that the subsequent growth of NiO(100) crystallites by annealing proceeds from the step edges.

In Fig. 2 we demonstrate how, for further oxygen exposure (50 L), these NiO nuclei show the tendency to form two-dimensional (2D) islands on the Ni(100) substrate. The resolution is not very good on the CCT. This means that the islands visible on the image do not yet correspond to NiO(100) crystallites but have been interpreted as a loose agglomeration of interacting NiO nuclei. Very often the islands of NiO nuclei are ligned up on the step edges of the substrate which supports the model of Ref. [10] that the growth of the NiO(100) crystallites

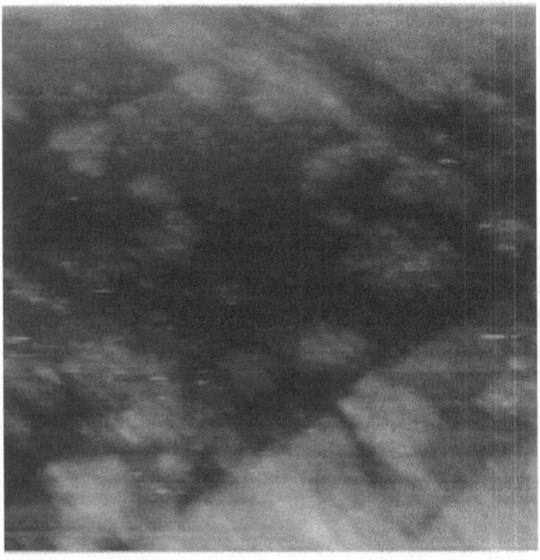

Fig. 2: CCT from Ni(100) which has been exposed to 50 L of O_2 at RT. U = - 0.5 V (sample negative), I = 0.5 nA and area 1000x1000 $Å^2$. The weak line structure visible in the direction of the diagonals is resulting from monoatomic steps.

proceeds from step edges of Ni(100). The LEED pattern of the surface shown in Fig. 2 still exhibits the c(2x2) symmetry.

After an exposure of 5000 L at RT and subsequent annealing to 270 °C, the NiO(100) layer has been fully developed. LEED at this stage normally shows the sharp substrate spots and those from NiO(100) which are relatively broad and tilted against the surface normal. CCT's (see Ref. [11]) exhibit a more or less regular arrangement of small islands with a typical lateral extension of 50 Å, which are sometimes aligned along [110]-like directions. However, the overall corrugation of such surfaces is in the order of 100 Å and therefore much larger than that found on the clean substrate. We were unable to resolve periodic atomic structures on the NiO crystallites.

To overcome the problem of the large lattice mismatch of NiO and the Ni substrate, we tried growing NiO by deposition of Ni in a O_2 atmosphere onto Ag(100) and Au(111) where the lattice mismatch of NiO(100) and NiO(111), respectively, is only about 2%. From photoemission and LEED results, it could be concluded that NiO(100) and NiO(111) layers with good order and which cover the metal completely can be grown by this method [7]. However, in case of the Au(111) substrate, the same experiment performed in the STM apparatus mostly led to three domains of NiO(100) instead of NiO(111) as was concluded from the LEED pattern. We are presently investigating the possible influence of the residual gas atmosphere on the growth mode of NiO. Freund et al. have recently discussed the presence of OH species on NiO(111)/Ni(111) and their role for stabilization of NiO(111) surfaces [12].

In Fig. 3 we show a three-domain NiO(100) structure on Au(111) for small coverages. Nominally 0.45 ML of Ni (referred to a dense-packed layer of Ni atoms) has been deposited under a simultaneous exposure with 60 L O_2, the substrate being at RT. NiO islands are formed, and they cover approximately 1/2 of the substrate surface where the "herringbone" 22x3 structure is still present on the uncovered parts (barely visible in the reproduction). The islands are aligned through the "herringbone" pattern, and their arrangement (but not their shape) is very similar to that of Ni islands on reconstructed Au(111) [13]. LEED of the surface showed in addition to the 1x1 spots of Au(111) the characteristic three-domain NiO(100) structure. Accordingly, on the NiO islands three systems of

Fig. 3: CCT from 0.45 ML Ni and 60 L O_2 on Au(111). U = -1 V, I = 0.5 nA and area 500x300 $Å^2$.

lines are recognized. Their separation of nearly 15 Å and orientation are consistent with the growth of three domains of NiO(100) on Au(111).

For RT deposition the NiO(100) islands seem to coalesce with increasing coverage. The overall structure remains quite rough, however, probably due to the lateral size of the domains which is tied to the 22x3 reconstruction of the substrate. LEED of 2.25 ML Ni and 500 L O_2 on Au(111) (RT) shows only weak substrate spots but a distinct three-domain NiO(100) pattern. Annealing such a sample slightly (approximately 200 °C for 5 min) leads to essential changes of the surface morphology. The surface becomes rather smooth and the oxide layer essentially follows the original terrace structure of the substrate. In Fig. 4 an overview image is reproduced where in the oxide film three levels may be recognized: the main terraces, holes in these terraces, and additional protrusions. Note that an unusually high sample bias voltage of 9 V had to be used for this measurement. Since it was impossible to obtain stable tunnelling conditions for small voltages, this might be an indication that a reduction of the defect-related conductivity has resulted from the annealing process.

4. Results: Al_2O_3

Recent work by using LEED, electron energy loss spectroscopy, and photoelectron spectroscopy has shown that a well ordered thin Al oxide layer may be grown on a clean NiAl(110) surface [14]. The procedure used was to anneal the clean surface in an O_2 atmosphere ($3 \cdot 10^{-6}$ mbar with an exposure of typically 1000 L) at 350 °C and with further annealing at approximately 900 °C in UHV. A complicated LEED pattern with sharp spots is obtained. This may be explained by the superposition of two nearly rectangular (88.7°) reciprocal lattices due to two possible domains of unit cells with a size of 10.6x17.9 Å^2 which are rotated by about 24° against the [110]-like directions. In Fig. 5 a CCT is reproduced, which has been measured on a surface not yet fully covered with the Al_2O_3 film. The presence of two different surface structures on the image is apparent. The left

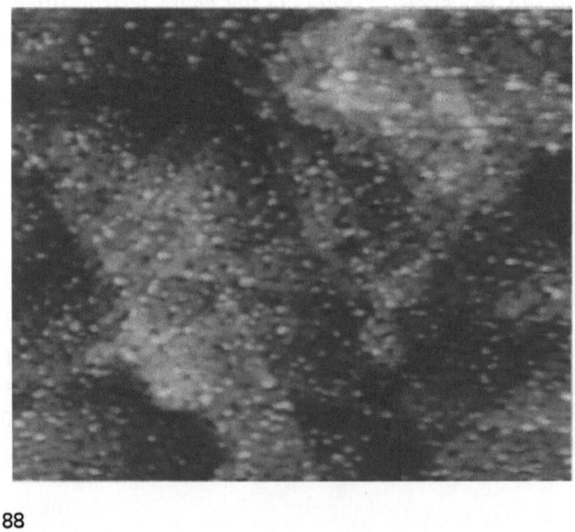

Fig. 4: CCT from 2.25 ML Ni and 500 L O_2 on Au(111) with subsequent annealing to 200° C for 5 min. U = 9 V, I = 2 nA and area 2500x2000 Å^2.

Fig. 5: CCT from partly oxidized NiAl(110). U = 4 V, I = 2 nA and area 500x500 Å^2. The CCT is shown in a three-dimensional representation, which uses illumination by light sources. The dark band in the upper part of the terrace corresponds to a monoatomic step with a height of 2 Å.

lower part of the sample showing a line structure is covered with the thin ordered oxide layer, while the rest corresponds to the clean NiAl(110) substrate. The 2D growth mode of Al_2O_3 is clearly established by these measurements. The step height (as seen by the tunnelling tip) between the clean (right top corner) and oxide covered part (left bottom corner) of the surface is strongly dependent on the sample bias U. While for small U (a few 0.1 V) the step height is approximately 0 Å, for U = 4 V (Fig. 5) a step height of 3 Å is found which may be explained by the additional contributions from the electronic states of the Al oxide which are only accessible for higher sample bias voltage.

Although for small U the presence of the oxide film does not lead to a general change of the magnitude of the average tunnelling current, the oxide film essentially affects the local corrugation. In Fig. 6 a CCT is displayed which con-

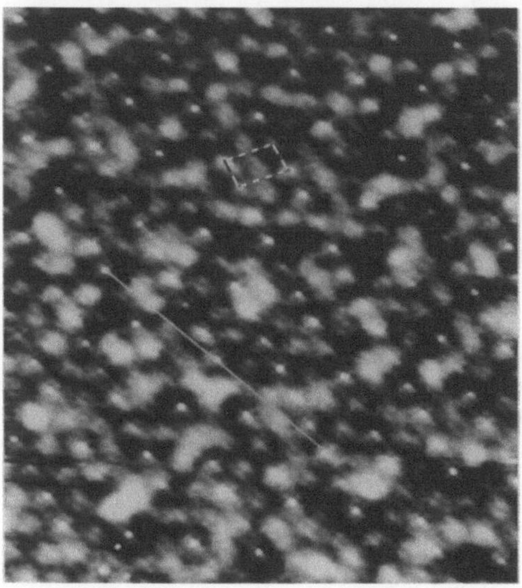

Fig. 6: CCT from Al_2O_3 film on NiAl(110). U = 30 mV, I = 4 nA, area 200x200 Å^2.

tains periodic structures corresponding to the oxide unit cells as indicated by the dashed lines. In particular, a regular arrangement of sharp protrusions reflects the surface mesh of the oxide. However, although the pattern does show atomic resolution, it cannot be explained in a simple or straightforward way by the atomic arrangement in the uppermost layer of the oxide. One reason lies in the fact that for small U tunnelling is accomplished by the substrate electrons, whose spatial distribution is determined by the presence of the oxide surface layer. Therefore, the CCT contains both information on the interface structure of the metallic substrate and the spatial variations of the oxide film. More details of our STM measurements on the Al_2O_3 film will be published elsewhere [15].

Acknowledgement

This work has been supported by the Deutsche Forschungsgemeinschaft through the Forschergruppe "Modellkat".

References

[1] N.M.D. Brown, Hong-Xing You: Surf. Sci. **233**, 317 (1990).
[2] W. Hanrieder, A. Scholz, H. Meixner: Surf. Sci. **243**, 166 (1991).
[3] G.S. Rohrer, V.E. Henrich, D.A. Bonnelle: Surf. Sci. **278**, 146 (1992).
[4] B.A. Kwetkus, K. Sattler: Ultramicroscopy **42-44**, 749 (1992).
[5] G. Tarrach: PhD Thesis, University of Basel (1993); G. Tarrach et al., to be published in Surf. Sci.
[6] Th. Berghaus, A. Brodde, H. Neddermeyer, St. Tosch: Surf. Sci. **184**, 273 (1987).
[7] K. Marre, H. Neddermeyer: Surf. Sci. (to be published).
[8] R.S. Saiki, A.P. Kaduwela, J. Osterwalder, C.S. Fadley: Phys. Rev. **B 40**, 1586 (1989).
[9] W.-D. Wang, N.J. Wu, P.A. Thiel: J. Chem. Phys. **92**, 2025 (1990).
[10] M. Bäumer, D. Cappus, H. Kuhlenbeck, H.-J. Freund, G. Wilhelmi, A. Brodde, H. Neddermeyer: Surf. Sci. **253**, 116 (1991).
[11] G. Wilhelmi, A. Brodde, D. Badt, H. Wengelnik, H. Neddermeyer in The Structure of Surfaces III", eds. S.Y. Tong, M.A. Van Hove, X. Xide, K. Takayanagi, Springer, Berlin (1991) p. 448.
[12] H.-J. Freund et al. (private communication); see also H.-J. Freund, H.Kuhlenbeck, M. Neumann, this volume.
[13] D.D. Chambliss, R.J. Wilson, S. Chiang: J. Vac. Sci. Technol. B **9**, 933 (1991).
[14] R.M. Jaeger, H. Kuhlenbeck, H.-J. Freund, M. Wuttig, W. Hoffmann, R. Franchy, H. Ibach: Surf. Sci. **259**, 235 (1991).
[15] Th. Bertrams, A. Brodde, H. Neddermeyer (to be published).

Partial Valence Densities-of-States of TiO_2 and TiO_{2-x} from Resonant Photoemission and Photoelectron Diffraction

R. Heise and R. Courths

Universität Duisburg, Laboratorium für Festkörperphysik, Lotharstr. 1
D - 47048 Duisburg, Germany

Abstract. The use of resonant photoemission (ResPE), X-ray excited photoelectron diffraction (XPED) and inverse photoemission from $TiO_2(110)$ (rutile) surfaces for the determination of partial (atom-resolved) densities-of-states of valence and conduction electrons is reported. The hybridization between O 2p and Ti orbitals (3d and 4s) extends over the whole valence band. Positions and splitting of the empty Ti 3d states are given. Three occupied defect states with titanium character are detected in oxygen deficient surfaces.

1. Introduction

Because of its optical and electronic properties, the transition metal (TM) oxide TiO_2 (rutile) is used as a substrate in catalytic and electrochemical processes and has applications in gas-sensing devices, in solar cells and in photochemical energy-conversion processes. A great deal of information about the bulk and surface electronic and geometric structure for stoichiometric and defective TiO_2 has already been gathered [1,2], but a number of details have still to be clarified.

In stoichiometric TiO_2 the metal atom is octahedrally coordinated and has, in an ionic model, a nominally $3d^0$ (Ti^{4+}) electronic configuration. The surface can be reduced by ion bombardment or heat treatments to create oxygen-vacancy-defects (TiO_{2-x}) and partially populate Ti 3d states (V_O-Ti^{3+}-complexes) [3-8]. These defects play a decisive role in charge transfer reactions (chemisorption, photodecomposition of water and heterogeneous catalysis). However, TiO_2 (as other TM oxides too) cannot simply be understood as an ionic solid, but has rather strong covalent character [9,10]. A ~ 40% covalent character in the Ti-O bond can be estimated [10] (using Pauling's electronegativity concept), corresponding to ~ 1.6 Ti-like valence electrons (per formula unit). The determination of the amount, the energetic positions and the orbital characters of the cation contributions to the valence states is one of the demands to the experimentator.

We report on the determination of titanium and oxygen partial (atom-resolved) densities-of-states in the valence band energy region (PVBDOS) of stoichiometric and defective $TiO_2(110)$ surfaces with the use of resonant photoemission (ResPE) from Ti valence states (Ti 3p-3d, 4s resonances) [11-15] and of X-ray photoelectron-diffraction (XPED) [16-18] of valence electrons. We also present information about the unoccupied conduction band states as revealed by inverse photoemission (IPE).

2. Experimental

All experiments were carried out with initially slightly reduced (heating in vacuum) $TiO_2(110)$ crystals. A heavily reduced surface was prepared by Ar-ion bombardment (500 eV ion beam for several minutes) and subsequent annealing in UHV (surface R). The "stoichiometric" surface was obtained by subsequent annealing in 10^{-6} Torr O_2 until a good (1x1) LEED pattern was observed (surface S). Residual oxygen vacancies could be diminished by additional oxygen exposure (surface S'). The defect concentration in an 1x1 surface was increased by additional heating the stoichiometric surfaces in UHV (surface R'). Similar preparations are described in literature [3-8]. The PE spectra were taken in an angle-resolved mode and details of the normal-emission ResPE experiments with synchrotron radiation from BESSY and of the XPED experiments with AlK_α can be found in Refs. 15 and 18. Normal incidence IPE spectra were measured using a bandpass detector working at $h\nu$ = 9.7 eV (CaF_2 window and iodine filled Geiger-Müller-counter).

3. Photoelectron Diffraction (XPED) from $TiO_2(110)$

Due to the scattering of the primarily excited photoelectron wave (emitter) on neighbouring atoms (scatterers), the elastic differential scattering cross section for electrons - with energies above some 100 eV - has a strong zeroth-order interference peak in forward direction, which is accompanied by weaker higher-order diffraction peaks (Fig. 1). X-ray excited photoelectron diffraction (XPED) has found useful applications in surface physics because of its sensitivity to short range geometric structures and bond directions of adsorbed molecules [16]. Combining XPED with XPE spectra of valence and shallow core electrons of the constituents of the crystal (Fig. 1) and exploiting the differences in the angular dependences of the individual PE cross-sections it is possible to extract the atom-resolved VBDOS's [17,18].

For $TiO_2(110)$ the diffraction patterns of core electrons depend markedly on the kind of atom [15,18-20]. Figure 1 shows as an example polar scans along the $[\bar{1}10]$ azimuth. The difference in the lattice sites are best seen around 45° polar angle, where Ti 3p exhibits a strong peak which is not present for O 2s. The VB

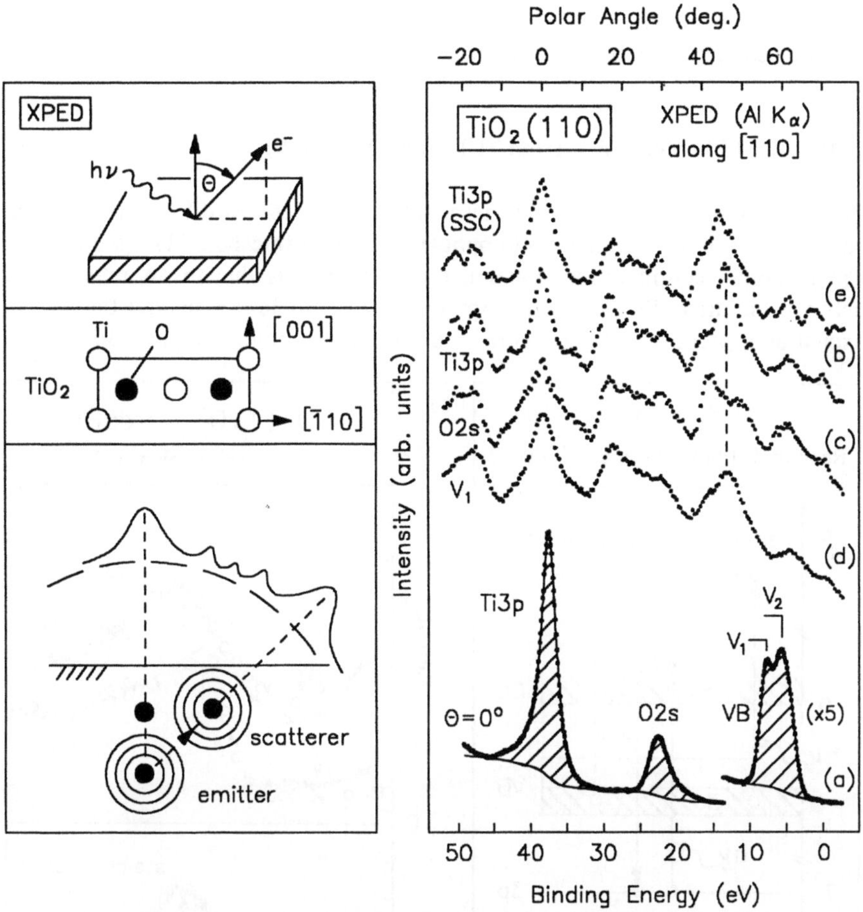

Fig. 1: Left: Experimental geometry, $TiO_2(110)$ surface and XPED scheme. Right: a: XPE spectrum excited along the surface normal of $TiO_2(110)$ with Ti 3p, O 2s and valence band emissions. b-d: Polar scans along the $[\bar{1}10]$ azimuth for the Ti 3p, O 2s and VB peak intensities. e: Theoretical Ti 3p XPED pattern within the SSC model.

emission shows this peak too, which immediately demonstrates the presence of occupied valence orbitals on Ti sites. The determination of PVBDOS's is described in Refs. 17 and 18, and our results for $TiO_2(110)$ [15,18] are presented below.

Figure 1 also shows the result of a single-scattering-cluster (SSC) calculation (this model is described in Ref. 21) for TiO_2 with a bulk truncated surface. The difference between theory and experiment is so small that severe geometrical distortions in the surface region such as shear planes [7,22,23] can be, in agreement with Ref. 19, excluded. Because XPED works with rather long electron escape depths (~ 15 Å) and light elements are weak scatterers, oxygen reconstructions are probably not visible.

4. Resonant Photoemission from TiO_2 and $TiO_{2-x}(110)$

Photoemission from 3d states is resonantly enhanced (ResPE) if the photon energy is swept through the 3p-3d optical absorption threshold [11] (see Fig. 2). Using atomic configuration notation the resonance is due to an interference between the direct PE process (1) $3p^6 3d^n + h\nu \rightarrow 3p^6 3d^{n-1} + e^-$, and the optical excitation process of a core 3p electron into empty 3d states

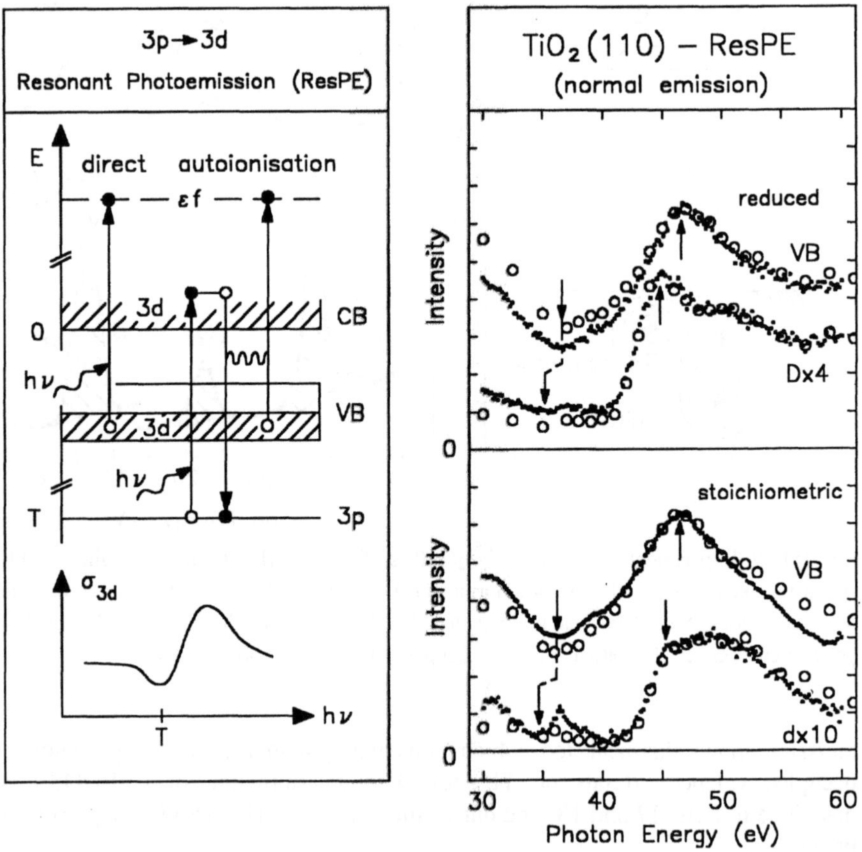

Fig. 2: Left: Energy level scheme for 3p-3d resonant photoemission. Right: Photon energy dependence of the total VB emission intensity from stoichiometric and reduced $TiO_2(110)$ along the surface normal and corresponding data for the defect-state emissions in the band gap (see Fig. 3). Circles represent an energy-integration of the spectra after subtraction of an inelastic background (see Fig. 3). The points represent averages of constant-initial-state curves.

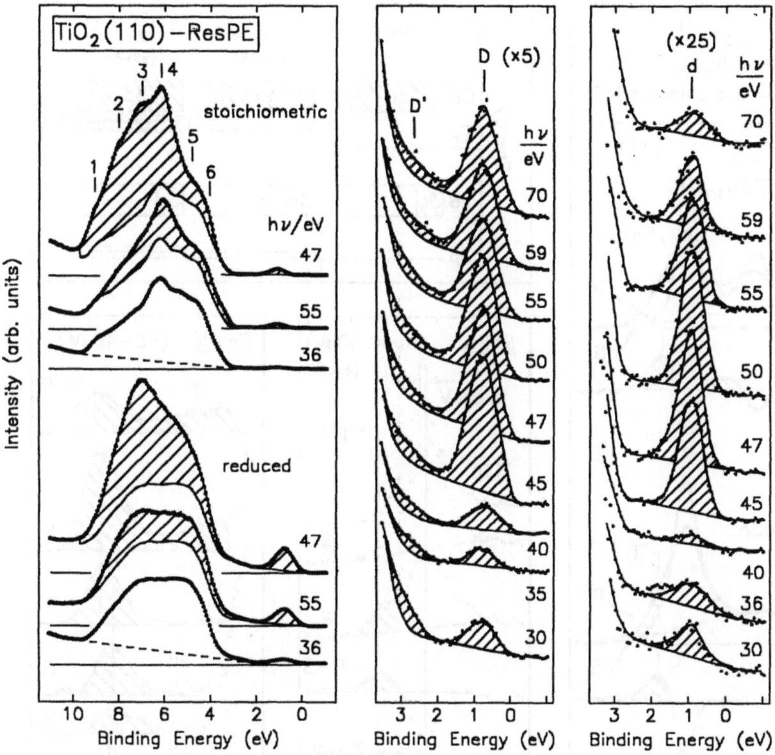

Fig. 3: Left: Normal emission PE spectra from stoichiometric and reduced $TiO_2(110)$ taken for photon energies hν = 36 eV ("off-resonance"), 47 eV ("in-resonance") and 55 eV. For details of the data evaluation see Ref. 15. Right panels: Series of spectra from the oxygen-induced defect states in the band gap for photon energies between 30 and 70 eV.

(2) $3p^6 3d^n + h\nu \rightarrow (3p^5 3d^{n+1})^*$, which autoionizes into a continuum state (3) $(3p^5 3d^{n+1})^* \rightarrow 3p^6 3d^{n-1} + e^-$, which is degenerate with the final state of the direct process. The variation of the 3d cross-section with photon energy shows a Fano-like profile [12] with a minimum at the onset of the absorption process (threshold T) and a delayed maximum. Evidence for the failure of the ionic model ($3d^0$ configuration) for TiO_2 is given by the observation of 3p-3d ResPE [11-15, and references therein]. The occupation number n in Eqs. 1-3 represents the amount of Ti 3d (or 4s) orbital contribution to the valence states [15].

Figure 2 shows the photon energy dependencies of the energy-integrated valence band intensities from stoichiometric (surface S) and reduced (surface R) $TiO_2(110)$. These resonance curves are characterized by a strong maximum at hν = 47 eV (Ti 3p-3d resonance [14,15]), an onset at about hν = 36 eV, which

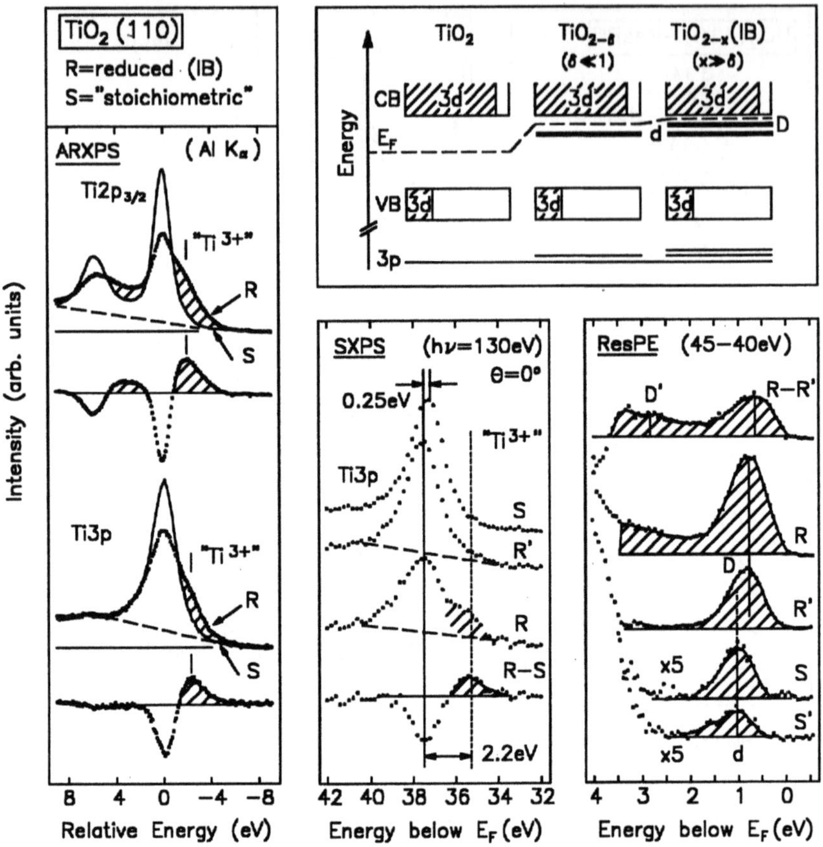

Fig. 4: Normal emission PE spectra from stoichiometric (S and S') and reduced (R and R') TiO_2(110) surfaces. Left: Ti 2p and 3p excited with AlK_α radiation. Center: Ti 3p emissions excited with hν = 130 eV. Right: Resonant photoemission from band-gap defect states. Shown are the differences between in-resonance (hν = 45 eV) and off-resonance (hν = 40 eV) spectra. Above: Energy level scheme.

agrees with the binding energy of the Ti 3p electrons (see Fig. 4), and a shoulder-like structure between 50 and 55 eV (Ti 3p-4s resonance [14]). Selected onset and in-resonance PE spectra are presented in Fig. 3 (see Ref. 14 for spectra recorded in an angle-integrated mode). The 47 eV in-resonance spectrum shows six Ti-derived structures 1-6. The dominant peaks 1-4 at higher binding energy resonate at photon energies between 45 and 47 eV [15], and thus reflect Ti 3d valence states. The weaker low-energy structures 5 and 6 resonate at about 55 eV [15] (in agreement with Ref. 14), thus showing a strong Ti 4s character.

Also shown in Figs. 2 and 3 are the resonances from the oxygen-vacancy induced Ti-like defect states d and D in the band-gap on the surface S and R, respectively. Their resonance profiles show marked differences as compared to those of the valence bands and mutually. This indicates that one is dealing with two different Ti-like states, although their energetic position at 0.8 eV below the Fermi level is identical. Their strong intensity above $h\nu = 50$ eV shows that their orbital character is not pure 3d but has 4s mixed in. We mention that the PE cross section is strongly enhanced between 36 and 47 eV by ~ 14 for d and ~ 8 for D. These values indicate a different orbital mixing in these states.

Reduced Ti species are visible as shoulders at the low binding energy side of core emissions (XPE) from the unreduced species (Fig. 4). A detailed inspection of such core spectra [6] has revealed three Ti defects with binding energies comparable to Ti in Ti_2O_3 ("Ti^{3+}"), TiO ("Ti^{2+}") and metallic Ti ("Ti^{0}"). Henrich et al. [3] have already given evidence for three distinct phases upon surface-defect-concentration. Recent scanning tunneling microscope studies [22,23] have postulated shear-plane-like geometrical defects due to Magneli phases (proposed in Ref. 7). High-resolution Ti 3p spectra excited with soft X-rays ($h\nu = 130$ eV, Fig. 4) show a shift of the "stoichiometric" Ti 3p level (surface S) to higher binding energy by ~ 0.25 eV upon increasing defect concentration (surfaces R' and R). This upward shift of the Fermi level (see insert in Fig. 4) shows that the defect state D has 0.25 eV higher energy than defect state d, which is obvious after correction of the VB spectra for this core level shift. Upon ion bombardment a third defect D' with a more continuous energy distribution is created. The lowered Ti 3p binding energy of the reduced Ti species is also reflected in shifted resonance onset and maximum energies (Fig. 2). The concentration of the defect d on surface S' is estimated to $\sim 5 \cdot 10^{-3}$ (using the resonance curves in Fig. 2) and d is thus probably due to point defects [7].

As representative for the Ti PVBDOS in stoichiometric TiO_2 we take the average of the differences between spectra in resonance ($h\nu = 37 - 60$ eV) and the off-resonance spectrum at 36 eV, which in turn represent the O PVBDOS in good approximation. These data are shown in Fig. 5 and will be discussed in chapter 5.

5. Densities of States

An octahedral ligand field splits the five Ti 3d states into e_g and t_{2g} symmetry states (this symmetry classification is convenient because the distortion of the octahedron in TiO_2 is weak), which interact with the O 2p orbitals through (pdσ) and (pdπ) interaction (see Fig. 5). Electronic structure calculations [9,10] show that the Ti contribution is strongest in the high binding energy region and is mainly of e_g type due to the stronger (pdσ) interaction. The upper part of the low energy region contains only a small cation contribution and is dominated by non-bonding oxygen states. However, it does not seem appropriate to simply

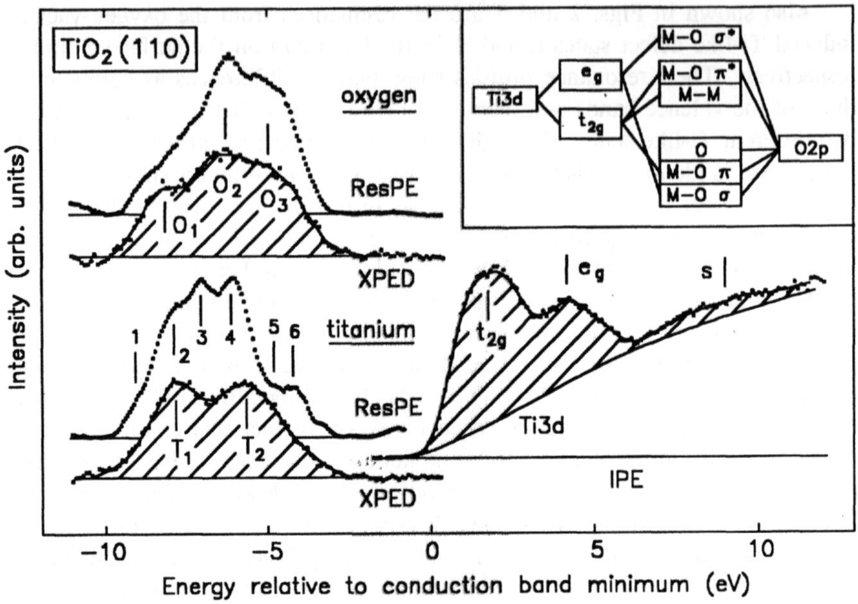

Fig. 5: Valence and conduction band density-of-states for $TiO_2(110)$. Left: Comparison of experimental atom-resolved valence-band densities-of-states derived from resonant photemission (ResPE) and X-ray photoelectron diffraction (XPED). All curves are normalized to the same integral value. Right: Normal-incidence inverse photoemission spectrum representing the unoccupied state-density. The insert shows a molecular-orbital scheme.

interprete the valence bands into energetically separated σ- and π-bonding, overlap and nonbonding regions.

The characteristic properties of the experimental state densities in stoichiometric $TiO_2(110)$ (Fig. 5) are as follows: (i) The hybridization between O and Ti states extends over the whole valence band, which has a width of about 6.5 eV. (ii) The Ti DOS as derived from XPED exhibits two pronounced structures T_1 and T_2 at 7.6 eV and 5.5 eV. (iii) The corresponding ResPE DOS shows six structures 1-6 which partly differ in energy and intensity from XPED. The most striking discrepancies as compared with XPED are the existence of structure 3, which falls into the minimum between the XPED peaks, the large intensity of structure 4, which coincides with T_2, and the pronounced structure 6. From the resonance profiles (Fig. 2 and 3) it is concluded that the dominant structures of the Ti DOS at high BE have 3d character, whereas the structures at lower binding energy have 4s states mixed in (see chapter 4). This orbital mixing might be the reason for the mentioned discrepancies, because XPED and ResPE have different orbital cross sections. In addition, direct transitions and surface effects [24] may also influence ResPE due to the low kinetic energy used. (iv) The XPED and ResPE derived densities for oxygen O 2p are very similar and show a three peak structure (O_1, O_2 and O_3 at energies 8.0 eV, 6.2 eV and 4.9 eV, respectively), and the low BE structure is more pronounced than for the Ti DOS probably due to nonbonding oxygen.

Experiment and theory [9,10] agree in the width of the valence band (~ 6.5 eV) and that the Ti PVBDOS is non-negligible. However, most of the calculations give the Ti 3d contribution to be dominant in the lower half of the VB. From the resonant enhancement of the valence band and the Ti 3d like defect d (chapter 4) the Ti orbital count in the VB is estimated to be about 1.3 [15], and this value is in rather good agreement with the estimate of the covalent character of the Ti-O bond given in the introduction [10]. A detailed comparison between ground-state theory and photoemission derived state-densities demands caution, because in principle it requires the calculation of PE spectra in a one-step model to check whether state-densities in the usual sense are measured or wether orbital effects (e.g. angular and quantum number dependent cross sections) and effects due to the localization of the hole left behind distort the PE spectra.

Information about the densities of unoccupied states is obtained from methods such as inverse photoemission (IPE), electron-energy-loss spectroscopy (EELS) and X-ray absorption spectroscopy (XAS). The normal incidence IPE spectrum displayed in Fig. 5 (compare also with Ref. 25) is dominated by two strong peaks at 2.7 and 4.0 eV above the conduction band minimum, which are obviously associated with the Ti 3d t_{2g} and e_g states, and a low intensity structure at 8.6 eV corresponding to Ti 4s. This interpretation is in line with theory [9,10] and with core level EELS [26] and XAS [27] studies.

Acknowledgements

This work was supported by the Deutsche Forschungsgemeinschaft (DFG Grant No. Co 110/2-1) and by the Bundesministerium für Forschung und Technologie (BMFT Grant No. 05436AAB).

References

[1] V.E. Henrich: Rep. Progr. Phys. **48**, 1481 (1985)

[2] W. Göpel: Progr. Surf. Sci. **20**, 9 (1985)

[3] V.E. Henrich, G. Dresselhaus, H.J. Zeiger: Phys. Rev. Lett. **36**, 1335 (1976)

[4] R.H. Tait, R.V. Kasowski: Phys. Rev. **B20**, 5178 (1979)

[5] V.E. Henrich, R. Kurtz: Phys. Rev. **B23**, 6280 (1981)

[6] W. Göpel, J.A. Anderson, D. Frankel, M. Jaehnig, K. Phillips, J.A. Schäfer, G. Rocker: Surf. Sci. **13**, 333 (1984); G. Rocker, W. Göpel: Surf. Sci. **181**, 530 (1987)
[7] R.G. Egdell, S. Eriksen, W.R. Flavell: Solid State Commun. **60**, 835 (1986)
[8] R.L. Kurtz, R. Stockbauer, T.E. Madey, E. Roman, J. De Segovia: Surf. Sci. **218**, 178 (1989)
[9] P.I. Sorantin, K.H. Schwarz: Inorg. Chem. **31**, 567 (1992)
[10] K.M. Glassford, J.R. Chelikowsky: Phys. Rev. **B46**, 1284 (1992)
[11] L.C. Davis: J. Appl. Phys. **59**, R25 (1986)
[12] K.E. Smith, V.E. Henrich: Phys. Rev. **B38**, 9571 (1988)
[13] J.M. McKay, M.H. Mohamed, V.E. Henrich: Phys. Rev. **B35**, 4304 (1987)
[14] Z. Zhang, S.-P. Jeng, V.E. Henrich: Phys. Rev. **B43**, 12004 (1991)
[15] R. Heise, R. Courths: Solid State Commun. **84**, 599 (1992)
[16] S.A. Chambers: Surf. Sci. Repts. **16**, 261 (1992); Adv. Phys. **40**, 357 (1991); and references therein
[17] A. Stuck, J. Osterwalder, T. Greber, S. Hüfner, L. Schlapbach: Phys. Rev. Lett. **65**, 3029 (1990)
[18] R. Heise, R. Courths: Surf. Sci. (1993), in press
[19] B.L. Maschoff, J.-M. Pan, T.E. Madey: Surf. Sci. **259**, 190 (1991)
[20] U. Bardi, K. Tamura, M. Owari, Y. Nihei: Appl. Surf. Sci. **32**, 352 (1988)
[21] S. Kono, S.M. Goldberg, N.F.T. Hall, C.S. Fadley: Phys. Rev. **B22**, 6085 (1980)
[22] Q. Zhong, J.M. Vohs, D.A. Bonnell: Surf. Sci. **274**, 35 (1992)
[23] G.S. Rohrer, V.E. Henrich, D.A. Bonnell: Surf. Sci. **278**, 146 (1992)
[24] P.J. Moller, M.-C. Wu: Surf. Sci. **224**, 265 (1989)
[25] A.E. Taverner, P.C. Hollamby, P.S. Aldridge, R.G. Egdell, W.C. Mackrodt: Surf. Sci. (1993), in press
[26] R.D. Leapman, L.A. Grunes, P.L. Fejes: Phys. Rev. **B26**, 614 (1982); R. Brydson, B.G. Williams, W. Engel, H. Sauer, E. Zeitler, J. Thomas: Solid State Commun. **64**, 609 (1987)
[27] F.M.F. de Groot, M. Grioni, J.C. Fuggle, J. Ghijsen, G.A. Sawatzky, H. Petersen: Phys. Rev. **B40**, 5715 (1989); L.A. Grunes: Phys. Rev. **B27**, 2111 (1982)

Part III

Adsorption on Oxide Surfaces: Molecules and Ultrathin Metal Films

Clean ZnO Surfaces and their Interaction with Gases

K. Jacobi

Fritz-Haber-Institut der Max-Planck-Gesellschaft,
Faradayweg 4-6, D-14195 Berlin, Germany

Abstract. Low-index surfaces of ZnO have been studied through ARUPS, LEED and TDS. The electronic properties of the clean ZnO (0001)Zn, $(000\bar{1})$O and $(10\bar{1}0)$ surfaces will be reviewed. Time-dependent changes of electron affinity and work function will be explained through oxygen diffusion from the surface into the bulk. Adsorption of D_2O, NO, CH_3OH, C_6H_6 and Xe has been studied in detail. C_6H_6 forms an ordered monolayer of a complicated orientational order. The polar molecules exhibit the strongest bonding to surface Zn sites.

1. Introduction

The electronic and structural properties of ZnO single-crystal surfaces have been reviewed some years ago [1]. The main, strongly correlated, problems are the existence and nature of electronic surface states or resonances, the reconstruction and the preparation of stoichiometric surfaces. From UV photoemission [2-7] it is known that there are no filled surface states in the band gap region of a density high enough to be detectable by this method. Surface states of densities below 10^{12} cm^{-2} have been inferred from surface photoconductivity measurements for surfaces prepared by cleavage [8]. Using the same preparation method Fermi-level pinning was found at 0.3 eV below the conduction band edge for the prism face and at 0.4 eV for the $(000\bar{1})$O face by means of work function (Kelvin method) and surface photoconductivity measurements [9,10]. This pinning can be explained by assuming surface states in agreement with the surface conductivity measurements. About the chemical or structural nature of these low density surface states nothing is known at present [1].

With respect to the structure of the surfaces and the question of a supposed reconstruction the most elaborate study was performed for the $(10\bar{1}0)$ face. From a LEED intensity analysis Duke et al. [11] concluded that the first-layer Zn atoms are vertically displaced downwards by 0.45 ± 0.1 Å, whereas the first-layer O atoms remain on their bulk positions within the accuracy of this method. This was confirmed through model calculations recently [12].

Undoped ZnO crystals exhibit n-type conductivity with the Fermi level lying 0.2 eV below the conduction band minimum. Whereas there is agreement that the bulk n-type conductivity is caused by some excess Zn, it is not known whether the main donor is an oxygen vacancy or an interstitial zinc. To achieve and prove

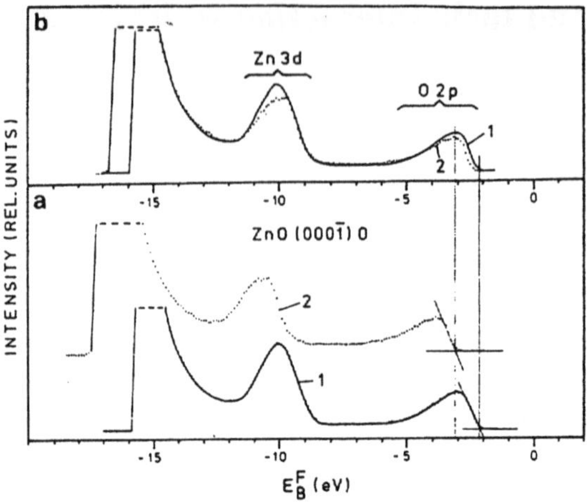

Fig. 1: Angle-resolved photoelectron spectra for the ZnO(000$\bar{1}$)O face at normal emission with a photon energy of $\hbar\omega$ = 21.2 eV. The binding energy $E_B{}^F$ is related to the Fermi level. Curve 1 is measured shortly after annealing at 700 K and curve 2 about 4 h later with a substrate temperature of 300 K. In the lower part (a) curve 2 is displaced only vertically. In the upper part (b) curve 2 is shifted to the right by 0.9 eV. From [6].

stoichiometric ZnO surfaces within an UHV environment is very difficult so far. Many studies therefore start with a cleaved surface using the fact that the nonpolar (10$\bar{1}$0) as well as the polar (0001)Zn and (000$\bar{1}$)O surfaces can be prepared by cleavage. On the other hand, cleaved surfaces seem not to be as ideal as one may anticipate. First, a cleaved surface contains O and Zn fragments which are subliming during the first weak annealing [7], and second, the cleaved surface is undergoing temperature-dependent changes of its electronic properties [10] which are not completely understood.

In the following, we will summarize our own work on low-index ZnO surfaces prepared through Ar-ion bombardment and annealing. From the observation of strong surface resonances in photoemission and several face-specific adsorption properties we conclude that the imperfections left at the surface are not numerous enough to obscure the main face-specific properties of the low-index ZnO surfaces.

2. The Clean Low-Index ZnO Surfaces

2.1 The Oxygen Diffusion Hypothesis

In this chapter we will discuss some observations for the clean low-index surfaces which have been derived from combined ARUPS and LEED measurements [6]. The ZnO samples were cut from crystals grown in vapor phase. It was In-doped in the case of the (10$\bar{1}$0) face. The two polar (0001)Zn and (000$\bar{1}$)O faces were prepared and studied together on a different sample which was annealed in Zn vapor at 1125 K for 26 h to improve the electric conductance. Only samples of high conductance could be cooled to 100 K (40 K) by liquid N_2 (He) and heated by direct current. All samples were cleaned by repeated Ar-ion sputtering

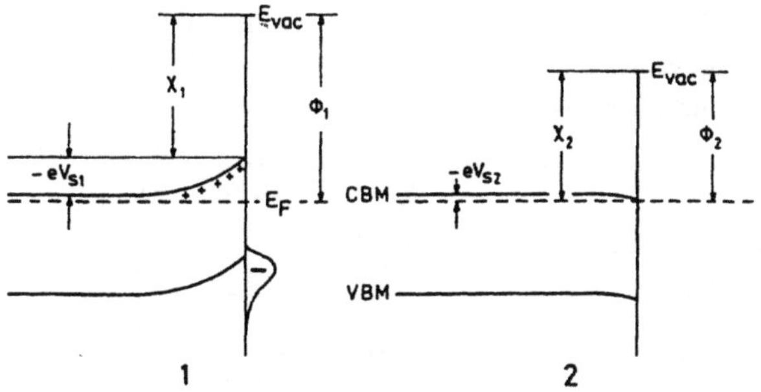

Fig. 2: Schematic sketch for the band bending eV_s of the conduction (CB) and valence band edge (VB) shortly after annealing (1) and 300 min later (2) for the ZnO(000$\bar{1}$)O face. Work function ϕ electron affinity χ are also indicated. From [6].

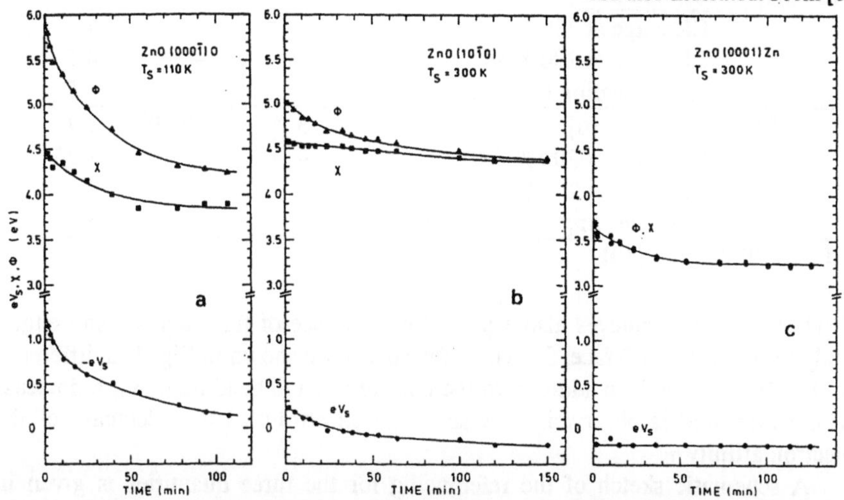

Fig. 3: Change of work function ϕ, electron affinity χ and band bending eV_s with time after annealing for 1 min at 850 K for the (10$\bar{1}$0) face (b) and at 700 K for the polar (000$\bar{1}$)O (a) and (0001)Zn faces (c). Ts is the temperature of the sample holder. From [6].

(500 eV, 3 μA) at 825 K and subsequent annealing at 825 K [(10$\bar{1}$0)] or by Ar-ion sputtering (500 eV, 2 μA) at room temperature and subsequent annealing at 700 K [(0001) and (000$\bar{1}$)].

From ARUPS the absolute value of the work function ϕ can be evaluated. E_{VBM}, the valence band maximum, can be determined by extrapolation as indicated in Fig. 1 which also elucidates how the electron affinity χ and the band bending eV_s can be derived. The binding energy $E_B{}^F$ is referred to the Fermi level. One recognizes a peak at about -3 eV which is due to oxygen 2p derived states at the upper valence band edge. The peak at around -10 eV is due to the Zn 3d states. Curve 2, which was measured about 4 h later, shows that the threshold energy is shifted to the left by 1.6 eV. But also the spectral features from Zn 3d and O 2p and the whole energy distribution curve are shifted to the left. Both curves are plotted together in part b of Fig. 1 with curve 2 shifted to the right by 0.9 eV so that the Zn 3d peaks of both curves coincide. For the (0001)Zn and the

Table 1: Electron affinity χ, band bending eV_s and work function ϕ from this work and from Moormann et al. [9,10]; the values from this work are extrapolated to time zero after ion bombardment (IB) and annealing (see Fig. 3). From [6].

Face		χ (eV)	$-eV_s$ (eV)	ϕ (eV)
$(000\bar{1})$O	IB, 700 K [6]	4.5	1.3	6.0
	Cleavage at 300 K [a]	–	–	4.95
	Cleavage and annealing to 800 K [b]	4.19	0.22	4.64
$(10\bar{1}0)$	IB, 825 K [6]	4.60	0.25	5.05
	Cleavage at 300 K [a]	–	–	4.64
	Cleavage at 290 K [b]	–	–	4.7
	and annealing to 750 K [b]	–	–	4.5
(0001)Zn	IB, 700 K [6]	3.7	–0.20	3.7
	Cleavage at 300 K [a]	–	–	4.25
	Cleavage at 200 K [b]	–	–	4.4
	and annealing to 400 - 700 K [b]	–	–	3.9

[a] From [9]; [b] from [10].

$(10\bar{1}0)$ face one achieves also a perfect coincidence of the valence band edges. Only for the $(000\bar{1})$O face, for which the curves are shown in Fig. 1, a difference of 0.2 eV was found. In addition to the change in band bending $e\Delta V_s$, a decrease of the threshold is observed. This shift can be attributed to a decrease of the electron affinity χ.

A schematic sketch of the referencing for the three quantities is given in Fig. 2. For the absolute positioning it is assumed that the Fermi level is 0.2 eV below the conduction band minimum in the bulk. The results of these measurements are summarized in Fig. 3 and Table 1. There are several arguments which support the assumption that the annealing leads to more perfect, i.e. more stoichiometric and better ordered, surfaces: The structure of the UP spectra gets sharper in a very reproducible way. The maximum work functions ϕ_m are ordered in the sequence $\phi_m(000\bar{1}) > \phi_m(10\bar{1}0) > \phi_m(0001)$. This order is expected for oxygen-containing surfaces under the assumption that the overall charge is negative for the surface oxygen and positive for the surface zinc. Furthermore, the absolute values found are equal or even larger than those found for cleaved surfaces. Finally, in adsorption studies (see below) of H_2O, Xe and methanol as well as isopropanol specific differences are found between the three surfaces. Therefore, the values for the ideal surfaces are evaluated by extrapolating the curves in Fig. 3 to time zero. These values are listed in Table 1 and compared with the data of Moormann et al. [9,10]. For the O face, Moormann et al. [9] have found an upward band bending by 0.22 eV. They argued that this was caused by pinning from surface states lying about 0.45 eV below the conduction band minimum. They found this in good agreement with IR surface photoconductivity measurements of Lüth [8]. From the ARUPS measurements it was concluded that

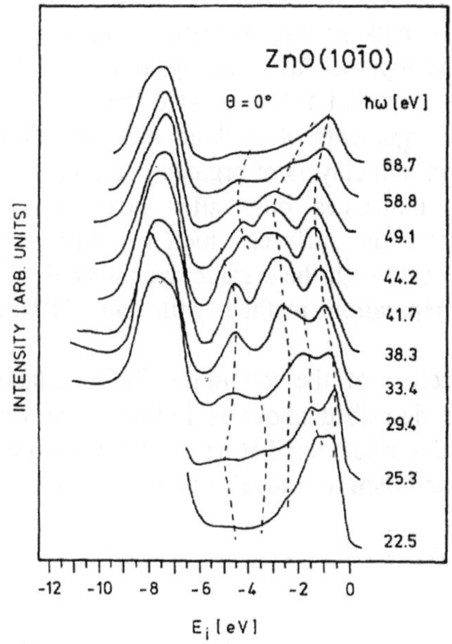

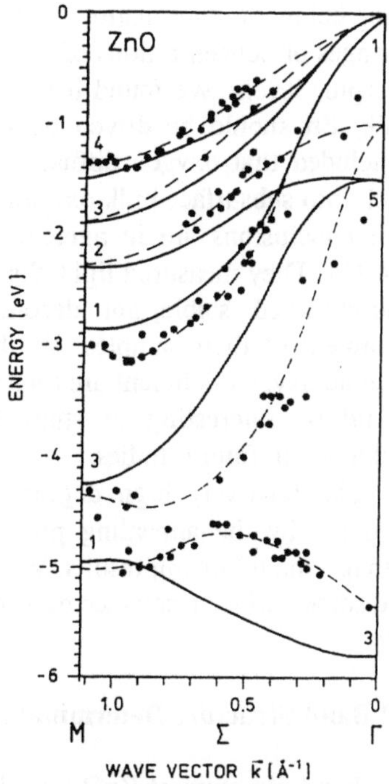

Fig. 4: Angle-resolved UV photoelectron spectra for normal emission (Θ = 0°, i.e. along ΓM). The initial state energy E_i is referenced to the valence band maximum. Parameter is the photon energy ħω as indicated. From [7].

Fig. 5: Band structure for ZnO along the ΓΣM line. The energy is referenced to the valence band maximum. The dots connected by the broken lines are the experimental values. The heavy lines are from the calculation of Chelikowsky [16]. From [7].

these states are not intrinsic surface states but may have been introduced by the cleavage procedure. For the surface photoconductivity measurements it was stated already by Lüth [8] that one cannot clearly distinguish between surface states and near-surface donors. The latter explanation becomes even more likely considering that Lüth has not found any difference between the two polar faces.

With respect to the cleaved surfaces there is another important result to be noted. Kohl et al. [13] have found sublimation of Zn and O within a first heating cycle after cleavage which was detectable at 650 K for the Zn face and 820 K for the O face. During the later annealing cycles only the exponential rise of sublimation rates for both kinds of atoms is observed for T > 1000 K. Thus, in conclusion, it is likely that these fragments left at the surface after cleavage are modifying the χ, eV_s and ϕ values. This explains also the main difference (see Table 1) between the cleaved and annealed surface in case of the (0001)Zn face.

With respect to the time-dependent changes, first it was checked that they are not caused by interaction with residual gases. The annealing temperatures are lying below the sublimation temperatures for Zn and O, and changes of χ indicate changes of concentrations of surface species and it was assumed not of donor or acceptor levels. We found it unlikely that, under conditions of excess Zn in the bulk, Zn should be driven back into the bulk during annealing. Instead, we concluded that oxygen diffuses out to the topmost layer during annealing and back into subsurface bulk oxygen vacancies created at the elevated temperatures. Our conclusions are in agreement with arguments of Heiland and coworkers [14,15]. They measured light absorption in monolayers of organic dye molecules and observed stable, not decomposing, molecules only after heavy oxygen treatment of their samples. We believe that the surfaces which we observe to become oxygen-deficient are those which react with the organic dye molecules. It would be interesting to study their preparation method with our ARUPS technique in future studies.

The relatively high degree of perfection of the low-index ZnO surfaces prepared by the annealing procedure as described above is indirectly proven through the following results. We briefly describe a band structure determination and some surface resonance observations and then turn to adsorption studies.

2.2 Band Structure Determination

The band structure of ZnO was determined at normal emission for the $(10\bar{1}0)$ surface, i.e. along ΓM [7]. Figure 4 exhibits a representative set of spectra. The peaks between -6 and -9.5 eV are due to the Zn 3d states. It was shown recently that the Zn 3d states have to be included in band structure calculations [17]. The spectra have been fitted using Gaussian line shapes. The peaks show strong dispersion as indicated through broken lines in Fig. 4. Using free electron final states, an inner potential of V_0 = 9.8 eV was evaluated and the position of the initial state bands was determined as shown in Fig. 5. There are five different

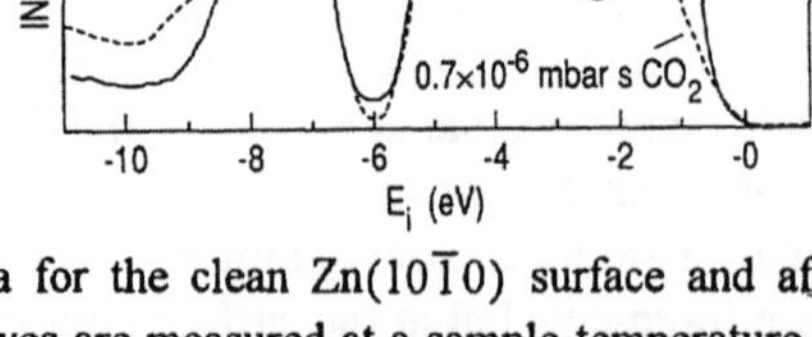

Fig. 6: ARUP spectra for the clean Zn$(10\bar{1}0)$ surface and after dosing with $0.7\cdot10^{-6}$ mbar s CO_2. Both curves are measured at a sample temperature of 100 K. Photon energy was 34.3 eV and angle of emission was 22.9° relative to surface normal. These parameters were chosen to meet the M point of the bulk Brillouin zone and $\overline{X}$ of the surface Brillouin zone [18].

Table 2: Desorption temperatures (in K) for D_2O on ZnO($10\bar{1}0$), (0001) and ($000\bar{1}$)

Interpretation	D_2O on:			H_2O on:	
	ZnO($10\bar{1}0$)	ZnO(0001)Zn	ZnO($000\bar{1}$)O	Ag(110)O [b]	Cu(110)/O [c]
D_2O on interstitial Zn	–	500	500		
D_2O on Zn sites	–	320	(320) [a]	320	283
(two-dimensional	–	260	–	240	232
D_2O cluster)	(220) [a]	220	(220) [a]	200	198
D_2O on oxygen sites	190	–	190		
Two-dimensional ice	168	168	168	170	175
Three-dimensional ice	152	152	152		160

[a] Weak; [b] ref. [21]; [c] ref. [22].

bands resolved in the ΓM direction.

Figure 6 exhibits two spectra measured off-specular at $\hbar\omega = 34.3$ eV [18], parameters which have been chosen to hit the $\bar{X}$ point in the 2D Brillouin zone. According to Ivanov and Pollmann [19] strong surface resonances are expected at this point. After adsorption of some CO_2 these resonances are quenched completely. Figure 7 indicates a very high degree of perfection of the uppermost surface layer considering the high intensity of surface emission relative to the bulk one.

3. Adsorption on Low-Index ZnO Surfaces

3.1 Adsorption of Water (D_2O)

Typical TD spectra are shown in Fig. 7 and summarized in Table 2 [20]. The interaction of D_2O with ZnO is found to be highly site-specific. The ($000\bar{1}$)O face interacts only weakly with water. The main chemisorption state at 190 K is separated from the first ice peak by only 22 K. This peak is due to an interaction of water species with oxygen sites. Consistent with this assumption, this state is not found on the Zn-terminated (0001)Zn face, whereas it is seen also strongly on the prism face. On both the (0001)Zn and the ($10\bar{1}0$) prism face the most intensive chemisorption state is found at 320 to 340 K. Quite obviously, this state has to be correlated with Zn sites. On the ($10\bar{1}0$) face mainly one strong peak develops for the interaction with the Zn sites, whereas a triple-peak structure is found in the (0001)Zn face. A similar triple-peak structure is found for water adsorbed on metal surfaces precovered by a small amount of oxygen [21,22]. These oxygen-precovered metal surfaces and the (0001)Zn face seem to establish similar conditions to allow the buildup of these special two-dimensional ice clusters which show the triple-peak structure in TDS.

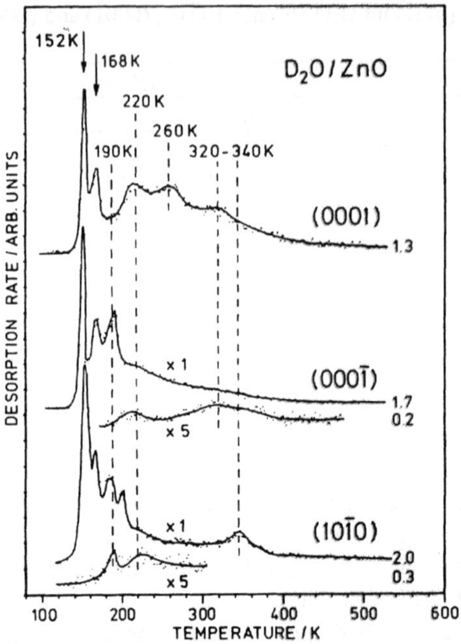

Fig. 7: Comparison of characteristic thermal desorption spectra for the three investigated crystal faces. Exposures are given in units of 10^{-6} mbar s. From [20].

Beside these main chemisorption species there is also a weak desorption signal in the Zn-site region on the $(000\bar{1})$O face. This can be attributed to some defect sites built by diffusion of oxygen into the bulk, giving rise to some Zn sites in the first substrate layer. Finally, what we call two-dimensional ice and what we believe is water in the second layer is somewhat more strongly bound than the three-dimensional ice species for the third and higher-order layers. This shows that the binding energy of the weakly interacting water molecules is greatly dependent on the relative orientation of the molecules. Any special ordering in the first layer implies a two-dimensional ice state in the second layer different from three-dimensional ice.

3.2 Adsorption of NO on ZnO$(10\bar{1}0)$

NO exposure onto ZnO$(10\bar{1}0)$ at 90 K leads to two adsorbed NO species, one desorbing at 110 K and one desorbing at ~420 K [23]. The latter splits and shifts with increasing exposure into three distinguishable structures (440 K, 390 K and ~320 K after 12 L exposure). Whereas the weakly bound NO is quickly adsorbed and saturates at ~2 L, the strongly bound NO adsorption is activated and saturates only after ~50 L exposure. It cannot be excluded that the strongly bound NO species is dissociatively adsorbed. During adsorption at 90 K the NO converts partially to N_2O which desorbs at 140 K, somewhat shifting at high exposures. The oxygen left in this reaction is not desorbed and it is concluded that it finally diffuses into the bulk. After desorption of the weakly bound NO and the N_2O by annealing only to 190 K, readsorption of weakly bound NO is strongly reduced and formation of N_2O is finally stopped after a few adsorption-desorption cycles. This is due either to the accumulation of oxygen near the surface hindering the reaction to N_2O or to the interference by the accumulated strongly bound NO. The strongly bound NO increases, the weakly bound NO decreases the electron affinity. The influence of N_2O on the electron affinity could not be separated from the simultaneous change due to adsorbed NO. Arguments can be given that the oxygen is not adsorbed on top of the surface but instead penetrates so deep into the substrate that it is not seen in ARUPS even at 90 K.

3.3 Adsorption of Methanol and Isopropanol

The salient features of TDS from CH_3OH/ZnO(0001)Zn faces are summarized in Fig. 8 [24]. After higher exposures larger fluxes were observed from condensed methanol, growing to multilayer coverage and desorbing at $T_{(max)}$ = 120 K. Part B of Fig. 8 identifies one contributory factor to the relatively low flux of unchanged CH_3OH desorbing from the first monolayer; since it shows that simultaneous monitoring at m/z = 28 and 44 revealed significant fluxes of CO and CO_2

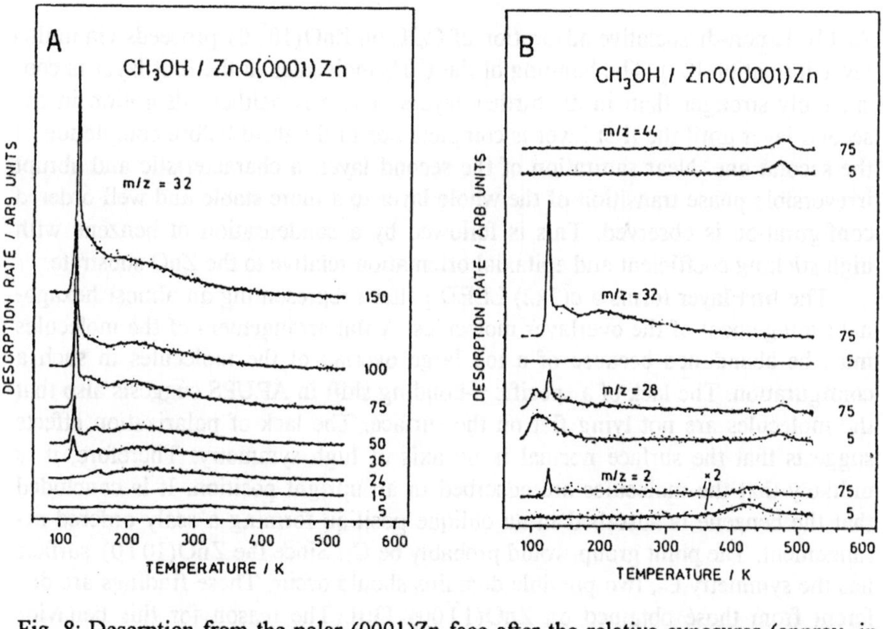

Fig. 8: Desorption from the polar (0001)Zn face after the relative exposures (approx. in 10^{-6} mbar s) to CH_3OH indicated to the right of each TPDS. A. TPDS of the unchanged molecule after progressively increased exposures to CH_3OH; B. Comparison of TPDS of unchanged alcohol (m/z = 32) with that of decomposition products CO_2, CO and H_2 (at m/z = 44, 28, and 2), attributable to dissociation of formate-type surface intermediates. From [24].

desorbing with $T_{(max)}$ = 480 K. This would be consistent with many reports claiming the formation of surface formate intermediates, HCO_2, from CH_3O fragments at CH_3OH/ZnO interfaces and their dissociation to CO and CO_2 at comparable temperatures during TDS [25-27].

The most striking difference to emerge from comparison of the foregoing results from the "Zinc" face with TDS from CH_3OH/ZnO($000\bar{1}$) systems was the absence of any significant TDS peaks attributable to CO_2, CO or H_2 from the oxygen face. The important inference which may be drawn from this contrast between TDS from the two faces in identical experimental conditions is that close approach of CH_3OH adsorbate to surface zinc - which would be guaranteed at the (0001) zinc face, but strongly inhibited by bulky O_s^{2-} ions at ($000\bar{1}$) oxygen faces - was a necessary condition for initiating the production of formate-like surface intermediates, e.g. via precursors such as (CH_3O^- ... Zn^+). The observations upon TDS from isopropanol/ZnO systems followed a similar overall pattern to those described for CH_3O. These results were all confirmed later by Vohs and Barteau [28].

3.4 Adsorption of Benzene on ZnO($10\bar{1}0$)

At 110 K non-dissociative adsorption of C_6H_6 on ZnO($10\bar{1}0$) proceeds via multilayer formation [29]. The bonding of the C_6H_6 molecules in the first layer is considerably stronger than in all further layers. There is neither adsorption in the second layer until the first layer is complete nor in the third before completion of the second one. Near saturation of the second layer, a characteristic and abrupt irreversible phase transition of the whole layer to a more stable and well ordered configuration is observed. This is followed by a condensation of benzene with high sticking coefficient and epitaxial orientation relative to the ZnO substrate.

The first layer forms a c(2x2) LEED pattern representing an almost hexagonal arrangement of the overlayer molecules. A flat arrangement of the molecules must be abandoned because of a too large overlap of the molecules in such a configuration. The lack of a specific π-bonding shift in ARUPS suggests also that the molecules are not lying flat on the surface. The lack of polarization effects suggests that the surface normal is no axis of high symmetry. Therefore, it is unlikely that the molecules are adsorbed in an upright position. It is concluded that the benzene is adsorbed in an oblique position forming a scaly ordered arrangement. The point group would probably be C_1. Since the ZnO($10\bar{1}0$) surface has the symmetry C_s, two possible domains should occur. These findings are different from those obtained on ZnO($1\bar{1}00$) [30]. The reason for this behavior could not be elucidated by our experiments.

The arrangement of the C_6H_6 molecules in the ordered second and all further layers is different from the first layer as indicated by the c(4x3) LEED pattern. It can be supposed that the formation of an apparent or true c(4x3) superstructure in the ordered second and all further layers is caused by an accidental common periodicity between the meshes of the ZnO($10\bar{1}0$) substrate and the benzene

(111) plane. Unlike the first layer where the arrangement of the benzene molecules is determined by the predominant interaction with the substrate the monolayer structure represents a suitable support for the multilayer growth of benzene with its proper bulk structure.

3.5 Adsorption of Xe

It is well known that the interaction of Xe with sp-materials (both metals and semiconductors) is weaker than with d-metals. Therefore, the ZnO samples had to be cooled using liquid He in order to adsorb Xe. For the measurements reported here a lowest temperature of about 40 K was reached. In TDS multilayer and monolayer peaks are observed [31] demonstrating that layer growth occurred despite the weak interaction with the substrate. The ZnO$(10\bar{1}0)$ surface shows a special rich structure in the monolayer regime and it was concluded that the adsorption starts in the troughs. The TD spectra from (0001)O and $(000\bar{1})$Zn are similar, providing no evidence for a preferential interaction with oxygen or zinc surface atoms.

LEED indicates epitaxial growth of Xe on Zn$(10\bar{1}0)$ but disordered growth on the polar faces. Xe deposition onto the polar faces leaves their work function unchanged. The work function of the $(10\bar{1}0)$ decreases by about 0.5 eV during the first half monolayer deposition. We interpret this work function change as a change of electron affinity caused by a reduction of the known reconstruction during filling of the surface troughs. Thus, in contrast to the observation on metals the Xe atoms in the monolayer are not polarized on all ZnO surfaces. Interestingly, the method failed to indicate different adsorption sites or surface inhomogeneities other than the trough adsorption on $(10\bar{1}0)$. This seems to be generally true for semiconductor and insulator surfaces. Although layer growth is demonstrated through TDS, no clear distance-dependent changes in electronic binding energies are observed in ARUPS. For (0001)Zn no effect at all is observed. The effects for $(10\bar{1}0)$ and $(000\bar{1})$O are opposite and are attributed to different shielding effects in the final state. Redistribution processes in the O 2p band emission were observed with all ZnO faces shifting part of the emission from a region close to the valence band maximum to binding energies higher by about 1 eV.

References

[1] G. Heil, H. Lüth: in The Chemical Physics of Solid Surfaces and Heterogeneous Catalysis, Vol. 3, eds. D.A. King , D.P. Woodruff, Elsevier, Amsterdam 1984

[2] H. Lüth, G.W. Rubloff, W.D. Grobman: Solid State Commun. **18**, 1427 (1976)

[3] W. Ranke: Solid State Commun. **19**, 685 (1976)

[4] W. Gaebler, K. Jacobi, W. Ranke: Surf. Sci. **75**, 355 (1978)

[5] W. Göpel, J. Pollmann, I. Ivanov, B. Reihl: Phys. Rev. **B 26**, 3144 (1982)

[6] K. Jacobi, G. Zwicker, A. Gutmann: Surf. Sci. **141**, 109 (1984)
[7] G. Zwicker, K. Jacobi: Solid State Commun. **54**, 701 (1985)
[8] H. Lüth: Surf. Sci. **37**, 90 (1973)
[9] H. Moormann, D. Kohl, G. Heiland: Surf. Sci. **80**, 261 (1979)
[10] H. Moormann, D. Kohl, G. Heiland: Surf. Sci. **100**, 302 (1980)
[11] C.B. Duke, R.J. Meyer, A. Paton, P. Mark: Phys. Rev. **B 18**, 4225 (1978)
[12] Y.R. Wang, C.B. Duke: Surf. Sci. **192**, 309 (1987)
[13] D. Kohl, M. Henzler, G. Heiland: Surf. Sci. **41**, 403 (1974)
[14] H. Grüning, G. Heiland: Surf. Sci. **122**, 569 (1982)
[15] C. Beneking, G. Heiland: J. Lumin. **43**, 9 (1989)
[16] J.R. Chelikowski: Solid State Commun. **22**, 351 (1977)
[17] P. Schröer, P. Krüger, J. Pollmann: Phys. Rev. **B** (in press)
[18] K. Jacobi, G. Zwicker: (unpublished)
[19] I. Ivanov, J. Pollmann: Phys. Rev. **B 24**, 8285 (1981)
[20] G. Zwicker, K. Jacobi: Surf. Sci. **131**, 179 (1983)
[21] E.M. Stuve, R.J. Madix, B.A. Sexton: Surf. Sci. **111**, 11 (1981)
[22] K. Bange, D. Grider, J.K. Sass: Surf. Sci. **126**, 437 (1983)
[23] G. Zwicker, W. Ranke, K. Jacobi, D. Pöss: Ber. Bunsenges. Phys. Chem. **88**, 364 (1984)
[24] G. Zwicker, K. Jacobi, J. Cunningham: Int. J. Mass Spectr. and Ion Proc. **60**, 213 (1984)
[25] U. Ueno, T. Onishi, K. Tamaru: Trans. Faraday Soc. **66**, 756 (1970)
[26] Y. Kikuzono, S. Kagami, S. Naito, T. Onishi, K. Tamaru: Faraday Disc. Chem. Soc. **72**, 135, 194 and 196 (1981)
[27] J. Cunningham, B.K. Hodnett: J. Chem. Soc. Faraday I **77**, 2777 (1981)
[28] J.M. Vohs, M.A. Barteau: Surf. Sci. **201**, 481 (1988)
[29] D. Pöss, W. Ranke, K. Jacobi: Surf. Sci. **105**, 77 (1981)
[30] G.W. Rubloff, H. Lüth, W.D. Grobman: Chem. Phys. Lett. **39**, 493 (1976)
[31] A. Gutmann, G. Zwicker, D. Schmeisser, K. Jacobi: Surf. Sci. **137**, 211 (1984)

Molecular Adsorption on TiO_2 and ZnO Surfaces

G. Thornton

Interdisciplinary Research Centre in Surface Science and Chemistry Department, Manchester University, Manchester M13 9PL, UK

Abstract. Synchrotron radiation measurements have been used to explore the interaction of model oxide surfaces with molecular adsorbates. The (100) and (110) faces of rutile TiO_2 and the $(10\bar{1}0)$ plane of ZnO are chosen as examples of the substrates studied. In the case of $TiO_2(100)$, the substrate structure has been characterised by STM in an attempt to understand the reactivity to H_2O. The latter involves molecular adsorption at $T \leq 130$ K, with dissociation at higher temperatures; a behaviour which is apparently independent of O-vacancy and step density. Finally, structural studies of benzene, pyridine, CO_2 and formate on $ZnO(10\bar{1}0)$ are described. All four species bond with substrate-induced alignment.

1. Introduction

One of the key issues addressed in this volume is the difference in behaviour of oxide and other ionic surfaces compared with the much more widely studied surfaces of metals. Although it is difficult to generalise about a class of materials with such diverse properties, a major difference between metal and oxide surface reactivity lies in the role of minority sites such as vacancies and steps. This can be understood by remembering that a metal ion at an oxide-surface step site will have an associated electrostatic potential similar in many respects to that associated with a 'promoter' alkali metal atom adsorbed on a metal surface. Of course we expect other, much more fundamental differences in the reactivity, for instance there is no metal-surface counterpart to complex anion formation involving substrate atoms. These are the type of interactions explored in this article.

There has of late been a marked increase in the number of single crystal studies aimed at elucidating the detailed mechanism of oxide surface reactions, including the role of specific minority sites in the reactivity [1]. The approach we have adopted involves the study of model reactions using synchrotron radiation techniques, several examples of which are described below. The first is an investigation of the effect of oxygen vacancies and steps on the reaction of H_2O with $TiO_2(100)$. This follows from an earlier study of water adsorption on $SrTiO_3(100)$, where catalytic dissociation by step sites was evidenced [2]. In the case of $TiO_2(100)$, there is little effect of either steps or O-vacancies, which prompted us to carry out a detailed study of the surface morphology by scanning tunnelling microscopy and spectroscopy STM(S).

Earlier work using photoemission has explored the bonding of molecules such as CO_2 and formate to $ZnO(10\bar{1}0)$ surfaces [3], motivated by the possible involvement of these reactions in commercial methanol synthesis [4]. In the case of CO_2, evidence was adduced for carbonate formation [3]. The third example described here is a NEXAFS study of the carbonate and formate species on $ZnO(10\bar{1}0)$.

The final example also involves the study of adsorption on $ZnO(10\bar{1}0)$ by NEXAFS. In this case the adsorbates are benzene and pyridine. The aim here is investigate the bonding of large aromatic molecules to an oxide surface for comparison with their behaviour on metal surfaces. As for the methanol synthesis intermediates, this structural work follows earlier photoemission studies [1].

2. Experimental

STM(S) studies employed an Omicron UHV instrument, which is designed for operation at room temperature. The synchrotron radiation measurements were all carried out at the SRS, Daresbury Laboratory. C and N K-edge NEXAFS studies used a hemispherical analyser on station U5.1 to record the C KLL and N KLL Auger yield as a monitor of the surface absorption coefficient. NEXAFS measurements were carried out as a function of polar angle with the ***E***-vector in the [0001] and the $[11\bar{2}0]$ azimuths of $ZnO(10\bar{1}0)$. Photoemission measurements used the VG ADES 400 system on station 6.2, where the angular resolution was $\pm 2°$ and the combined energy resolution was ca. 0.15 eV [FWHM] at a photon energy of 33 eV. The base pressure of the sample chambers was $\leq 1 \cdot 10^{-10}$ mbar.

Preparation of the stoichiometric (1x1) and oxygen depleted (1x3) $TiO_2(100)$ surfaces on planar and 2.6°-vicinal samples employed methods described elsewhere [5]. The $ZnO(10\bar{1}0)$ 1x1 substrate was cleaned by cycles of Ar^+ bombardment and annealing to 800 K. The final anneal was carried out in $1 \cdot 10^{-6}$ mbar oxygen, with the temperature reduced to 550 K at this pressure. This procedure was carried out in an attempt to restore the stoichiometry. LEED was used to check the surface order, with either AES or photoemission used to monitor the surface contamination level, being ≤ 0.02 ML (monolayer) in all cases. Exposure to H_2O was carried out using the vapour of doubly distilled water which was thoroughly degassed before use. The formate adlayer on $ZnO(10\bar{1}0)$ was formed by exposing the clean sample to 10 L (1 Langmuir=$1.32 \cdot 10^{-6}$ mbar·s) of formic acid at 293 K, while CO_2 exposure (60 L) was carried out with the sample at 180 K. The pyridine (benzene) adlayer was formed by exposing the sample to 1(5) L at 293 K (180 K), with measurements being made at the adsorption temperatures.

3. Results and Discussion

$TiO_2(100)$-H_2O. Two types of photoemission experiments were performed to examine the effect on the reactivity of O-vacancies and steps. These involved adsorption at 130 K with subsequent annealing of the sample to higher temperature, and room temperature adsorption. Results obtained from our stepped $TiO_2(100)$1x1 surface recorded in the former experiment are shown in Fig. 1 in the form of difference spectra. Data obtained from the stepped 1x3 and the planar 1x1 and 1x3 surfaces, which will be presented in full elsewhere [6], are similar. Room temperature H_2O adsorption of about 10^3 L on all surfaces yields difference spectra similar to those obtained by low temperature adsorption and annealing to room temperature.

The difference spectra shown in Fig. 1 evidence molecular adsorption at 130 K, the three features corresponding to the $1b_2$, $3a_1$ and $1b_1$ orbitals of H_2O. When aligned with a gas phase spectrum of H_2O at the $1b_2$ energy a $3a_1$ bonding shift of 0.9 eV is observed in the difference spectrum, similar to that observed by Brookes et al. [2] for $SrTiO_3(001)$. Annealing the 130 K dosed surface results in the attenuation of the $1b_2$ feature, the apparent shift to higher BE of the $3a_1$ feature, and the appearance of a new peak at about 5.6 eV BE. We assign the spectrum of the 290 K annealed surface by comparison with the 130 K data, as well as with photoemission data from H_2O-dosed $SrTiO_3(001)$ [2] and Si(100)2x1 [7] surfaces. This leads us to associate the 10.7 eV BE feature with the Ti-O bonding orbital, the 7.7 eV BE peak with the O-H bonding orbital, and the 5.6 eV feature

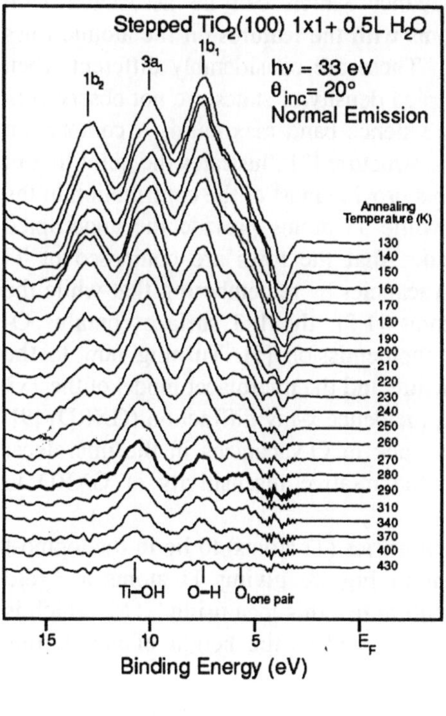

Fig. 1: Photoemission difference spectra showing the effect of sequentially annealing our stepped $TiO_2(100)$1x1 surface exposed to 0.5 L H_2O at 130 K to the temperatures indicated. Difference spectra were obtained by subtracting a clean surface spectrum recorded at 130 K from dosed surface spectra recorded at the anneal temperature.

Fig. 2: A 30 Å x 40 Å constant height image of planar $TiO_2(100)1x3$ recorded at a bias voltage of +2V, and a tunnelling current of 0.2 nA before disengaging the feedback loop. The 1x3 unit cell is indicated.

with the O 2p lone pair orbital. In other words, we associate changes to the spectra on annealing the surface from 130 K to 290 K to activated dissociation of H_2O.

These results indicate that neither the steps present on our vicinal surface nor the O-vacancies on the 1x3 surface markedly affect the reactivity to water. While the lack of an effect of O-vacancies is consistent with $SrTiO_3(001)$ results, the steps created by the vicinal off-cut were expected to produce sites similar to those which give rise to catalytic dissociation of H_2O on $SrTiO_3(001)$. To explore the reasons for this discrepancy we have examined the structure of the $TiO_2(100)$ substrates using STM(S).

Figure 2 shows a constant height image of the 1x3 vicinal surface obtained at positive sample bias, i.e. tunnelling into unoccupied states of the substrate. Individual atoms can be distinguished on the rows, which are parallel to [001]. Current profiles between and along the rows indicate a surface unit cell of dimensions 3.3 Å x 15.4 Å, to be compared with those expected for a 1x3 unit cell of 2.96 Å x 13.77 Å, a discrepancy which can be associated with instrumental factors. We identify Ti atoms with the features on the atomic rows in Fig. 2 by reference to STS data [8]. These are considerably different when recorded on and between the rows; occupied density of states are not observed in the former down to ca. 1 eV below the valence band maximum. A comparison with calculations of the surface electronic structure [8] illustrates the point that in this relatively ionic material oxygen atoms provide most of the contribution to the occupied states near the Fermi level, while Ti atoms provide the unoccupied states. This provides compelling evidence that the rows are composed of Ti atoms. These can be associated with O-vacancies by remembering that while the 1x1 surface should terminate in O atoms [1,5], the 1x3 surface contains O-vacancies, as previously evidenced by photoemission [5]. Turning now to the relationship between the O-vacancy structure and the microfacet model of the 1x3 surface proposed on the basis of grazing incidence x-ray diffraction (GXRD) [9]. An obvious discrepancy arises from the lack of O-vacancies in the microfacet model, which presumably arises from the relative insensitivity of GXRD to oxygen atoms.

We might reasonably expect the O-vacancies (Ti atoms) to lie in the topmost layer of a microfacet model, as shown in Fig. 3, giving Ti atoms a 3-fold coordination to oxygen. While it is difficult to test this idea using STM, which is sensitive to the local electronic structure as well as the height of atoms, this

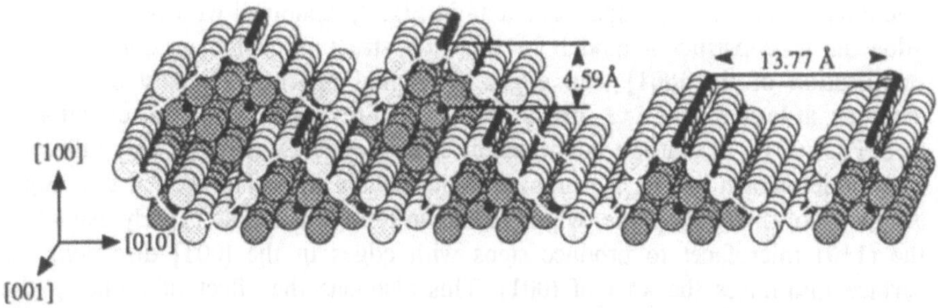

Fig. 3: Three dimensional microfacet model of the 1x3 surface modified to include oxygen vacancies and the proposed step models for the [010] and [001] directions. Ti atoms are represented by the small black spheres, while O atoms are represented by the larger spheres. The 1x3 unit cell is indicated.

model has been confirmed using surface core shift photoelectron diffraction [10]. As noted above, the lack of an effect of O-vacancies on the reaction to water was also observed in $SrTiO_3(001)$ measurements. Nevertheless, the O-vacancy geometry derived here, with a low coordination Ti site, might have been expected to promote reactivity. An explanation for this phenomenon awaits detailed calculations of the local electronic structure.

Of more interest here is the lack of promotion by steps, which are active on $SrTiO_3(001)$ [2]. STM images of the 1x3 vicinal surface, cut 2.6° off (100) towards [001], contain the [001] rows observed in Fig. 2, which typically run along wide (ca. 500 Å) terraces to terminate in irregular step edges parallel to [010]. These are the steps introduced by the vicinal off-cut, which we had arranged to produce steps similar to those on $SrTiO_3(001)$. However, step edges are also observed in the [001] direction, which are more regular, giving rise to spot-splitting of LEED beams [5]. Both types of step can be observed in the image shown in Fig. 4. In this image impurity atoms decorate the bottom of

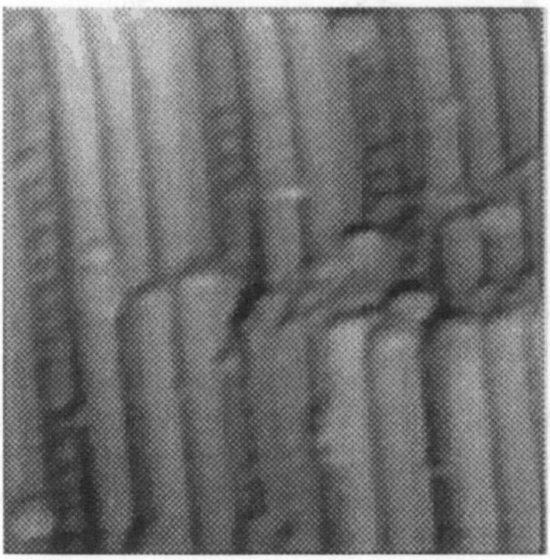

Fig. 4: A 150 Å x 150 Å constant current image of 2.6° vicinal $TiO_2(100)1x3$ recorded at +2 V, 0.3 nA. The rows lie parallel to [001].

the [001] direction step edges. The data in Fig. 4, combined with height profiles allow us to construct a model for the step structure which accounts for the stabilisation of the [001] step edges by vicinal off-cut-derived steps. It also explains, at least for the 1x3 surface, why we fail to observe step-site activation as in $SrTiO_3(001)$, in that the environment of step edge Ti atoms is very different. The model shown in Fig. 3 summarises our conclusions. The vicinal step edges are facets of the (001) plane, which is known to be metastable [1]. By extending the (110) microfacet to produce steps with edges in the [001] direction, the surface minimises the area of (001). This also has the effect of reducing the concentration of the step-edge sites with which we expected to mimic the reactivity of $SrTiO_3(001)$. However, this is probably not the major difference introduced by surface relaxation; it is more likely the very different environment of the step edge site introduced by the 1x3 reconstruction.

ZnO(10$\bar{1}$0)-formate, CO_2. The C K-edge NEXAFS spectra of formate and CO_2 on ZnO(10$\bar{1}$0), recorded with the X-ray ***E***-vector in the [11$\bar{2}$0] azimuth are shown in Fig. 5. Both sets of data indicate orientation of the molecules on the surface, because the relative intensities of the NEXAFS peaks vary with angle. The formate data show a peak at an energy of 288.7±0.2 eV, with a smaller

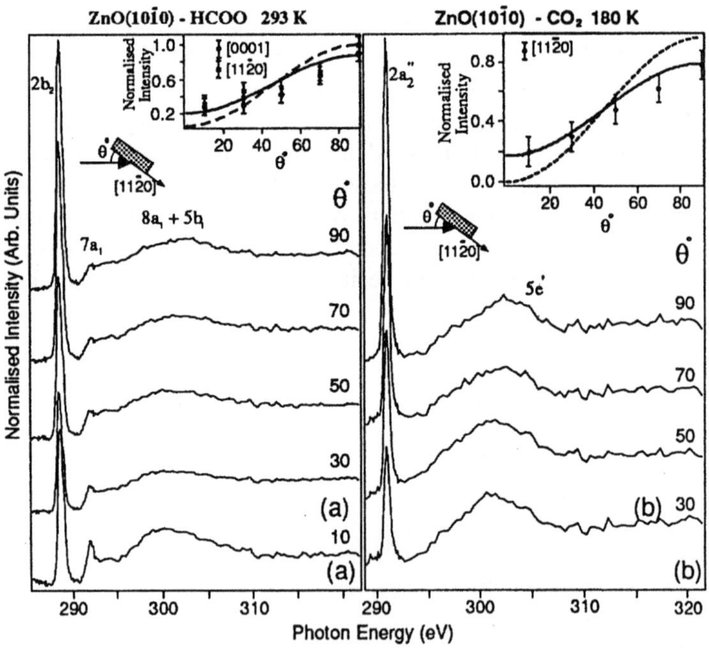

Fig. 5: C K-edge NEXAFS spectra of formate (a) and CO_2 (b) adsorbed on ZnO(10$\bar{1}$0). The insets show the intensity variation of the lowest energy resonances, compared with those expected on the basis of an upright geometry (dashed line) and with a tilt of the HCOO- (CO_3^{2-}) plane of 20° (25°) (solid line).

feature at 292.0±0.2 eV. The positions of these features and that at ca. 300 eV are comparable with those observed from formate on Cu(110) and Ag(110) [11,12]. Here we focus on the 288.7 eV and 292.0 eV peaks, which can be assigned to excitation from C 1s to the $2b_2$ and $7a_1$ unoccupied molecular orbitals, respectively, assuming the C_{2v} symmetry of the isolated formate ion [11]. The $2b_2$ orbital is polarised about the plane of the molecule, i.e. the 288.7 eV peak is a π-type resonance. The $7a_1$ orbital is totally symmetric and has mostly C-H character. Using dipole selection rules [13], the $2b_2$ resonance should be excited when the ***E***-vector is perpendicular to the plane of the molecule, and the $7a_1$ should be excited when the ***E***-vector is parallel to the C_2 axis.

The data in Fig. 5(a) and those recorded with the ***E***-vector in the [0001] azimuth show that the $7a_1$ peak has maximum intensity in both azimuths when the beam is at grazing incidence. We can conclude from this that the formate molecular plane is oriented more or less vertically. That the data recorded with the ***E***-vector in the [0001] azimuth show similar behaviour in the intensity variation of the $2b_2$ resonance indicates that there is no azimuthal ordering of formate, as observed in the case of Ag(110) [12]. Hence, formate adopts a geometry like that shown in Fig. 6. In this and other adsorption-site models we assume bulk termination, since we have no means of gauging the effect of adsorption on the clean substrate relaxation [1].

Without temperature dependent measurements it is not possible to determine whether the azimuthal disorder is static or dynamic. On the basis of no azimuthal ordering, we can analyse the $2b_2$ intensity variation in a quantitative fashion using the expression [13]:

$$I(2b_2) \propto P(\cos^2\Theta_E \cos^2\Theta_t + 0.5 \sin^2\Theta_E \sin^2\Theta_t) + (1-P)\, 0.5 \sin^2\Theta_t,$$

where Θ_E is the angle of incidence, Θ_t is the angle of tilt of the molecule relative to the surface normal, and P is the degree of polarisation (0.95). The results in the inset of Fig. 5(a) indicate that the C_2 axis is tilted by up to 20° from the surface normal. It is also possible that the formate molecule is tilted in only the [0001] azimuth, with rotation about the C_2 axis giving rise to the azimuthal independence of the NEXAFS.

The ZnO(10$\bar{1}$0)-CO_2 spectra shown in Fig. 5(b) are very similar to those of

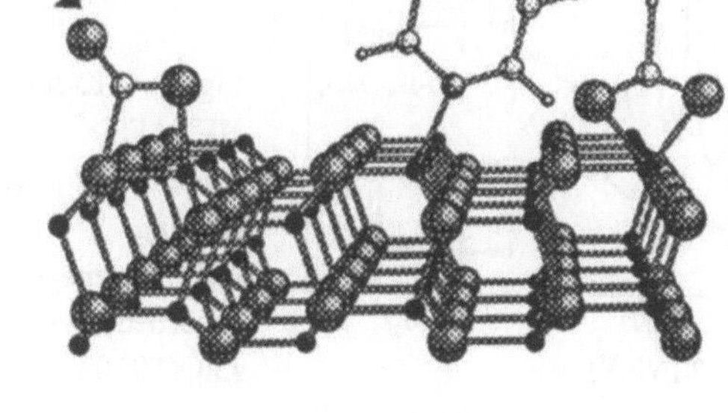

Fig. 6: Proposed structural models for (from left to right) CO_2 (carbonate), pyridine and formate on ZnO[10$\bar{1}$0]. The Zn and O atoms of the substrate are depicted as small and large spheres, respectively, scaled to their ionic radii.

carbonate on Ag(110) [14]. Assuming the D_{3h} symmetry of the isolated ion [14], the peak at 290.5±0.02 eV can be assigned to the $2a_2''$ unoccupied molecular orbital which has C $2p_z$ and O $2p_z$ character, and is therefore polarised about the molecular plane. When the *E*-vector is in the [11$\bar{2}$0] azimuth, the intensity of the $2a_2''$ resonance is greatest at normal incidence (see Fig. 5(b)). This, coupled with the invariance of the $2a_2''$ intensity with incidence angle in the [0001] azimuth indicates that the molecular plane lies close to the surface normal in the [0001] azimuth, as shown in Fig. 6(b). The results of a quantitative analysis of the $2a_2''$ intensity dependence seen in Fig. 5(b), based on the expression [13]:

$$I(2a_2'') \propto P(\cos^2\Theta_E \cos^2\Theta_t + \sin^2\Theta_E \sin^2\Theta_t),$$

are shown in the inset of Fig. 5(b). These indicate that the carbonate molecular plane is tilted by up to 25° from the surface normal towards the [11$\bar{2}$0].

The geometry shown for the carbonate species differs from that on Ag(110), where it lies flat on the surface [14]. However, a similar geometry has been suggested for carbonate on Cr_2O_3(111) [15]. In Fig. 6 we suggest a plausible model which allows one of the O atoms from CO_2 to bond to an adjacent Zn atom at the bulk Zn-O distance, while maintaining the bulk carbonate C-O distance of 1.29 Å. This is close to the geometry proposed on the basis of cluster calculations [16].

ZnO(10$\bar{1}$0)-pyridine, benzene. The C K-edge NEXAFS data in Fig. 7 are significantly easier to interpret than the formate and carbonate results, because the polarisation dependence is rather clear cut. Both the pyridine and benzene spectra are comparable with those obtained from the molecules adsorbed on metal

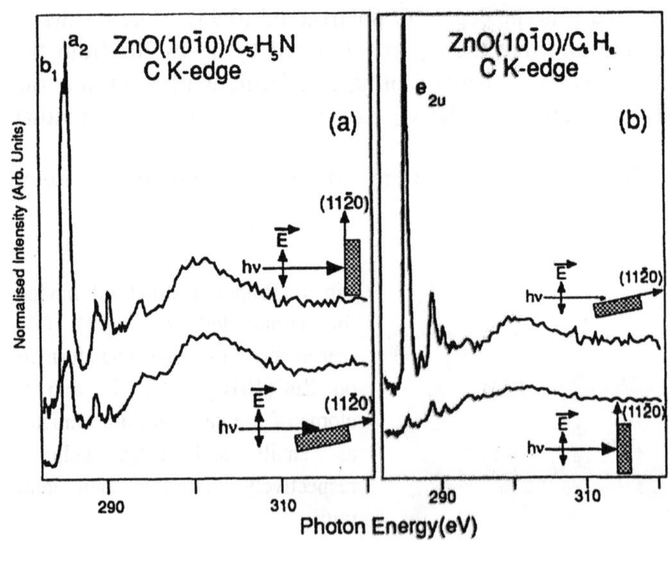

Fig. 7: C K-edge NEXAFS spectra of pyridine (a) and benzene (b) adsorbed on ZnO(10$\bar{1}$0).

surface [13]. Here we will simply focus on the lowest-energy resonances which are polarised about the molecular planes. In the pyridine data these lie at 285.0±0.2 eV and 285.5±0.2 eV, corresponding to excitation into b_1 and a_2 orbitals, respectively. In the benzene data in Fig. 7(b), it lies at 285.5±0.2 eV, corresponding to excitation into an e_{2u} orbital. For both adsorbates, these π-like resonances are excited by a component of the X-ray ***E***-vector lying perpendicular to the molecular plane. By inspection of the data in Fig. 7(a), we can conclude that pyridine stands more or less upright on the surface, with the molecular plane inclined towards the [0001] azimuth. The results recorded with the ***E***-vector in the [0001] azimuth are consistent, showing little polarisation dependence. The data in Fig. 7(b) indicate that benzene is tilted about 10° away from the idealised geometry in which the molecule lies flat on the surface, consistent with previous LEED work [17]. This is evidenced in the data as an incomplete attenuation of the e_{2u} resonance at normal incidence.

In the case of pyridine, the upright geometry strongly suggests that the molecule bonds via the N lone pair to substrate Zn atoms. This Lewis base behaviour of pyridine has been observed in metal surface chemistry [13]. It might be expected to give rise to a geometry close to that shown in Fig. 6, where the N-Zn bond lies along the tetrahedral angle for Zn coordination.

4. Conclusions

In this work we have examined the effect of surface defects on the reaction of H_2O with $TiO_2(100)$ surfaces, finding little modification. Further progress awaits calculations which explore the effect of the local geometry of Ti atoms on the reactivity. We have also examined the adsorption geometry of formate, CO_2, pyridine and benzene on $ZnO(10\bar{1}0)$. Only in the case of CO_2 is there a marked difference between its behaviour compared with that on metal surfaces. On $ZnO(10\bar{1}0)$ a carbonate species is formed in which the plane of the molecule lies within 25° of the surface normal in the [0001] azimuth.

Acknowledgements

The work described above involved a collaboration with R. Davis, J.F. Walsh, C.A. Muryn, V.R. Dhanak, K.C. Prince, P.J. Hardman, J.J. Crouch, G.N. Raikar, D.S-L. Law, P.W. Murray, F.M. Leibsle, H.J. Fisher and C.F.J. Flipse. It was funded by the UK Science and Engineering Research Council.

References

[1] P.A. Cox, V.E. Henrich: Surface Properties of Oxide Materials, Cambridge University Press 1993

[2] N.B. Brookes, F.M. Quinn, G. Thornton: Catalysis Today **2,** 547 (1988)
[3] C.T. Au, W. Hirsch, W. Hirschwald: Surf. Sci. **199**, 507 (1988)
[4] K. C. Waugh: Catalysis Today **15**, 51 (1992)
[5] C.A. Muryn, P.J. Hardman, J.J. Crouch, G.N. Raikar, G. Thornton, D.S.-L. Law: Surf. Sci. **251/252**, 747 (1991)
[6] C.A. Muryn, P.J. Hardman, J.J. Crouch, G.N. Raikar, G. Thornton, D.S.-L. Law: Surf. Sci. to be submitted for publication
[7] K. Fives, R. McGrath, C. Stephens, I.T. McGovern, R. Cimino, D.S.-L. Law, G. Thornton: J. Phys. Condensed Matter **1**, SB105 (1989)
[8] P.W. Murray, F.M. Leibsle, H.J. Fisher, C.F.J. Flipse, C.A. Muryn, G. Thornton: Phys. Rev. **B 46**, 12877 (1992)
[9] P. Zschack, J.B. Cohen, Y.W. Chung: Surf. Sci. **262**, 395 (1991)
[10] P.J. Hardman, N.S. Prakash, C.A. Muryn, G.N. Raikar, G. Thornton, R.J. Blake: Phys. Rev. B, submitted for publication
[11] J. Somers, A.W. Robinson, Th. Lindner, D. Ricken, A.M. Bradshaw: Phys. Rev. **B 40**, 2053 (1989)
[12] P.A. Stevens, R.J. Madix, J. Stöhr: Surf. Sci. **230**, 1 (1990)
[13] J. Stöhr: NEXAFS Spectroscopy, Springer, Berlin 1992
[14] M. Bader, B. Hillert, A. Puschmann, J. Haase, A.M. Bradshaw: Europhys. Lett. **5**, 443 (1988)
[15] H. Kuhlenbeck, C. Xu, B. Dillmann, M. Haßel, B. Adam, D. Ehrlich, S. Wohlrab, H.-J. Freund, U.A. Ditzinger, H. Neddermeyer, M. Neuber, M. Neumann: Ber. Bunsenges Phys. Chem. **96**, 15 (1992)
[16] W. Göpel: Progr. Surf. Sci. **20**, 9 (1985)
[17] D. Pöss, W. Ranke, K. Jacobi: Surf. Sci. **105**, 77 (1981)

Single-Crystal Studies of Molecular and Atomic Adsorption on Transition-Metal Oxides

V. E. Henrich

Surface Science Laboratory, Department of Applied Physics
Yale University, New Haven, CT 06517, USA

Abstract. Photoemission spectroscopy has been used to study the interaction of inorganic molecules and metals with well characterized single-crystal surfaces of transition-metal oxides. Stoichiometric, nearly perfect surfaces of maximal valency oxides are generally inert, with most adsorption activity resulting from the presence of surface defects. Stoichiometric surfaces of lower oxides range from weakly adsorbing (e.g., O_2 or H_2O on Ti_2O_3), where defects greatly enhance reaction, to highly reactive (e.g., SO_2 on Ti_2O_3), where defects play little if any role. The reaction of metals with oxide surfaces is governed largely by the relative oxygen affinities of the respective oxides, although point defects also play an important role, particularly for relatively inert metals.

1. Introduction

In order to study the fundamental interactions that occur when atoms or molecules are adsorbed on metal oxides, adsorption studies have been conducted on well characterized single-crystal samples of a number of different transition-metal oxides, including both maximal valency oxides (i.e., oxides in which the metal cation is in its highest oxidation state) and lower oxides (i.e., oxides whose cations are partially reduced) [1]. Interest in these basic studies has arisen largely from the catalysis community, where molecule/oxide interactions are important in heterogeneous catalysis on oxides, and metal/oxide interactions are of interest in oxide-supported metal catalysts. Experiments have been conducted primarily on oxides of the 3d-transition metals. That system contains both isostructural families of oxides whose cation valence states differ (e.g., FeO, MnO, CoO and NiO; or Ti_2O_3 and V_2O_3), and oxides in which the same cation has different valence states (e.g., VO, V_2O_3, VO_2, V_6O_{13} and V_2O_5). It is therefore possible to vary geometric and electronic structure of the oxide independently, so that the role of each in adsorption can be addressed.

In the experiments described here, samples were prepared by cleaving or fracturing in ultrahigh vacuum (UHV). The exception is TiO_2 (110), for which a well ordered, stoichiometric surface can be prepared by ion bombardment and

Springer Series in Surface Sciences, Vol. 33 Eds.: E. Umbach · H.-J. Freund
Adsorption on Ordered Surfaces of Ionic Solids and Thin Films

subsequent annealing in an O_2 ambient [2]. Some results have also been obtained on Ti_2O_3 $(10\bar{1}2)$ surfaces that were prepared by ion bombardment followed by a specific annealing procedure [3]. The exact nature of a "stoichiometric, well ordered" oxide surface is only now being determined through the use of atomic force microscopy (AFM) and scanning tunneling microscopy (STM), but there is no question that such surfaces are in fact more complex structurally than simple models based upon truncation of bulk structure would suggest. More work is necessary to clarify the state of the single-crystal oxide surfaces that are used in adsorption studies.

2. Role of Surface Defects in Adsorption

Stoichiometric, nearly perfect surfaces of most maximal valency transition-metal oxides are quite inert [1]. Lower oxides of the transition metals exhibit varied behavior toward molecular adsorption, but even for those the stoichiometric surfaces are generally less reactive than surfaces that contain defects. The defects that are most important in increasing the chemical activity of oxide surfaces are oxygen vacancies. These are also the predominant types of defects that are present on metal oxides, since thermal annealing, bombardment by ions, electrons or photons, etc. generally tend to reduce oxide surfaces. Extended defects such as steps are also present in large numbers on most oxide surfaces, but they tend to be much less active for adsorption than are O-vacancy point defects, since the electronic structure of stepedge atoms is very similar to that of atoms in perfect terraces.

To understand the importance of O-vacancy defects in adsorption, consider the idealized model of the corundum $(10\bar{1}2)$ surface in Fig. 1. This is the natural cleavage face of Ti_2O_3, V_2O_3 and Cr_2O_3 (although corundum itself -- Al_2O_3 -- does not cleave). The large circles represent the O^{2-} ions, with shading increasing with depth below the surface plane, and the small solid circles are the metal cations. (Since the electronic structure of transition-metal oxides is dominated by the valence state of the cation, with the anions having essentially a closed shell, O^{2-} configuration, attention will be focussed on the cations.) In the bulk of corundum, and all other octahedrally coordinated metal oxides, the metal cations are surrounded by six near-neighbor O ions. On atomically flat terraces of the $(10\bar{1}2)$ surface, that ligand coordination has been reduced to five-fold. One atomic-height step between two $(10\bar{1}2)$ terraces is shown in Fig. 1, as is one O-vacancy point defect. The cations at the step edge have their ligand coordination reduced from five to four, as do two of the cations adjacent to the O-vacancy point defect. But there is a fundamental difference between the electronic structure of the four-fold cations at steps and those at point defects. At steps both the cations and anions can still have the same valence that they had in the bulk or on terraces, since both cations and anions are removed in creating a step. Experimental measurements on heavily stepped oxide surfaces show that the

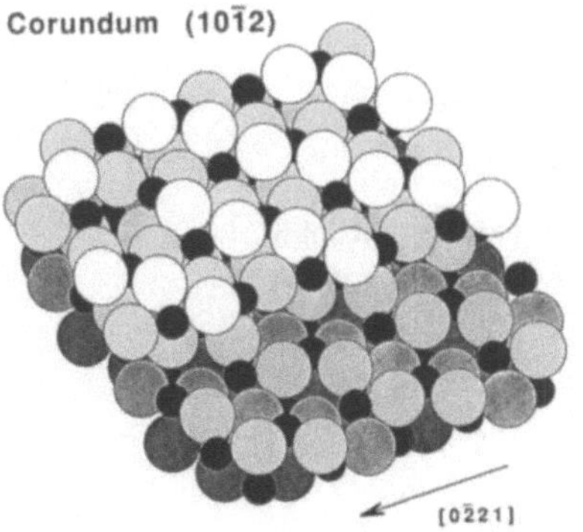

Fig. 1: Idealized model of the geometric structure of the corundum (10$\bar{1}$2) surface. See text for details.

electronic structure [at least at the level of resolution that can be obtained in ultraviolet photoemission spectroscopy (UPS)] is essentially the same as that of the bulk. But at O-vacancy point defects the local stoichiometry has changed. Removal of an O^{2-} ion from the surface must be accompanied by the localization of two electrons at the defect site in order to maintain local charge neutrality. (In other words, the oxygen is removed as an O atom, with its two electrons remaining at the defect site.) This creates totally new electronic states, often occurring in the bulk bandgap, that are not present on the stoichiometric surface, and it is those states that appear to be responsible for most of the chemisorption properties of metal oxides.

By far the most thoroughly studied transition-metal oxide in surface science is TiO_2. However, since there are several other papers in this volume that deal explicitly with TiO_2, we will primarily consider other transition-metal oxides here. We will first consider maximal valency oxides, where stoichiometric surfaces exhibit little or no chemical activity at room temperature.

3. Defects on Maximal Valency Oxides

In many respects the chemisorption behavior of NiO is characteristic of that of most other maximal valency transition-metal oxides (although the adsorption of H_2O on NiO is unique) [4,5]. Since it has the rocksalt crystal structure, the only stable crystal face for NiO is (100), and it cleaves extremely well along that plane; cleaved samples exhibit among the best LEED patterns of any cleaved transition-metal oxide. At room temperature the cleaved surface is totally inert to H_2, O_2, SO_2 and H_2O, and adsorption of CO is only seen for very high exposures. When NiO(100) is reduced by inert gas ion bombardment, a fraction of a monolayer of O-vacancy defects is produced. The defective surface interacts strongly with O_2, as indicated by the UPS spectra in Fig. 2. Most striking is the depopulation the Ni 3d defect state (the shoulder at the upper edge of the valence band); that is the emission which arises from electrons localized at O-vacancy defects. Although the UPS difference spectra do not provide unique

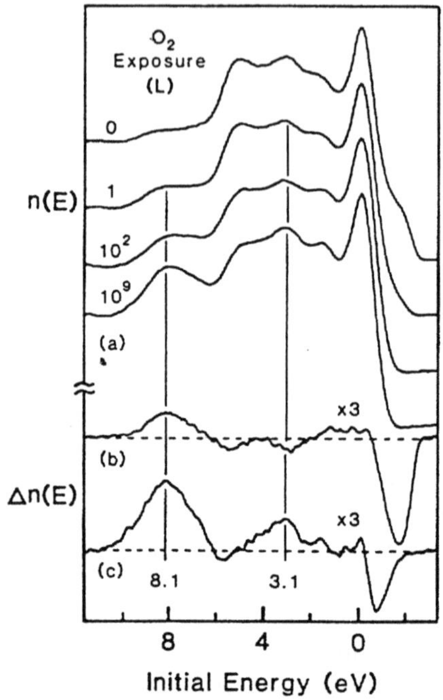

Fig. 2: (a) Ultraviolet photoemission (UPS) spectra for Ar^+-ion-bombarded NiO(100) exposed to the indicated amounts of O_2. (b) Difference spectrum between 0 and 1 L O_2. (c) Difference spectrum between 100 and 10^9 L O_2. hν = 21.2 eV.

identification, the phase that adsorbs during the first 100 L of exposure is thought to be atomic O that has an electronic configuration close to O^{2-}. This is typical of oxidation/reduction interactions on oxide surfaces, where electrons on surface cations play a major role in chemisorption and are transferred to the adsorbed species [1]. For exposures above 100 L, a second adsorbed phase is seen, which has been tentatively identified as adsorbed O_2^{2-}. This behavior is similar to that on most other 3d-transition-metal oxides.

Studies of the interaction of SO_2 with NiO surfaces (motivated by the importance of sulfur as a catalyst poison) show only a weak interaction even on defective surfaces [6]. X-ray photoelectron (XPS) spectra, including some taken using synchrotron radiation in order to maximize the sensitivity of the S 2p emission, indicate that adsorbed S is present both in sulfide bonds to the surface cations and as an S-O complex, probably resulting from incomplete dissociation on the surface.

The interaction of H_2O with NiO(100) surfaces is very interesting [4]. The cleaved surface is inert to H_2O at room temperature, but, surprisingly, the reduced surface exhibits only very weak dissociative adsorption. This is in contrast to all other 3d-transition-metal oxides that have been studied, where H_2O readily adsorbs dissociatively at O-vacancy defects [1]. However, pre-exposure of the defect surface to O_2 results in greatly enhanced dissociative adsorption of H_2O. Exposure to O_2 produces some form of adsorbed atomic O that is different from the lattice O^{2-} ions. It thus appears that the presence of non-lattice O is necessary in order to dissociate H_2O and initiate adsorption. This adsorption

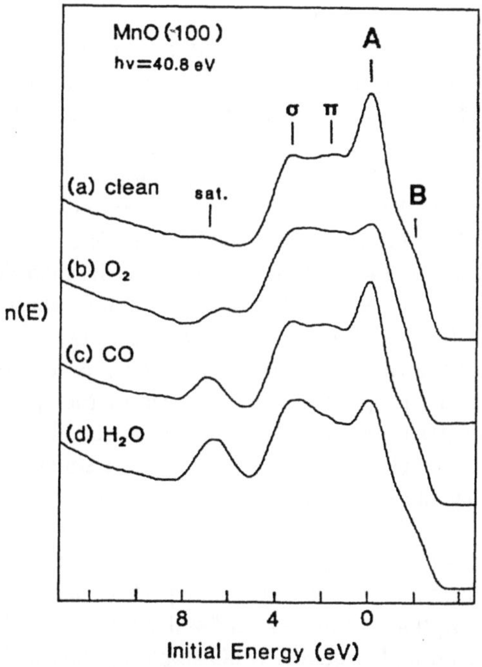

Fig. 3: UPS valence-band spectra of UHV-cleaved MnO (100) (a) clean, and then exposed to 10^6 L of (b) O_2, or (c) CO, or (d) H_2O. The energy scale is referenced to peak A. hν = 40.8 eV.

system deserves further study, since it should give further insight into the nature of adsorbed O species on oxide surfaces. CO is one molecule that will react with stoichiometric NiO, although the probability of interaction is very small [5]. On UHV-cleaved (100) surfaces, no reaction with CO is seen for small exposures, but for large exposures (≥ 10^8 L) CO abstracts lattice O ions from the surface, presumably through the formation of CO_2, which subsequently desorbs. (The same behavior is seen for CoO [7]). By reducing the surface, CO molecules generate O-vacancy defect sites at which other CO molecules can adsorb. Thus the O-vacancy site is the active one for the oxidation of CO on NiO(100). CO is observed to adsorb molecularly at room temperature on NiO(100) surfaces that had been pre-reduced by ion bombardment, with UPS spectra exhibiting emission from the 4σ and unresolved (1π, 5σ) molecular orbitals.

4. Adsorption on Lower Oxides

Lower oxides of the 3d-transition metals exhibit varied chemisorption behavior. MnO has been observed to behave differently from the other rocksalt transition-metal monoxides in adsorption [8]. Both the cleaved and the subsequently ion-bombarded MnO(100) surfaces interact strongly with O_2, H_2O and CO. The changes in the He II UPS spectra produced by exposure of the cleaved surface to

10^6 L of those gases are shown in Fig. 3; the changes observed for the ion-bombarded surface are generally similar. The changes produced by O_2 exposure occur rapidly; within the first 10 L, peak B (largely of Mn 3d origin) almost disappears and peak A (also Mn 3d) is attenuated. This is believed to result from dissociative chemisorption, with adsorbed O^{2-} species removing electrons from the Mn^{2+} ions, raising them to higher valence states. Interaction of the MnO (100) surface with H_2O is also strong. The initial sticking coefficient for H_2O is comparable to that for O_2, but, as for other transition-metal oxides, H_2O exhibits a much weaker interaction with the Mn 3d electrons (peaks A and B) than does O_2. UPS difference spectra indicate that the adsorbed species for low exposures is OH^-. CO interacts with cleaved MnO(100) less strongly than does O_2 or H_2O; there is a smaller interaction with the Mn 3d electrons than there is for H_2O, although additional features, such as the peak at 7 eV, appear in the spectra. Difference spectra suggest that the adsorption is molecular. However, there are questions concerning the quality of the cleaved surface used in these studies. X-ray diffraction analysis showed that the sample contained approximately 7 at.% α–Mn_3O_4. Diffuse MnO(100)-(1x1) LEED patterns were observed, but their quality was significantly poorer than the patterns obtained for NiO. Thus the surface most likely contained a high density of steps, and perhaps also point defects. The surface used for adsorption is thus not as "well characterized" as one would like, and measurements on higher quality MnO single crystals will be necessary in order to clarify the adsorption behavior of MnO.

Ti_2O_3 behaves very differently when exposed to O_2 than to SO_2, although both are oxidizing molecules. Since the oxidation of Ti_2O_3 to TiO_2 is thermodynamically very favorable, Ti_2O_3 is expected to interact strongly with O_2. Somewhat surprisingly, that interaction is relatively weak at room temperature [9]. Figure 4 presents a series of UPS spectra for room temperature O_2 exposure of the UHV-cleaved $(10\bar{1}2)$ surface. The decrease in amplitude of the Ti 3d emission indicates oxidation of the surface Ti ions, as does the disappearance of the shoulder on the high-binding-energy side of the O 2p valence band, which corresponds to the largest Ti 3d admixture to the valence-band wavefunctions [8]. As for NiO, two adsorbed phases occur as a function of exposure, with the initial phase adsorbed O^{2-}. However, even after 10^6 L of O_2, Ti^{3+} ions remain in the near-surface region, indicating that chemisorption has effectively stopped. The presence of O-vacancy defects on the Ti_2O_3 surface significantly enhances its interaction with O_2.

When the stoichiometric Ti_2O_3 $(10\bar{1}2)$ surface is exposed to SO_2, a much stronger reaction ensues, as shown in Fig. 5 [10]. In fact, the interaction of SO_2 with Ti_2O_3 is the strongest adsorbate/substrate interaction that has been observed for any molecule on any metal oxide! In this case, defects are clearly *not* necessary for adsorption. The interaction occurs for such low SO_2 exposures that Fig. 5 shows two different plots of the UPS spectra for SO_2 on stoichiometric Ti_2O_3 $(10\bar{1}2)$ at room-temperature: (a) shows the first 1 L of exposure, and (b) shows exponential exposures up to 10^4 L. The Ti 3d orbitals are rapidly

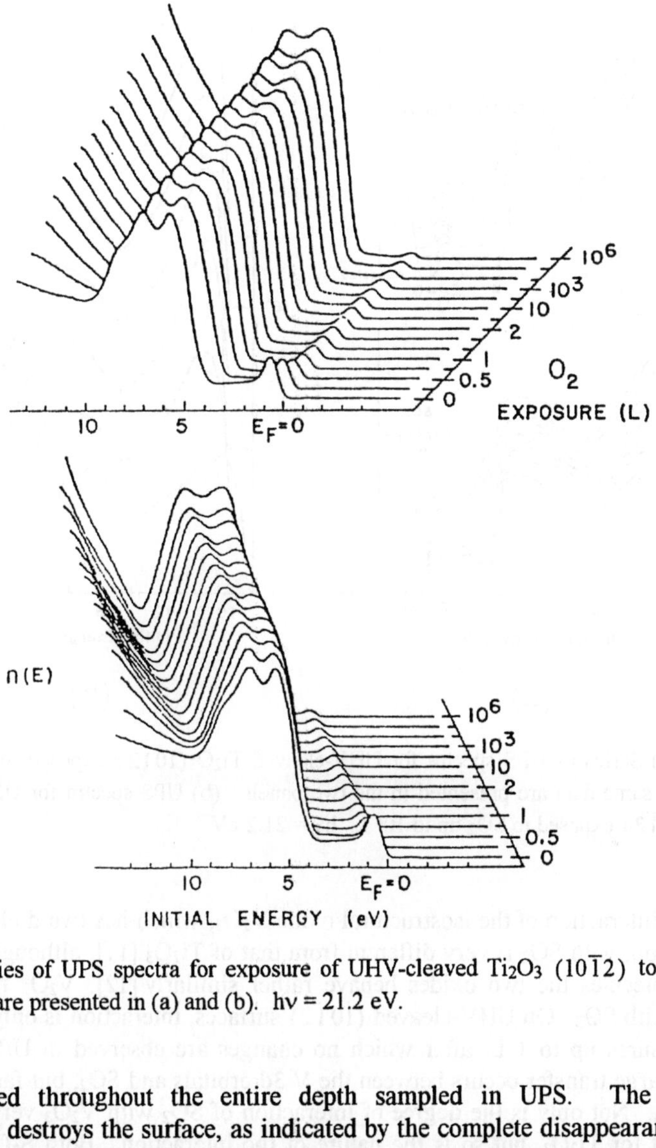

Fig. 4: Series of UPS spectra for exposure of UHV-cleaved Ti_2O_3 $(10\bar{1}2)$ to O_2. The same data are presented in (a) and (b). $h\nu = 21.2$ eV.

depopulated throughout the entire depth sampled in UPS. The reaction physically destroys the surface, as indicated by the complete disappearance of all LEED patterns (which are excellent for the clean surface) by 0.5 L exposure. Careful examination of the UPS spectra shows that changes in the spectra continue to occur up to 10^4 L, the highest exposure used in the study. XPS measurements of the S 2p binding energy indicate that SO_2 completely dissociates and bonds to Ti ions in the surface region, forming a mixture of TiO_2 and TiS_2. This reaction is so strong that it is essentially corrosion rather than chemisorption.

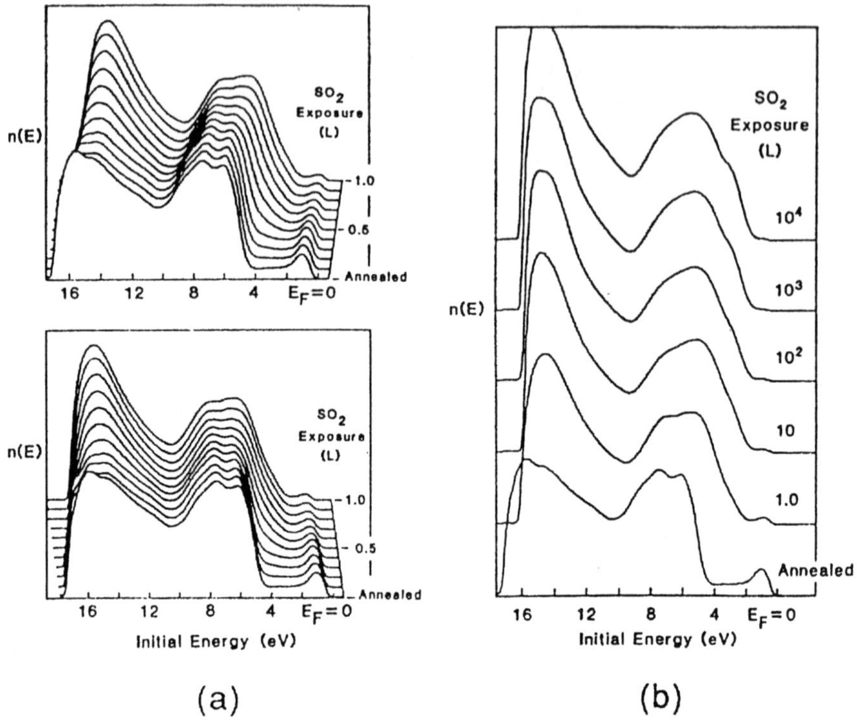

Fig. 5: (a) Series of UPS spectra for UHV-cleaved $Ti_2O_3(10\overline{1}2)$ exposed to SO_2 up to 1 L. The same data are presented in the two panels. (b) UPS spectra for UHV-cleaved $Ti_2O_3(10\overline{1}2)$ exposed to SO_2 up to 10^4 L. $h\nu = 21.2$ eV.

The interaction of the isostructural oxide V_2O_3, which has two d-electrons on each cation, with SO_2 is very different from that of Ti_2O_3 [11], although for most other molecules the two oxides behave rather similarly [12]. V_2O_3 reacts only weakly with SO_2. On UHV-cleaved $(10\overline{1}2)$ surfaces, interaction is only observed for exposures up to 1 L, after which no changes are observed in UPS spectra. Some charge transfer occurs between the V 3d orbitals and SO_2, but far less than for Ti_2O_3. Not only is the degree of interaction of SO_2 with V_2O_3 very different from that for Ti_2O_3, but so is the nature of the interaction. Both S-O and S-V bonds are present, with the former predominating. Thus the $V_2O_3(10\overline{1}2)$ surface does not completely dissociate SO_2, as Ti_2O_3 did. Also unusual is the fact that defective, reduced V_2O_3 surfaces react only slightly more with SO_2 than do stoichiometric ones.

The interaction of H_2O with Ti_2O_3 was studied in order to separate the role of surface cation d-orbital population from geometric effects in the interaction of

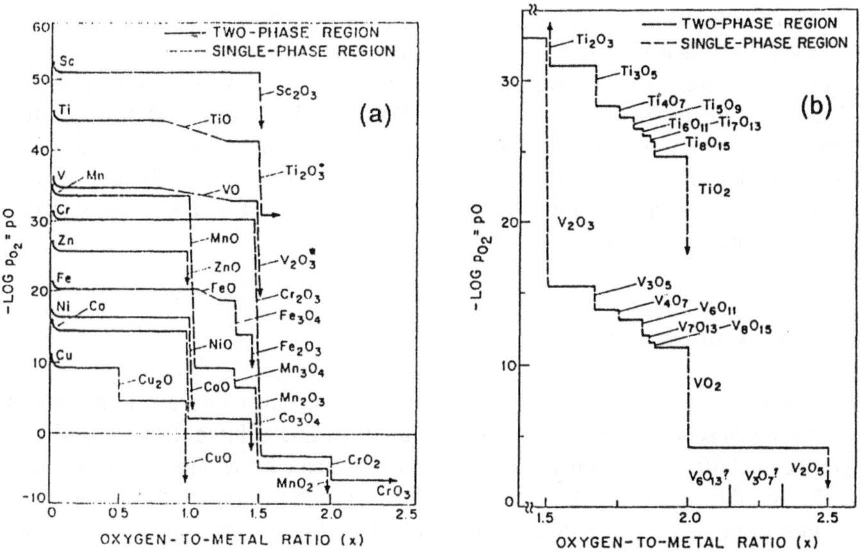

Fig. 6: Oxygen affinities of the 3d-transition metals and their oxides; (b) shows a detail for the oxides of Ti and V only [From Ref. 15].

H_2O with titanias [13]. The perfect $TiO_2(110)$ surface neither adsorbs nor dissociates H_2O to any appreciable extent, but H_2O does dissociate readily at O-vacancy point defects on reduced TiO_2 surfaces. Since O-vacancy point defects are always accompanied by partial population of the 3d orbitals of adjacent Ti cations, it was assumed for several years that the presence of Ti 3d electrons was the necessary and sufficient prerequisite for dissociative adsorption. However, exposure of the UHV-cleaved $Ti_2O_3(10\bar{1}2)$ surface, where *all* Ti cations have a $3d^1$ configuration, to H_2O at room temperature showed only *molecular* adsorption. As for H_2O on reduced TiO_2, there is only a weak interaction with the Ti 3d electrons, as expected for a donor/acceptor interaction [1]. Although the initial sticking coefficient for H_2O on $Ti_2O_3(10\bar{1}2)$ is quite large, saturation surface coverage is less than one-half monolayer. For reduced Ti_2O_3 surfaces, adsorption of H_2O is initially dissociative, as on reduced TiO_2 surfaces. These results show that the presence of Ti 3d electrons on titania surfaces is not sufficient for the dissociation of H_2O, although their presence may well be necessary. Some indirect indication that d-electrons are in fact necessary in order to dissociate H_2O comes from the fact that UHV-fractured $TiO_2(110)$ surfaces, which contain a large density of steps, interact only weakly with H_2O at room temperature. To the extent that the local atomic geometry at steps resembles that at point defects, it appears that *both* structural defects *and* d-electrons are necessary in order for titania to dissociate H_2O.

Some unusual results have been obtained from experiments that are currently underway on adsorption on V_6O_{13} [14]. That oxide has a quasi-two-dimensional layered structure, with easy cleavage along the (001) basal plane. The interaction of O_2 at room temperature with both UHV-cleaved surfaces and surfaces that were subsequently reduced by bombardment with Ar^+ ions was studied by UPS. Although the interaction was weak in both cases, it was *weaker* for the reduced surface than for the stoichiometric one. Such a result, while not unique, is rare among the transition-metal oxides [1].

5. Metal / Oxide Interfaces

The properties of the interface between a metal and an oxide are important in catalysis, gas sensors, etc. Several other papers in this volume discuss results on specific metal/oxide systems. Such interactions can be complex, but a useful guide to the expected behavior can be obtained simply from the bulk thermodynamics of the system. Figure 6 shows the oxygen affinity of the 3d-transition metals and their oxides, in units of the log of the partial pressure of O_2 at 1000 K required for equilibrium between the different phases [15]. Consider the interaction of metallic vanadium with titania surfaces. This system has been studied by using both stoichiometric $TiO_2(110)$ substrates and TiO_2 surfaces that had been pre-reduced by ion bombardment [16]. When V metal is deposited by thermal evaporation onto stoichiometric TiO_2 in UHV, it will compete with the titania for O ions. The oxygen affinity of V metal is 35 from Fig. 6 (a), while the affinity necessary for the reduction of TiO_2 is about 25 [Fig. 6 (b)]. V would therefore be expected to reduce the TiO_2 surface, which is precisely what is found experimentally; changes in the valence band electronic structure in UPS and the appearance of low-binding-energy shoulders on the Ti^{4+} core level peaks in XPS indicate reduction of the TiO_2 substrate. However, if the TiO_2 crystal is first reduced by ion bombardment to a composition close to Ti_2O_3, then the oxygen affinity necessary for reduction of Ti_2O_3, which is about 42 from Fig. 6 (a), is *greater* than that of V metal, and the V should not be able to remove O ions from the titania. UPS spectra exhibit far smaller changes in the valence band spectra for this case than for stoichiometric titania, and XPS spectra confirm that the Ti ions in the reduced titania are not further reduced by V deposition.

The situation is different when the metal deposited has a much lower O affinity than do the cations in the oxide. An example of this is Rh deposition onto TiO_2. Rh has an extremely low oxygen affinity, and when deposited onto stoichiometric TiO_2 surfaces in UHV it does not perturb the substrate electronic structure perceptibly [17]. However, when O-vacancy point defects are present on the TiO_2 surface, a very different type of interaction is observed. The adsorbed Rh *oxidizes* the surface defects, removing electrons from Ti cations at defect sites and transferring them onto the Rh atoms. The importance of defects in chemisorption is characteristic of metal/oxide systems in which the metal has a much lower O affinity than does the oxide.

Acknowledgements

This work was partially supported by NSF, Division of Materials Research Grant DMR-9015448; and DoE, Office of Basic Energy Sciences Grant DE-FG02-87ER13773.

References

[1] V.E. Henrich, P.A. Cox: The Surface Science of Metal Oxides, Cambridge University Press, Cambridge 1993

[2] Z. Zhang, S.-P. Jeng, V.E. Henrich: Phys. Rev. **B 43**, 12004 (1991)

[3] J.M. McKay, V.E. Henrich: Surf. Sci. **137**, 463 (1984)

[4] J.M. McKay, V.E. Henrich: Phys. Rev. **B 32**, 6764 (1985)

[5] J.M. McKay, V.E. Henrich: J. Vac. Sci. Technol. A **5**, 722 (1987)

[6] X. Li, V.E. Henrich: (unpublished)

[7] J.L. Mackay, V.E. Henrich: Phys. Rev. **B 39**, 6156 (1989)

[8] R.J. Lad, V.E. Henrich: J. Vac. Sci. Technol. A **6**, 781 (1988)

[9] R.L. Kurtz, V.E. Henrich: Phys. Rev. **B 25**, 3563 (1982)

[10] K.E. Smith, V.E. Henrich: Phys. Rev. **B 32**, 5384 (1985)

[11] K.E. Smith, V.E. Henrich: Surf. Sci. **225**, 47 (1990)

[12] R.L. Kurtz, V.E. Henrich: Phys. Rev. **B 28**, 6699 (1983)

[13] R.L. Kurtz, V.E. Henrich: Phys. Rev. **B 26**, 6682 (1982)

[14] L. Liu, V.E. Henrich: (unpublished)

[15] T.B. Reed: Free Energy of Formation of Binary Compounds, MIT Press, Cambridge 1971

[16] Z. Zhang, V.E. Henrich: Surf. Sci. **277**, 263 (1992)

[17] H.R. Sadeghi, V.E. Henrich: J. Catal. **109**, 1 (1988)

Molecular Adsorption on Thin Ordered Oxide Films and Single Crystal Oxide Surfaces

*H.-J. Freund, H. Kuhlenbeck, and M. Neumann**

Lehrstuhl für Physikalische Chemie I, Ruhr-Universität Bochum
Universitätsstraße 150, D-44780 Bochum, Germany

* Fachbereich Physik, Universität Osnabrück
Barbarastraße 7, D-49069 Osnabrück, Germany

Abstract. We report adsorption studies on clean and modified surfaces of NiO, Cr_2O_3 and Al_2O_3. Small molecules such as NO, CO and H_2O are adsorbed. The oxides are studied as bulk single crystals as well as as single crystal films. We are able to discriminate between adsorption on ideal terrace sites and defect sites by using EEL-spectra of surface states.

Employing various methods, e.g. ARUPS, HREELS and NEXAFS, we characterize the bonding between molecules and oxide substrates and show that concepts known from molecule metal interactions cannot directly be transferred to oxide surfaces.

1. Introduction

The electronic structure of metal surfaces is predominantly determined by the delocalized nature of the wavefunctions [1]. Localization of the electronic wavefunction in a metal, although present in certain cases, does not play the dominant role for a correct description of the electronic structure of a metal. Exceptions are, of course, the rare earths and actinides with their highly localized levels.

In oxides, on the other hand, localization is crucially important to even approximately understand the electronic structure. Often we are dealing with completely ionic or at least strongly ionic compounds, the constituents of which may be basically described as ions with their localized charges. Frequently, electron correlation plays a very important role to understand the electronic structure of the ions and to describe the transport properties of oxides. One case is NiO which is an example for a so called charge-transfer-insulator [2]. Compounds of this type have received a lot of attention in the past in particular with respect to photoemission and inverse photoemission [3].

The differences in electronic structure alluded to above between metals on one hand and oxides on the other hand are expected to have major consequences also for the interaction of the respective substrates with molecules from the gas phase, in the sense that concepts to describe molecule surface bonding developed

Springer Series in Surface Sciences, Vol. 33 Eds.: E. Umbach · H.-J. Freund
Adsorption on Ordered Surfaces of Ionic Solids and Thin Films

over the last two decades for metals cannot be simply transferred to try to understand bonding towards oxide surfaces.

Therefore several groups have started to investigate adsorption on oxide surfaces experimentally. We have chosen to use bulk oxide single crystal surfaces as well as thin epitaxial oxide films as substrates [4-7]. The main reasons to study both types of substrates are:

i) Bulk single crystals may be cleaved in vacuo and stable surfaces with very few defects are created.
ii) Thin oxide films with various crystallographic orientations including non-polar and polar surfaces may be grown on metallic substrates.
iii) Through variation of the preparation conditions of the films the number and type of defects on such films can be varied and thus their influence on the adsorption behaviour may be studied in comparison to bulk single crystals.
iv) Thin films on metal substrates can be cooled to liquid nitrogen or liquid helium temperatures without sacrifying the conductivity of the samples needed for application of electron spectroscopies.

Oxide surfaces play important roles in catalysis. Some transition metal oxides are catalytically active and others, mainly simple metal oxides are often used as support materials for catalysts. So far we have studied NiO(100) [8-14], NiO(111) [15-17], Cr_2O_3(111) [18-22], CoO(111) [23] and γ-Al_2O_3(111) [13,24-27]. Ni and Co ions prefer the oxidation state II+ and do not have the tendency to change oxidation states while Cr with oxidation state III+ in Cr_2O_3 may relatively easy change its oxidation state either to lower or higher values. In this sense NiO and CoO represent limiting cases as opposed to Cr_2O_3. Al_2O_3 is an example for a support oxide material.

2. Results and Discussion

2.1 NiO

The most thoroughly studied oxide surface in our laboratories is the non-polar surface of NiO, i.e. NiO(100), which we investigated both as a bulk single crystal surface and as a thin film.

We start our discussion by considering this surface. We shall, when possible, compare the properties of the NiO(100) surface with the polar NiO(111) surface. Figure 1 shows the ideal geometric arrangement for both surfaces. LEED investigations on the bulk NiO(100) surface reveal that the surface is rumpled by less than 0.02 Å with an inward relaxation of the surface layer by 0.04 Å which is about 2% [28]. In other words, the deviation from the bulk terminated structure is not dramatic. The quality of a cleaved NiO(100) surface, as indicated by the LEED pattern in Fig. 1, is quite comparable to the quality of metal single crystal surfaces, indicating that the number of point and line defects is rather low. While also NiO(100) [8,9] and NiO(111) [16,17] films exhibit distinct LEED patterns, the spot sizes are considerably larger and the background intensities are higher as compared with the bulk single crystal indicating a higher density of defects on the film surfaces. The comparison of the bulk single crystal and film systems thus allows us to study the influence of defects on the adsorption properties [10,11].

Let us first consider the ideal clean (100) oxide surface. The electronic properties of the clean surface document themselves, among other things, in the existence of surface states [12]. These surface states are localized in the surface layer. For metal surfaces we know that their energy position and intensity, for example, in a photoelectron spectrum strongly depends on the presence of adsorbates on the surface. In the case of an oxide surface we have recently detected such surface states, however, not via photoemission but rather via electron energy loss spectroscopy [12]. Figure 2 shows the EEL-spectra of the electronic excitations of a bulk single crystal surface [29] and of a thin film. In comparison with the clean surfaces we show two spectra after adsorption of water and additional adsorption of NO. Several conclusions can be drawn from the observations: Water adsorption leads to the appearance of OH losses at 460 meV. Coupling to the substrate Fuchs-Kliewer phonons causes the side bands at higher loss energies to appear. The electronic excitations between 0.57 eV and 2 eV are, however, not influenced by water adsorption. On the other hand NO adsorption which shows up by the NO vibrational loss does affect these electronic excitations. In particular, the

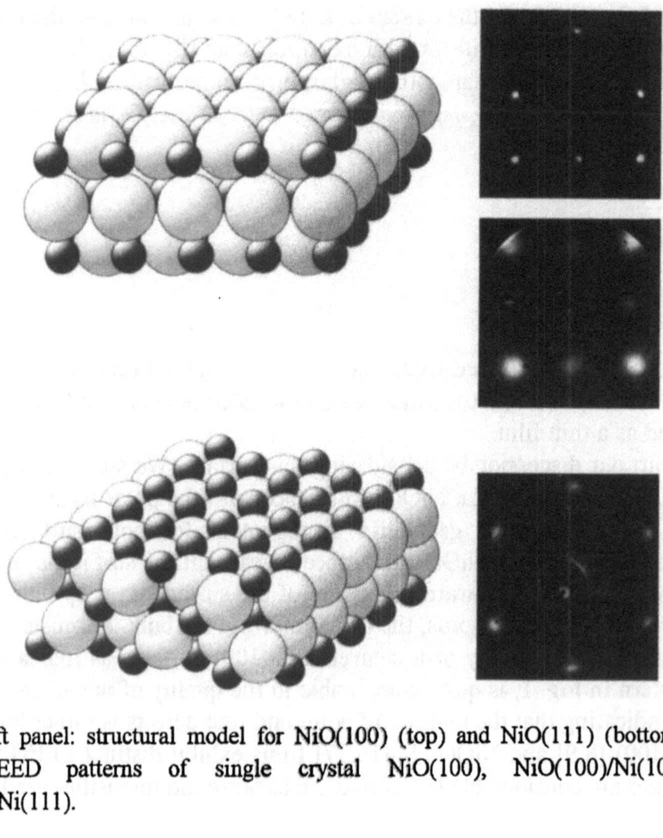

Fig. 1: Left panel: structural model for NiO(100) (top) and NiO(111) (bottom). Right panel: LEED patterns of single crystal NiO(100), NiO(100)/Ni(100), and NiO(111)/Ni(111).

bands at 0.57 eV and 1.62 eV of the clean surface are shifted to 0.9 eV and 1.75 eV, respectively.

These findings represent clear experimental evidence for the existence of surface states at oxide surfaces. The nature of these states has been explained through ab-initio cluster calculations [12]: The states between 0.57 eV and 2 eV are due to d → d transitions localized on the Ni ions. The lowest lying d → d transitions in bulk NiO occur at 1.05 eV and 1.79 eV and are due to excitations between the ground $^3A_{2g}$-state and the $^3T_{2g}$ and $^3T_{1g}$-states, respectively. They may be well described via electronic excitation within an octahedral NiO_6^{10-} cluster. The surface states can be derived from these bulk states by realizing that the Ni ions in the surface are coordinated only by five O^{2-} ions which reduces the ligand field strength with respect to the bulk and stabilizes the excited states. A particularly pronounced effect is observed for the lowest lying state shifting by 0.48 eV. It is this state that is strongly influenced by the presence of NO but not by OH. We believe that this is due to the preferential adsorption of OH on the defects at the surface while NO adsorbs preferentially on the regular NiO(100) terrace sites. As a consequence of the localization of the d → d excitations to the Ni^{2+} centers, the shift of the surface state proves that NO binds to the Ni^{2+} centers and not to the oxygen ions in the surface. This conclusion is in line with vibrational HREELS results where the stretching frequency is found close to the gas phase value which is compatible with this bonding configuration [8,9]. The NO axis turns out to be tilted by about 45° with respect to the NiO(100) surface

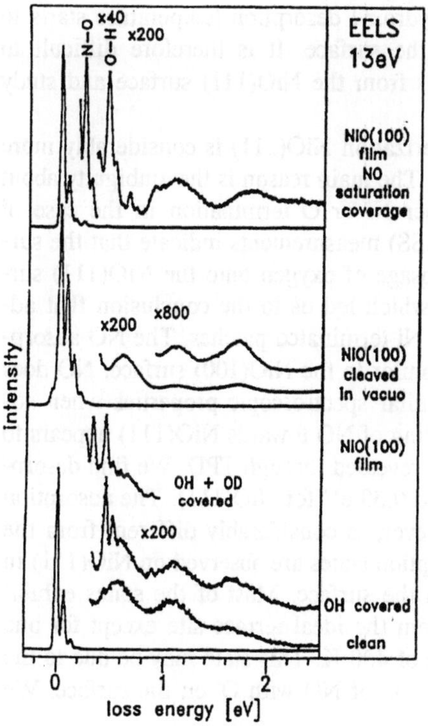

Fig. 2: ELS spectra in specular reflection at E_p = 13 eV of a) a cleaved NiO(100) surface, b) a clean thin NiO(100) film, c) a thin NiO(100) film after H_2O adsorption and d) a thin NiO(100) film after NO + H_2O adsorption.

normal as revealed through a NEXAFS analysis [8,9]. Together with the TPD data [8,9] which are compatible with an adsorption energy of 0.5 eV for both desorption from bulk and thin film NiO(100) surface the picture for the NO/NiO(100) system is rather complete. The experimental data on this system are complemented with theoretical analysis from several groups [30,31] which give a quite satisfactory rational both for the binding energy as well as for the adsorption geometry. This will be discussed in another paper in this volume [32].

The nature of the OH bond to the NiO(100) surface is considerably less clear as compared with the NO-NiO(100) bond. Firstly, we can conclusively show that the loss at 460 meV is due to OH groups by using isotopic exchange experiments. A spectrum, where part of the OH groups is exchanged by OD groups is included in Fig. 2 [16]. Secondly, it was already mentioned that the hydroxyl groups do not seem to adsorb on regular terrace sites but rather on defects. Experiments with water adsorption on bulk single crystals with very low point defect density (very low background intensity in LEED) show that dissociative sticking of water with hydroxyl formation is widely suppressed, corroborating the idea that defects are involved in the process. If a weakly sputtered bulk single crystal is exposed to water, surface hydroxyls are formed. In the case of a NiO(100) surface OH can be removed through thermal treatment. A surface not saturated with hydroxyl groups exhibits a modified behaviour with respect to exposure towards other adsorbates, e.g. NO. Hydroxyl groups sometimes also influence the geometric structure of the surface and its thermal stability. The NiO(111) surface represents such a case [16]. We have indications that NiO(111), formed via oxidation of a Ni(111) surface and heated above the hydroxyl desorption temperature starts to facet. NiO(100) facets are formed on the surface. It is therefore difficult to completely remove the surface hydroxyls from the NiO(111) surface and study the completely clean surface.

With respect to a structural characterization NiO(111) is considerably more complicated as compared with NiO(100). The main reason is the ambiguity about surface termination, which may be either Ni or O termination in the case of NiO(111). Ion scattering spectroscopic (ISS) measurements indicate that the surface exhibits a mixed termination. Predosage of oxygen onto the NiO(111) surface strongly attenuates NO adsorption which led us to the conclusion that adsorption takes place preferentially on the Ni terminated patches. The NO adsorption behaviour on this surface is quite similar to the NiO(100) surface. NO does not have exactly the same but rather similar spectroscopic properties when adsorbed on NiO(100) and NiO(111). Bonding of NO towards NiO(111) appears to be slightly stronger, i.e. by about 10%, as revealed through TPD. We find desorption energies of 0.52 eV for NiO(100) and 0.59 eV for NiO(111). The adsorption behaviour of the NiO(111) surface, however, is considerably different from the NiO(100) surface. Several new NO adsorption states are observed on NiO(111) in TPD spectra when OH is removed from the surface. Most of the states exhibit desorption energies not very different from the ideal terrace site except for one state which has a desorption temperature of 450 K. This state may be due to the formation of a NO_2^- species through reaction of NO with O^- on the surface. We do have experimental evidences from XPS data that such species may be present on the NiO surfaces of thin films [17]. This example illustrates how the presence of hydroxyl groups may influence the reactivity of an oxide surface. Bonding of CO on the NiO surfaces is considerably weaker than that of NO. Desorption temperatures are around 100 K. The experimental studies for CO adsorption have not yet been completed for NiO.

2.2 Cr_2O_3

$Cr_2O_3(111)$ may be prepared by oxidation of Cr(110) [19,20]. Similar to NiO there are surface states on the $Cr_2O_3(111)$ surface [19,20]. The surface is hexagonal and polar. Therefore the termination ambiguity discussed above for NiO(111) exists here as well. Electronic EEL-spectra show that the surface Cr ions adopt an oxidation state of 2+ as opposed to 3+ in the bulk. This means, the surface has a high reduction potential which is documented by the formation of oxygen adsorbates and simultaneous attenuation of the surface Cr^{2+} state. CO forms an ordered $(\sqrt{3} \times \sqrt{3})R30°$ structure on the $Cr_2O_3(111)$ surface. The orientation of the molecular axis is peculiar in this case: The CO axis is oriented parallel to the surface plane as revealed by angle resolved photoemission and NEXAFS [18,20]. It is evident from the discussion presented elsewhere [18-20] that the bonding of CO to $Cr_2O_3(111)$ is different from the bonding of CO to all metallic surfaces investigated so far.

2.3 Al_2O_3

As an example for a non-transition metal oxide we have investigated Al_2O_3. Aluminiumoxide is the prototype of a support material and particularly interesting in this respect. An Al_2O_3 layer with a quasi-hexagonal oxygen layer as the terminating surface may be grown via oxidation on top of a NiAl(110) alloy surface. The oxide layer is very thin, i.e. it exhibits a thickness of about 5 Å as revealed by XPS and AES. The structure of this film has been studied with various techniques including scanning tunnelling microscopy, LEED, EELS and ARUPS [24].

LEED indicates a well ordered structure with a large unit cell in real space. ISS measurements show that the oxide surface is oxygen terminated. The analysis of the intensity distribution among the LEED spots points to a hexagonal or quasihexagonal arrangement of oxygen atoms in the terminating layer.

This conclusion is corroborated by ARUPS measurements of the oxide bandstructure dispersion parallel to the surface [4-7]. The experimental dispersion curves are shown in Fig. 3 together with a computer simulation for a distorted hexagonal oxygen 2p band structure. The oxygen-oxygen distance in the quasihexagonal oxygen layer is by about 6% larger as compared with bulk Al_2O_3 but this is compatible with the limited thickness of the film. The dispersions have

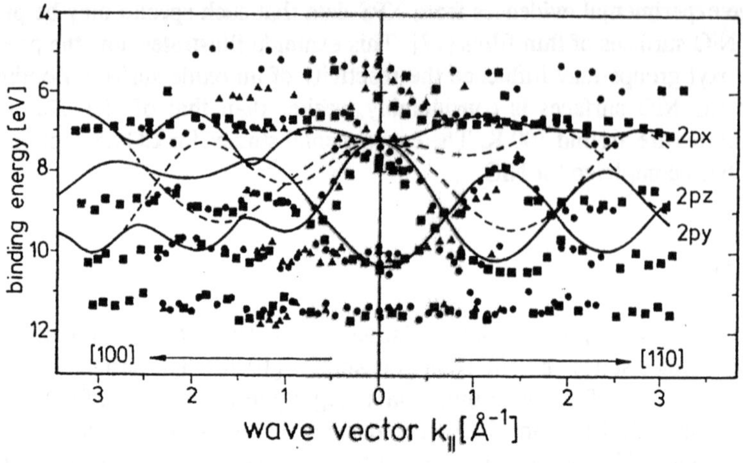

Fig. 3: Measured $k_{\|}$ dispersion of a γ-Al_2O_3(111) surface. The result of a tight binding band structure calculation of a two dimensional O^{2-} overlayer is compared with the experimental data.

been measured with respect to the two azimuths (110, 100) of the NiAl substrate and these are not high symmetry directions with respect to the oxide Brillouin zone. The calculations, which are described in detail elsewhere, reproduce the basic features of the bandstructure except for the two bands at the top and the bottom of the oxide band. These we assign to aluminium derived contributions indicating the aluminium-oxygen hybridization. What are the aluminium sites in the oxide? We believe from our HREELS data that this oxide film is related to a γ-Al_2O_3(111) surface, i.e. there are octahedral and tetrahedral sites.

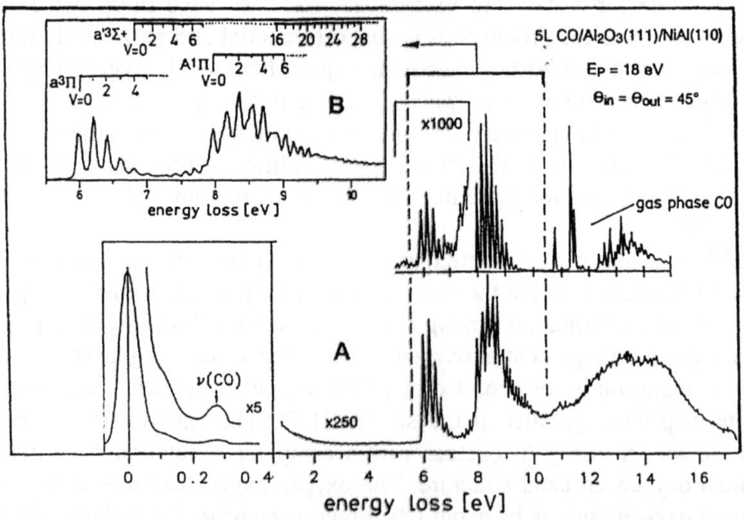

Fig. 4: EEL spectrum (E_p = 18 eV, $\theta_{in} = \theta_{out}$ = 45°) of 5 L CO on Al_2O_3/NiAl(110) at T = 35 K in comparison with details of the CO gas phase spectrum (E_p = 48 eV).

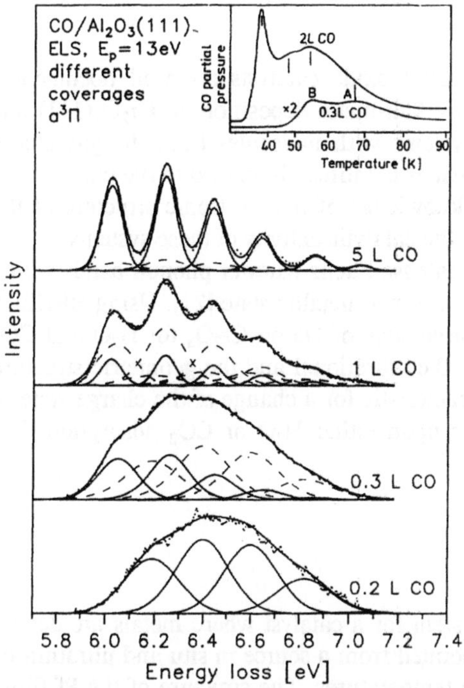

Fig. 5: EEL spectra of the $a^3\pi$ state of CO on Al_2O_3/NiAl(110) for different coverages. To the experimental data we have fitted vibrational progressions. The inset shows TPD spectra.

Most molecules interact with the γ-Al_2O_3(111) surface rather weakly. As an example we show in Fig. 4 ELS spectra of CO physisorbed on Al_2O_3 taken at 35 K [25]. The ELS spectra of the multilayer system show rather pronounced vibrational splitting of the electronically excited states. The comparison with gas phase spectra indicates the rather close correspondence. It is interesting to note that in the monolayer regime the line widths of the vibrational components increase and the bands shift to higher excitation energy which is demonstrated in Fig. 5 for the $a^3\pi$ excitation of CO. There are various CO states identifiable via their line positions and line widths in ELS and their TDS spectra (see inset in Fig. 5). On the basis of such results we can estimate the pressure one would have to apply in order to create a coverage similar to the low temperature situation at higher temperature [4-7].

2.4 Metal Deposits on Oxide Surfaces

One goal of our activities is to model a catalytic system by depositing metal on oxide surfaces. In the following we shortly refer to two examples for metal/metal-oxide systems:

Na/Cr_2O_3

Because alkali-metal promoters of metal oxide catalysts are well documented [33,34], it was near at hand to study alkali-metal deposition on $Cr_2O_3(111)$ and the interaction of such composite systems with molecules from the gas phase. Since the catalytic activity of metal oxides is intimately related to the charge state of the surface atoms [8,9,18-20], a knowledge of the electronic structure at the surface is fundamental to understand the catalytic activity of these systems.

Most oxide systems exhibit very intense Fuchs-Kliewer phonon bands which can be quenched by overlayers which are in a metallic state [21]. Using HREELS and ARUPS we have studied the charge state of Na on Cr_2O_3 for H_2O and CO_2 adsorbates. The changes observed in the vibrational and the geometric structure of the oxide and the Na film are characteristic for a change of the charge state of the Na atoms from metallic to ionic upon either H_2O or CO_2 adsorption. For details we refer to ref. [21].

Pt/Al_2O_3

$Pt/Al_2O_3(111)$ represents a model system for a catalyst where metals are deposited on a support material. Pt is evaporated from a source in situ and ultrathin Pt films are created at different surface temperatures. The structure of the Pt films and their interaction with the oxide surface is not completely understood yet but SPA-LEED and TDS of the CO covered films already show some characteristic features. The TDS data reveal in addition to a desorption signal between $350\,K < T < 500\,K$, which is likely due to desorption from metallic Pt, a low temperature signal. The latter desorption peak we attribute to desorption from Al_2O_3 which is possibly modified by Pt atoms incorporated into the oxide material [7]. This interpretation is corroborated by SPA-LEED data which reveal that deposition of Pt partly destroys the order within the oxide film [35].

3. Synopsis

Summarizing we have characterized the electronic properties of oxide surfaces including transition metal oxides and simple metal oxides. It is shown that the metal cations in the surface are characterized by a specific ELS spectrum indicating the existence of particular surface states which are very important to understand the adsorptive behaviour of these surfaces. The surface states allow us to differentiate ideal terrace sites from defect sites on oxide surfaces. We have discussed the difference between polar and non-polar surfaces considering the stabilization induced by adsorption at defects. Physisorption and chemisorption of small molecules at oxide surfaces lead to interesting adsorption geometries. The bonding involved is very different on oxide surfaces as compared to metal surfaces and the adsorption properties of the oxide surface are dramatically modified by depositing metal onto the surface.

Acknowledgements

Our work has been supported through the Deutsche Forschungsgemeinschaft (Forschergruppe "Modellkat"/Bochum, SFB 225/Osnabrück), the Bundesministerium für Forschung und Technologie (Verbundforschung), the Ministerium für Wissenschaft und Forschung des Landes Nordrhein-Westfalen and the Fonds der chemischen Industrie.

References

[1] N.F. Mott, H. Jones: The Theory of the Properties of Metals and Alloys, Dover Publications, New York

[2] S. Hüfner, J. Osterwalder, T. Riesterer, P. Hullinger: Solid State Commun. **52**, 793 (1984)

[3] S. Hüfner, P. Steiner, I. Sander, M. Neumann, S. Witzel: Z. Phys. B **83**, 185 (1991)

[4] H. Kuhlenbeck, H.-J. Freund in Metal-Ligand-Interactions: from atoms to clusters to surfaces, (Eds. D.R. Salahub, N. Russo), Nato advanced study institute, Series Vol. 378, p. 37 (1992), Kluwer Acad. Publishers

[5] H. Kuhlenbeck, H.-J. Freund: Adsorption and reaction of small molecules on oxide surfaces, Springer Proceedings in Physics, in press

[6] H.-J. Freund, Th. Mull, M. Menges, B. Baumeister, B. Adam, K. Al-Shamery, D. Weide, C. Fischer, P. Andresen: Low energy excitations and desorption ynamics from oxide surfaces, in: Desorption induced by electronic transitions (DIET V), Springer Series in Surface Science, in press

[7] H.-J. Freund, B. Dillmann, D. Ehrlich, M. Haßel, R.M. Jaeger, H. Kuhlenbeck, C.A. Ventrice Jr., F. Winkelmann, S. Wohlrab, C. Xu, Th. Bertram, A. Brodde, H. Neddermeyer: J. Mol. Catal., in press

[8] H. Kuhlenbeck, G. Odörfer, R.M. Jaeger, G. Illing, M. Menges, Th. Mull, H.-J. Freund, M. Pöhlchen, V. Staemmler, S. Witzel, C. Scharfschwerdt, K. Wennemann, T. Liedtke, M. Neumann: Phys. Rev. **B43**, 1969 (1991)

[9] St. Uhlenbrock, Chr. Scharfschwerdt, M. Neumann, G. Illing, H.-J. Freund: J. Phys. C **4**, 7973 (1992)

[10] M. Bäumer, D. Cappus, H. Kuhlenbeck, H.-J. Freund, G. Wilhelmi, A. Brodde, H. Neddermeyer: Surf. Sci. **253**, 116 (1991)

[11] M. Bäumer, D. Cappus, G. Illing, H. Kuhlenbeck, H.-J. Freund: J. Vac. Sci. Technol. A **10**, 2407 (1992)

[12] A. Freitag, V. Staemmler, D. Cappus, C.A. Ventrice Jr., K. Al-Shamery, H. Kuhlenbeck, H.-J. Freund: Chem. Phys. Lett., accepted

[13] U.J. Katter, H. Schlienz, M. Beckendorf, H.-J. Freund: Ber. Bunsenges. Phys. Chem. **97**, 340 (1993)
[14] Th. Mull, B. Baumeister, M. Menges, H.-J. Freund, D. Weide, C. Fischer, P. Andresen: J. Chem. Phys. **96**, 7108 (1992)
[15] D. Cappus, C.A. Ventrice Jr., K. Al-Shamery, C. Xu, D. Ehrlich, H. Kuhlenbeck, H.-J. Freund: Chem. Phys., submitted
[16] D. Cappus: Ph.D. thesis, Ruhr-Universität Bochum 1994
[17] M. Menges: Ph.D. thesis, Ruhr-Universität Bochum 1993
[18] C. Xu, B. Dillmann, H. Kuhlenbeck, H.-J. Freund: Phys. Rev. Lett. **67**, 3551 (1991)
[19] C. Xu, M. Haßel, H. Kuhlenbeck, H.-J. Freund: Surf. Sci. **258**, 23 (1991)
[20] H. Kuhlenbeck, B. Dillmann, M. Haßel, B. Adam, D. Ehrlich, S. Wohlrab, C. Xu, H.-J. Freund, U.A. Ditzinger, H. Neddermeyer, M. Neuber, M. Neumann: Ber. Bunsenges. Phys. Chem. **96**, 15 (1992)
[21] C.A. Ventrice Jr., D. Ehrlich, B. Dillmann, E.L. Garfunkel, H.-J. Freund: Phys. Rev. **B46** 12892 (1992)
[22] D. Ehrlich, Ph.D. thesis, Ruhr-Universität Bochum 1994
[23] M. Haßel, Ph.D. thesis, Ruhr-Universität Bochum 1994
[24] R.M. Jaeger, H. Kuhlenbeck, H.-J. Freund, M. Wuttig, W. Hoffmann, R. Franchy, H. Ibach: Surf. Sci. **259**, 235 (1991)
[25] R.M. Jaeger, H. Kuhlenbeck, H.-J. Freund: Chem. Phys. Lett. **203**, 41 (1993)
[26] S. Wohlrab: Ph. D. thesis, Ruhr-Universität Bochum 1993
[27] F. Winkelmann: Ph. D. thesis, Ruhr-Universität Bochum 1994
[28] C.G. Kinniburgh, J.A. Walker: Surf. Sci. **63**, 274 (1977)
[29] St. Uhlenbrock, M. Neumann, D. Cappus, B. Dillmann, H.-J. Freund: unpublished results
[30] M. Pöhlchen, V. Staemmler, J. Wasilewski: Surf. Sci., in press
[31] L.G.M. Pettersson: to be published
[32] V. Staemmler: this volume
[33] C. Lin, T. Ito, J. Wang, J.H. Lunsford: J. Am. Chem. Soc. **109**, 4808 (1987)
[34] X. Zhang, R.K. Ungar, R.M. Lambert: J. Chem. Soc. Chem. Commun. **8**, 473 (1989)
[35] J. Libuda, S. Wohlrab, F. Winkelmann, M. Bäumer, H.-J. Freund: to be published

The Growth and Structure of Ultrathin Metal Films on $TiO_2(110)$

Th.E. Madey, U. Diebold, and J.-M. Pan

Department of Physics and Laboratory for Surface Modification
Rutgers, The State University of New Jersey, Piscataway, NJ 08855-0849, USA

Abstract. We have studied the growth kinetics and structure of 3d transition metals (Cr, Fe, Cu) evaporated onto $TiO_2(110)$, a model oxide substrate. Using Low Energy Ion Scattering (LEIS) and X-ray Photoelectron Spectroscopy (XPS), we find that Cu grows in the Volmer-Weber (3-D island) mode. The mobility of Cu atoms is very high, and 3-dimensional clusters are formed even at 160 K for Cu thicknesses < 30 Å. In contrast, the Cr and Fe films completely cover the TiO_2 substrate for thicknesses of a few ML, indicating that Cr and Fe "wet" the surface better than Cu. The ability of the metal film to cover (wet) the oxide substrate correlates directly with the reactivity of the metal toward oxygen. There is no XPS evidence for a strong chemical interaction between the Cu film and the oxide substrate. However, for the Cr and Fe films, we observe a strong interaction between the films and the oxide substrate, i.e., oxidation of the metal overlayers and reduction of the substrate, at submonolayer thickness. A bcc(100) structure is observed for Cr and Fe films with thickness ≥ 10 Å, on both stoichiometric and sputtered substrates using Angle-Resolved XPS (ARXPS), Medium Energy backscattered Electron Diffraction (MEED) and LEED. The overlayer structure orientation is related to the substrate local structure.

1. Introduction

There is a large number of applications where the interfaces between metals and oxides are important. These include thin film technology, fiber optics, electronic devices and heterogeneous catalysis. Many of the properties of a metal oxide system depend on the physical and chemical interactions taking place at the interface.

An important scientific issue in studies of metal/oxide systems is the relation between the metal/oxide interaction at the interface and the growth modes of the metal thin films, i.e., the morphology of the films in the initial stages of growth [1]. In the present work, we have studied the growth of a series of 3d transition metals on $TiO_2(110)$ surfaces using Low Energy Ion Scattering (LEIS) together with X-ray, Photoelectron Spectroscopy (XPS) [2-4]. This is a powerful combination for studying ultrathin film growth [5]: because of the high sensitivity of

Springer Series in Surface Sciences, Vol. 33 Eds.: E. Umbach · H.-J. Freund
Adsorption on Ordered Surfaces of Ionic Solids and Thin Films

LEIS to the composition in the topmost layer, we can separate 3-D island (Volmer-Weber or VW growth) from the layer + island (Stranski-Krastanov or SK growth) and layer-by-layer (Frank-van der Merwe or FM growth) modes by identifying the coverage necessary to complete the first monolayer. There have been various reports [6-8] of the interaction of metals with TiO_2 surfaces that focus on the electronic properties of the interface; the focus in the present work is on the growth, reactivity and structure of ultrathin films.

TiO_2 has been studied extensively, mainly because of its interesting properties in catalysis, photocatalysis and other applications. The $TiO_2(110)$ surface is the charge neutral surface of rutile TiO_2, and is believed to have the bulk truncation structure [9,10]. An experimental advantage of TiO_2 is that the bulk can be reduced slightly by annealing so that the sample is free from charging problems. The surface region can be reoxidized to restore stoichiometry, as determined using photoelectron spectroscopy [11-13]. We observe in this work a clear correlation between the ability of the metal to wet or cover the oxide substrate in the initial stages of growth, and the reactivity of the metal towards oxygen. Of the three metals studied (Cr, Fe, Cu), the most reactive (Cr) wets or covers the TiO_2 surface best, while the least reactive (Cu) forms 3-dimensional clusters on the TiO_2 surface (VW growth) and is least effective in covering the substrate.

2. Experimental

The experiments are performed in an UHV system [10,14] with base pressure $2x10^{-10}$ Torr. It contains basic probes for surface characterization including X-ray Photoelectron Spectroscopy (XPS) and Low Energy Ion Scattering (LEIS). A concentric hemispherical energy analyzer with 100 mm radius is used for both XPS and LEIS. A retarding field optics with microchannel plates for image intensification is equipped for Low-Energy Electron Diffraction (LEED) and Medium Energy backscattered Electron Diffraction (MEED). Residual gas analysis is performed using a Quadrupole Mass Spectrometer (QMS). LEIS is performed at an incidence angle of 40° with a fixed scattering angle of 135°, and 1 keV He^+ and 1.5 keV Ne^+ are used as probe particles. The ion fluence for each measurement is controlled at $< 10^{14}$ ions/cm^2 to minimize possible beam damage [14]. Each film deposition is made on a freshly prepared substrate so that the accumulation of the beam damage from LEIS measurements is prevented. The substrate is a 5 x 10 x 1 mm^3 TiO_2 crystal with polished (110) surface and is annealed in UHV up to 1000 K for hours so that the bulk is reduced enough to become conductive. The stoichiometric surface is restored by annealing the sample in $2x10^{-6}$ Torr of oxygen [14]. The metal films are deposited from a liquid nitrogen cooled thermal evaporator source at a rate of 1 - 3 Å/min. The thicknesses of metal films are monitored by a Quartz Crystal Monitor (QCM). Rutherford backscattering (RBS) measurements are performed as a calibration of the true thickness. The thicknesses reported in this paper are average values of thickness in Angstroms (Å) by assuming uniformly thick films with the appropriate bulk density.

3. Results

3.1 LEIS Studies of the Growth of Reactive Metal Films on the TiO_2 Substrate

LEIS (0.5-5 keV) using rare gas ions is a very surface sensitive technique because only ions scattered from the topmost layer can survive due to the high neutralization probability of rare gas ions near a surface [15]. Figure 1A shows typical LEIS spectra of Fe (top) and Cr (bottom) on $TiO_2(110)$ using 1.0 keV $^4He^+$ [4]. The substrate oxygen peak is well separated from the metal (Ti, Cr, and Fe) peaks. For Fe on TiO_2, the mass separation of Fe and Ti is large enough that both peaks can be well resolved using He^+. To increase the mass resolution for Cr on TiO_2, $^{20}Ne^+$ with 1.5 keV primary energy is also used as a probe particle, as shown in the spectrum of Fig. 1B. Utilizing isotopically pure ^{20}Ne, satellite peaks from the naturally occurring ^{22}Ne isotope are eliminated.

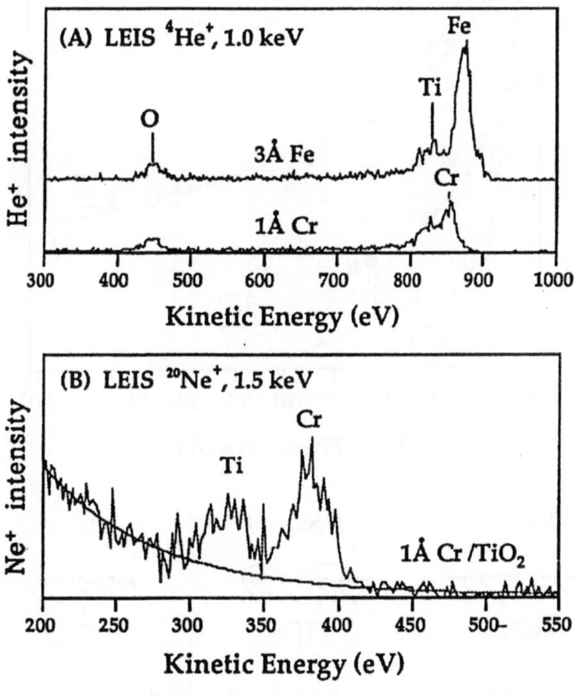

Fig. 1: LEIS spectra of (A) He^+ ion scattering from Fe or Cr deposited onto $TiO_2(110)$ surfaces, (B) $^{20}Ne^+$ ion scattering from Cr deposited on $TiO_2(110)$. The mass resolution is better using $^{20}Ne^+$.

Figure 2 shows the attenuation of the substrate Ti (LEIS) signal with increasing coverages of Cu, Fe or Cr deposited onto $TiO_2(110)$ at 300 K. The intensities are normalized to the value for the clean TiO_2 surface. The attenuation of the substrate signal is an upper limit for the covering of the surface by overlayer atoms. For complete covering (wetting) of the substrate by the metal overlayer one would expect a linear decrease with the signal approaching zero at monolayer coverage (~ 2-3 Å). Such behavior is observed for Fe on TiO_2 at 160 K [3] and in the initial stages of Cr deposition in Fig. 2, but is very different from the behavior for Fe and Cu shown in Fig. 2. In the case of Cu, the fact that a LEIS substrate signal is visible up to high average coverages is clear evidence for Volmer-Weber (3-dimensional islands) growth [2], in disagreement with the Stranski-Krastanov growth mode proposed formerly for $Cu/TiO_2(110)$ [6]. Data identical to those of Fig. 2 are found for Cu/TiO_2 even at 160 K [2]. In the case of Cr, the LEIS signal initially is attenuated linearly up to a point where ~ 80% of the substrate is covered. It appears that Cr grows in a quasi-two-dimensional

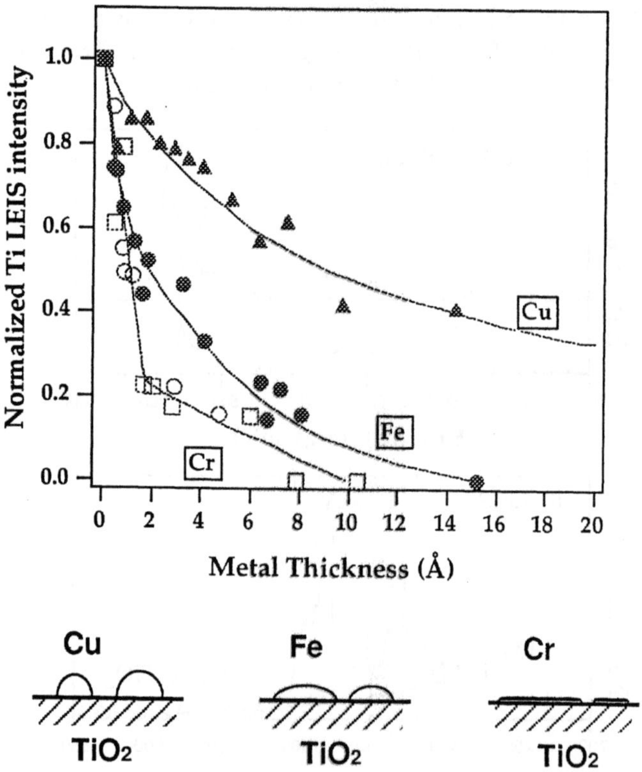

Fig. 2: Normalized LEIS Ti intensities as functions of metal thicknesses measured using a calibrated QCM. For Cr/TiO_2, both He^+ (open circles) and Ne^+ (open squares) scattering data are shown. The Fe and Cu data are for He^+ scattering. Schematics of the initial stages of film growth are shown.

fashion, where the first monolayer covers ~ 80% of the substrate. For higher Cr coverages, the growth of the Cr film deviates from 2-D growth and seems 3-dimensional [4]. The growth behavior of Cr/TiO_2 is similar to that reported recently for Cu/ZnO [16]. In the case of Fe at 300 K, the film grows clearly in 3-D islands initially, and its wetting ability is intermediate between that of Cr and Cu. Schematics of the initial growth modes are shown at the bottom of Fig. 2.

3.2 XPS of the Metal/Oxide Interface

The chemical interaction at the interface between metal and oxide involves oxidation of the overlayer metal and reduction of the TiO_2 substrate, and is characterized using XPS. Figure 3 compares the XPS Ti $2p_{3/2}$ signal of a clean TiO_2 surface and of the surfaces on which Cu, Fe and Cr have been deposited to an average thickness of 2 Å, respectively. For Cu, the Ti $2p_{3/2}$ lineshape remains unchanged, only a slight attenuation is observed, and no evidence for reduction of the substrate (or oxidation of the Cu) is found [2]. In sharp contrast, the Ti $2p_{3/2}$ peak is attenuated and broadened after deposition of 2 Å of Fe and Cr (Fig. 3a, 3b). The shoulder at low binding energy indicates that Ti^{4+} is reduced to Ti^{3+} and (in the case of Cr) to even lower oxidation states of Ti. The reduction is greatest for the most reactive of the three metals, Cr. The reduction of Ti cations is accompanied by oxidation of < 0.5 ML of Fe atoms and < 1 ML of Cr atoms [3,4].

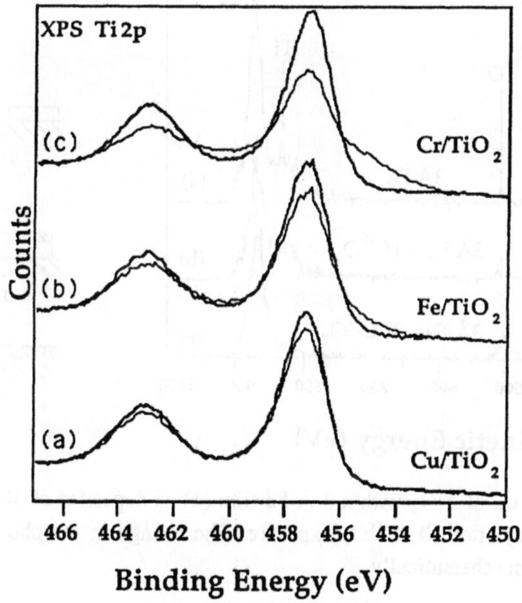

Fig. 3: Evidence for interface reactivity. XPS Ti 2p signals at 300 K for 2 Å (a) Cu, (b) Fe, (c) Cr deposited on $TiO_2(110)$. For comparison, the spectrum for the uncovered surface is shown in each case.

3.3 Oxidation of an Fe Film: Effect on Growth

For the reactive metals, it is of interest to understand the effects of oxidation on the growth of films on oxide substrates [3]. Figure 4 (a) shows LEIS spectra for a 3 Å Fe film on stoichiometric $TiO_2(110)$, and after exposure to 1 L (b) and 10 L (c) of $^{18}O_2$. Isotopic $^{18}O_2$ is used to separate adsorbed ^{18}O from substrate ^{16}O. Prior to the oxygen adsorption, 3 Å of Fe covers only about 50% of the substrate at 300 K. After 1 L of $^{18}O_2$, the Fe signal is attenuated due to preferential ^{18}O adsorption on the Fe clusters, but the Ti intensity is not changed significantly. After 10 L of $^{18}O_2$ exposure, the substrate ^{16}O and Ti signals are greatly suppressed. The dominant peaks in Fig. 4c are overlayer ^{18}O and Fe signals, suggesting that the overlayer has spread over the substrate. Parallel XPS measurements show that Fe is oxidized to Fe^{2+} [3]. We conclude that oxidation of Fe flattens the overlayer, causing it to spread and completely cover the substrate at 1 ML, as indicated schematically in Fig. 4. There was no evidence found for an oxygen-induced spreading of Cu on TiO_2.

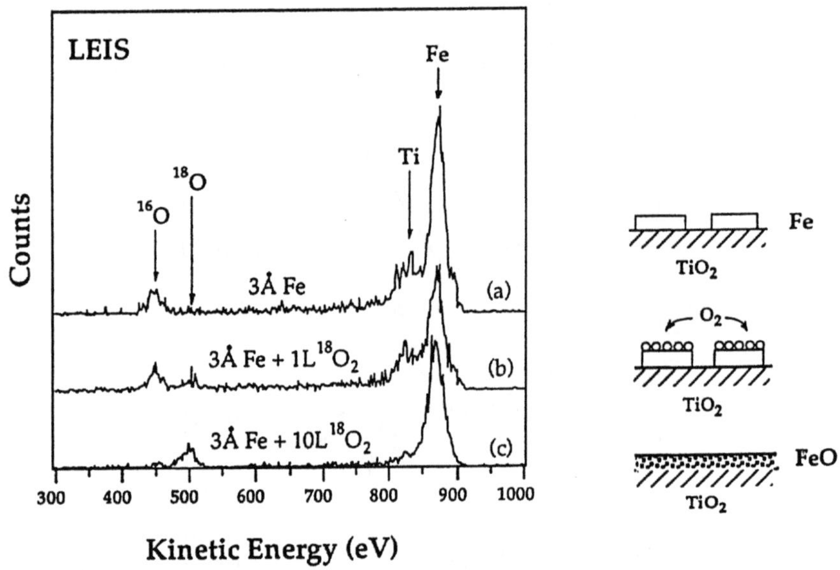

Fig: 4: LEIS spectra of TiO_2 with a 3 Å Fe film (a) as deposited at 300 K, (b) after 1 L $^{18}O_2$ exposure, (c) after 10 L $^{18}O_2$ exposure. The change in morphology caused by Fe oxidation is shown schematically.

3.4 Structure of Ultrathin Films

LEED studies of all three metals exhibit some evidence for long range order in thick (≥ 15 Å), annealed (≥ 300 °C) films. However, evidence for short range local order is seen even at 300 K in thinner films (~ 9 Å of Fe, Cr) using Angle Resolved XPS (ARXPS) and Medium Energy backscattered Electron Diffraction (MEED) [17]. Elastically backscattered electrons (E > 750 eV) or photoelectrons emitted from the TiO_2 surface show an enhancement of intensity along directions defined by atom rows [10]. The pattern symmetry is determined by forward-peaked electron-atom scattering (the "forward focusing" effect).

Figure 5 shows a comparison between MEED patterns for 750 eV electrons backscattered from clean TiO_2(110) (top), 10 Å Fe on TiO_2(110) (middle) and 100 Å of Cu on TiO_2(110) (bottom). The left column shows the data; the right column shows the results of a single scattering cluster calculation [10,17] that

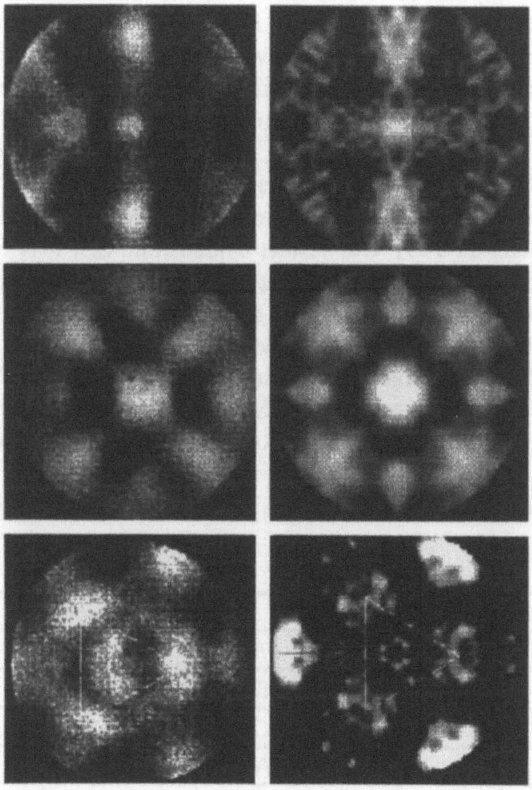

Fig. 5: MEED patterns using 750 eV electrons. Left column: data after background subtraction. Right column: results of single scattering cluster calculation. Top pair: clean, stoichiometric TiO_2(110); middle pair: Fe film on TiO_2(110); bottom pair: Cu film on TiO_2(110).

agrees well with the data. Note the distinct change of pattern symmetry, from twofold for clean $TiO_2(110)$ to fourfold for bcc Fe(100) on TiO_2, and threefold for fcc Cu(111) on TiO_2. A fourfold MEED pattern for bcc Cr(100) on TiO_2 is also found. Well defined MEED patterns are seen even for thin, unannealed films in the absence of long range LEED order.

4. Discussion

A correlation in which reactive metals show better wetting and bonding than less reactive metals is well known to ceramists, who have studied the macroscopic aspects of metal/oxide interfaces [18-20]. In a series of ultrahigh vacuum experiments based on thermal desorption spectroscopy, Peden et al. [1] observed a change of the growth mode of ultrathin metal films on an oxidized tungsten surface, from complete non-wetting (Volmer-Weber growth) for Cu to a layer growth mode for Cr, and correlated the wetting tendency with the metal-oxygen bond strength of the metal overlayer. Although the details are different, we observe a similar trend on TiO_2, and provide microscopic insights into the initial stages of ultrathin film growth. The ability of a metal film to cover or wet an oxide substrate is determined by the energy balance between the surface free energies for the overlayer and substrate, and the interface energy. As shown in Table 1, the surface free energy of TiO_2 is low compared to that of Cr, Fe and Cu; based on this fact one might predict a strong tendency for Volmer-Weber growth for all three metals on TiO_2. Our results demonstrate that it is not sufficient to consider only the surface free energies of overlayer and substrate for these systems. The fact that Cr and Fe spread more effectively than Cu is attributed to the strong metal/oxide bond at the interface. Simple thermochemical considerations predict a large negative interface energy for the Fe/TiO_2 system [3]. Thus, the different reactivity of the various metals towards the surface oxygen appears to be critical in determining the interaction between the metal

Table 1: Surface free energies of metals and TiO_2 from references [a,b]. Heat of formation from reference [b].

Surface free energy		Heats of formation [b]		
	erg/cm^2		kcal/mole	eV
TiO_2 [a]	280~380	O-Ti	- 120	- 5.2
Cr [b]	1700	O-Cr	- 71 ~ - 92	- 3.1 ~ - 4.0
Fe [b]	1800	O-Fe	- 65	- 2.8
Cu [b]	1300	O-Cu	- 40	- 1.7

[a] S.H. Overbury, P.A. Bertrand, and G.A. Somorjai, Chemical Reviews 75, 547 (1975).

[b] Robert C. Weast, Melvin J. Astle, and William H. Beyer, *CRC Handbook of Chemistry and Physics*, 1988-1989

overlayers and the TiO_2 surface. The reactivity itself is exhibited by electron transfer from the overlayer metal via the oxygen ion to the substrate metal ion. The larger the heat of formation of the metal oxide of the overlayer metal namely Cr > Fe > Cu, (cf. Table 1) the more effective is the covering of the surface atoms at the same average film thickness. This tendency is supported by recent measurements from our laboratory [21] for the highly reactive metal Hf on TiO_2(110), where a single monolayer completely covers the substrate at 300 K.

Acknowledgements

The authors thank Dr. B. Maschhoff for performing the single scattering cluster calculations for the MEED data, and Lizhong Zhang for assistance with the Cr measurements. This work has been supported in part by NSF Grant DMR-8907553.

References

[1] C.H.F. Peden, K.B. Kidd, N.D. Shinn: J. Vac. Sci. Technol. **A9**, 151 (1991)
[2] U. Diebold, J.-M. Pan, T.E. Madey: Phys. Rev. **B47**, 3868 (1993)
[3] J.-M. Pan, T.E. Madey: J. Vac. Sci. Technol. (1993) in press
[4] J.-M. Pan, U. Diebold, L. Zhang, T.E. Madey: Surf. Sci. (1993) in press
[5] U. Bardi: Appl. Surf. Sci. **51**, 89 (1991)
[6] P. J. Møller, M.-C. Wu: Surf. Sci. **224**, 265 (1989); M.-C. Wu and P. J. Møller: Surf. Sci. **224**, 250 (1989)
[7] Z. Zhang, V. E. Henrich: Surf. Sci. **277**, 263 (1992)
[8] J. Deng, D. Wang, X. Wei, R. Zhai, H. Wang: Surf. Sci. **249**, 213 (1991)
[9] V.E. Henrich: Rep. Progr. Phys. **48**, 1481 (1985)
[10] B.L. Maschhoff, J.-M. Pan, T.E. Madey: Surf. Sci. **259**, 190 (1991)
[11] W. Göpel, G. Rocker, R. Feierabend: Phys. Rev. **B28**, 3427 (1983)
[12] Y.W. Chung, W.J. Lo, G.A. Somorjai: Surf. Sci. **64**, 588 (1977)
[13] V. E. Henrich, R. L. Kurtz: Phys. Rev. **B23**, 6280 (1981)
[14] J.-M. Pan, B.L. Maschhoff, U. Diebold, T.E. Madey: J. Vac. Sci. Technol. **A10**, 2470 (1992)
[15] E. Taglauer, in: Ion Spectroscopies for Surface Analysis. eds. A. Czanderna and D. Hercules, Plenum Press, New York 1991, p. 363
[16] K.H. Ernst, A. Ludviksson, R. Zhang, J. Yoshihara, C.T. Campbell: Phys. Rev. B, in press
[17] J.-M. Pan, B.L. Maschhoff, U. Diebold, T.E. Madey: Surf. Sci., in press
[18] W.F. Egelhoff: Critical Reviews in Solid State and Material Sciences **16**, 213 (1990)
[19] S. A. Chambers: Surf. Sci. Reps. **16**, 261 (1992)
[20] K.C. Russell, S.-Y. Oh, A. Figueriedo: MRS Bulletin 46 (1991)
[21] L. Zhang, U. Diebold J.-M. Pan, T.E. Madey: in preparation

Structural and Electronic Properties of Ultrathin Cu-Layers on some Crystalline Metal-Oxides

P.J. Møller, S.A. Komolov, and E.F. Lazneva**

Department of Chemistry, University of Copenhagen, H.C. Ørsted Institute, Universitetsparken 5, DK-2100 Copenhagen Ø, Denmark
*Research Institute of Physics, St. Petersburg University, Uljanovskaya ul. 1, 198904-Petrodvorets, St. Petersburg, Russia.

Abstract. Recent results of surface electronic and geometrical properties of ultrathin copper layers on the crystalline metal-oxide substrates TiO_2(110), α-Al_2O_3(0001), CaO(001), ZnO(1120) and ZnO(0001) are reviewed. The basic experimental data of interest have been obtained by using LEED, AES, PES and EELS. New experimental results on laser-induced desorption (LID) from the system CO/Cu/ZnO(0001) are presented.

1. Introduction

The interest in copper/metal-oxide systems arises mainly from the large catalytical activity that several of these systems exhibit. Main questions concerning these metal-oxide supported metal catalysts are the respective roles of the metal deposits and the substrate and their interfacial relations. It is worthwhile to have detailed information on structural and electrical properties of Cu/metal-oxide systems on the atomic level and on the elementary chemisorption interactions in these systems. One can outline some reasons for the catalytic activity variations upon copper deposition on metal-oxide substrates. i) At the very initial sub-monolayer stage of the deposition the Cu atoms are often adsorbed weakly as dispersed Cu^+ cations, which may be considered as new active adsorption and reaction sites [1-3]. ii) Surfaces of copper islands (clusters) exhibit different adsorption and reaction properties in comparison with the metal-oxide substrate [4,5], therefore the activity of the catalysts may be considerably changed due to copper deposition. For instance, CO adsorbs most readily on the Cu-surface while CO_2 preferentially adsorbs on the zinc oxide substrate [6]. iii) Copper atom interaction with metal-oxide substrates can perturb the remaining oxide surface [2], in some cases by depositing initially on cation vacancy sites, e.g. on MgO [3]. It may cause variations of electron densities, of charge and energy relaxation and on transport properties on the open substrate area in the vicinity of Cu-adsorbed sites which lead to a change of the catalytic activity.

Springer Series in Surface Sciences, Vol. 33 Eds.: E. Umbach · H.-J. Freund
Adsorption on Ordered Surfaces of Ionic Solids and Thin Films

Most clear experimental data may be obtained using the modelling of Cu/metal-oxide systems. For this purpose one preferably uses single crystals as substrates and performs a copper deposition under ultrahigh-vacuum conditions [3,7-10]. Alternative approaches were realized by using copper crystals as a substrate [6,11]. At first, the system of interest should be characterized by surface crystallographic structure measurements and by deposited-layer thickness monitoring using low-energy electron diffraction (LEED), quartz crystal microgravimetry and Auger and energy-loss electron spectroscopy (AES and EELS) [3,9,10-18] or low-energy ion scattering [19]. Further electronic energetic structure of the system (binding energies of valence states, chemical shifts, charge transfer) is probed by a combination of photoelectron spectroscopy (PES) and EELS [20-23]. In order to recognize new active adsorption sites, to obtain chemi-sorption energies and to analyze directly a chemical composition of surface particles, the adsorption-desorption experiments were performed by using infrared spectroscopy [24,25], temperature-programmed desorption (TPD) [2,6,26,27], photo and laser-stimulated desorption [28,29] techniques. The laser desorption experiment gives us a possibility not only to measure mass spectra of desorbed particles, but also to obtain their velocity distributions [30-32].

In the present communication we brief some recent results of ultrathin copper-film growth on different metal-oxide substrates: α-Al_2O_3(0001), CaO(001), TiO_2(110), ZnO(1120), and ZnO(0001). Then we consider new experimental data of laser desorption from CO/Cu/ZnO(0001) surfaces.

2. Surface Geometrical Structure

2.1 Incommensurate Growth of Cu on a (1x1) TiO_2(110) Surface

The growth of an ultrathin Cu film on room temperature (1x1) surfaces of TiO_2(110) was studied by Møller and Wu for coverages ranging from 0 to 80 Å of deposited copper (the substrate surface may exhibit also a (1x2) missing-row structure) [13]. At deposition thicknesses d_{Cu} below 7 Å on the (1x1) surface, the integral diffraction spots are gradually attenuated with increasing thicknesses and the LEED background is relatively increased. A weak hexagonal diffraction pattern gradually appeared near $d_{Cu} \sim 7$ Å. It is interesting to notice that the Cu atoms are placed in registry with the substrate lattice along the $[1\bar{1}0]$ direction. The lattice constant of the Cu superlattice in **k** space is 3 times larger than that of the substrate lattice. This registry relation is confirmed by the fact that no splitting or distortion of the diffraction spots at coincident points in the LEED patterns are observed. No such relation is established along the [001] direction, however. In addition, a Cu hexagonal structure without distortion is recognized by measuring each edge of the unit mesh. The superstructure lattice constant is 2.50 Å, thus slightly smaller than the 2.55 Å Cu(111) value.

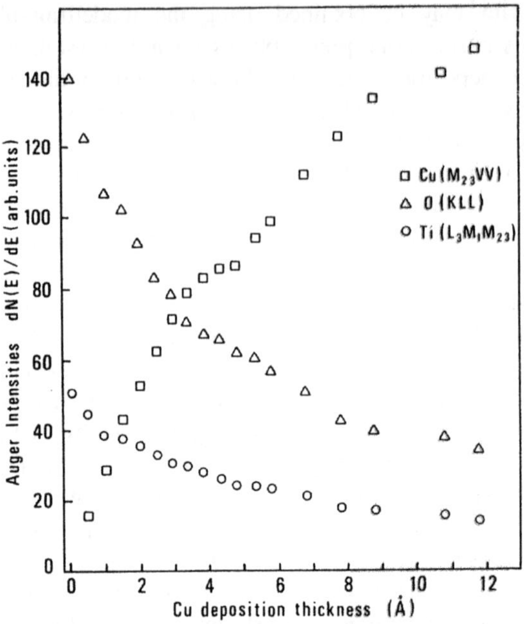

Fig. 1: $Cu(M_{2,3}VV)$, O(KLL) and $Ti(L_3M_1M_{2,3})$ Auger intensities versus thickness d_{Cu} of Cu deposited at 293 K upon a $TiO_2(110)$ surface. The first breakpoint corresponds to Cu monolayer-thick large-island coverage.

The growth has also been followed by AES. Both the $Cu(M_{2,3}VV)$ and O(KLL) Auger peak-to-peak intensity heights versus d_{Cu} exhibit a clear break point at 3.5 Å as read by a quartz crystal microbalance monitor (Fig. 1). This value corresponds to an apparent monolayer coverage. The attenuation of the $Ti(L_3M_1M_{2,3})$ signal is also inserted, but the monolayer breakpoint for this signal is not as sharp as those of O(KLL) and $Cu(M_{2,3}VV)$. The substrate LEED was still visible at a Cu-layer thickness of 40 Å [13]. In comparison with the Auger intensity change with d_{Cu} this indicates that the growth mode is island/crystallite formation on top of initial large monolayer islands [33].

2.2 Cu on an α-$Al_2O_3(0001)$ Surface

α-Al_2O_3 (the sapphire) is an insulator with a large band gap and charging-up of the surface easily happens under electron beam irradiation. A sharp (1x1) LEED pattern was seen from a clean α-$Al_2O_3(0001)$ surface only at primary energies E_p above 60 eV [9,15]. Within the 0.7 Å $< d_{Cu} < 10$ Å range the substrate LEED pattern was broadened and the background intensity was increased with d_{Cu}. A sharp LEED pattern from the copper overlayer, showing a Cu(111)R30°||α-$Al_2O_3(0001)$ epitaxy, was obtained in the 50-130 Å thickness range under heating to 150-160 °C [18] (the lattice constant of the Cu(111) layer is smaller than that of the alumina). AES measurements of Al(KLL), O(KLL) and Cu(LMM) indicated a change in growth mechanism at $d_{Cu} = 2$ Å. It was found that the Al-to-O intensity ratio of the $d_{Cu} < 2$ Å deposits was lowered more than for the thicker ($d_{Cu} > 2$ Å) deposits. This may indicate that the copper atoms are positioned in the a-top position of the aluminium sites during the initial stages of copper deposition. Thus, the deposited copper apparantly grows as two-dimensional monolayer islands for coverages up to about one monolayer. At increasing coverages, 2 Å < d_{Cu} < 50 Å, three-dimensional clusters are formed. Further deposition at room temperature results in full coverage of the substrate [9,15,18].

2.3 Cu on CaO(001), ZnO(11$\bar{2}$0) and ZnO(000$\bar{1}$) Surfaces

During Cu film growth on a well defined 1x1 CaO(100) surface an epitaxial behaviour was observed resulting in a c(1x1) Cu(100) LEED pattern formation [34,35]. The superstructure lattice constant was 2.41 Å, thus slightly smaller than the value of 2.55 Å for bulk copper. The Auger signal intensity-to-thickness dependencies showed breakpoints at $d_{Cu} \sim 2.2$ Å corresponding to the first layer formation. The following growth may be connected with the island formation mechanism [34,35].

No ordering structure was observed during copper deposition on the ZnO(1120) surface at room temperature. The mode of growth was found to be monolayer-simultaneous-multilayer up until about 40 Å average thickness. Annealing of such layers in the 200-500 °C interval did not cause an appearance of patterns related to diffraction from Cu. Moreover, heating of the substrate led to a considerable increase in the intensity of the substrate pattern, which suggests that agglomeration of the copper takes place [23]. In contrast to the former observation the thin copper-film growth on the polar ZnO(0001) (oxygen terminated) surface was characterized by a p(1x1) overlayer formation up to a monolayer limit. The following epitaxial growth is consistent with a model where 3-dimensional Cu (111) microcrystals begin to form after completion of the first p(1x1) monolayer [7].

3. Surface Electronic Structure

The electronic structure of Cu films deposited at room temperature upon a $TiO_2(110)$ surface have been studied by UPS and EELS [13]. The UPS results indicate continuous Cu 3d binding energy shifts with Cu coverage in the range 0.2 Å < d_{Cu} < 5 Å. The FWHM of the Cu 3d peaks gradually become broader, from 0.96 eV at the initial 0.2 Å Cu to 1.36 eV at a d_{Cu} of 5 Å. The narrow Cu 3d band and its behaviour lead to the proposal that the interactions between the d-band and substrate are very limited and that the Cu atoms therefore bonded to the surface through the delocalized 4s-p electrons. In the EELS measurements the Cu deposition attenuated the substrate losses and produced three new losses: A, B and C whose energy positions change versus copper thickness as it is shown in

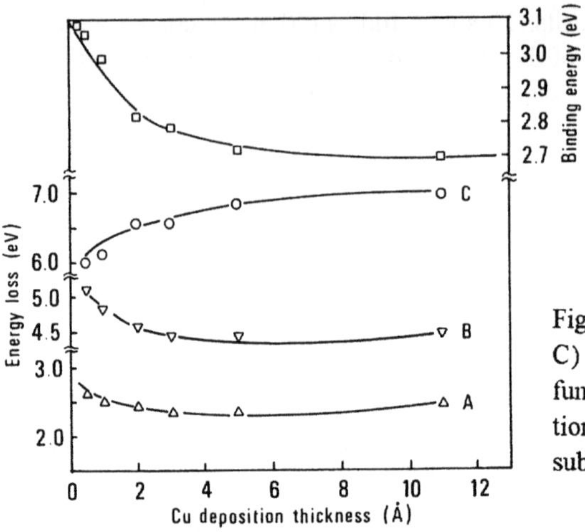

Fig. 2: The electron losses (A, B and C) and the Cu 3d binding energy as a function of the thickness of Cu deposition, d_{Cu}, on a (1x1)TiO_2(110) substrate at 293 K.

Fig. 2. The Cu 3d binding energy shift is also inserted. The losses A and B go through a weak minimum around $d_{Cu} \sim 3$ Å. They were associated with interband transitions from 3d-initial states.

Below 3 Å the shift of the Cu 3d band dominates the loss energy shifts, as seen in Fig. 2 the loss C exhibits a different behaviour, shifting monotonically towards a higher energy. In a previous study [21] a similar loss was assigned to a Cu-plasmon excitation and a model was developed in which the shift of the plasmon frequency was associated with the change of the free-electron density in the surface layer. Applying this model to the Cu/TiO_2(110) system it was possible to deduce a charge transfer from the plasmon-loss shift. There was a linear relation between the shift of plasmon frequency and Cu 3d binding energy [13]. This relation may be explained in terms of the charge potential model, in which the chemical shift of the core level is associated with the charge variation of the outer-shell electrons [36]. It is also interesting to point out that due to charge transfer the initially deposited Cu atoms probably exist as Cu^+ states on the TiO_2(110) surface. A similar behaviour of EEL-spectra was observed in the course of thin Cu-film deposition upon α-Al_2O_3(0001) [9,15] and ZnO(1120) [23] crystal surfaces. There was the difference that a Cu-surface plasmon did not change its energy during Cu deposition on the ZnO(1120) surface indicating no (or very little) charge transfer for the latter case at room temperature (contrary to the situation at higher temperatures) [23]. The same observation was made in a room-temperature Cu-deposition experiment onto CaO(001) surface [35]. One can see that the main electronic structure changes of Cu-deposits are taking place at submonolayer coverages. In the dependence on the type of a metal-oxide substrate and on the strength of adsorbent-adsorbate interaction there are variations to a different degree in chemical shifts of binding energies and in characteristic energy losses. The shift of a surface plasmon energy directly points out to a charge transfer between a deposited atom and a substrate. All electronic characteristics reach their bulk values at mean deposit thicknesses of 10-20 Å.

4. Desorption from a CO/Cu/Zn(0001) System under Pulse Laser Illumination

Laser-induced desorption (LID) is a valuable method in studies of the properties of adsorption systems and of elementary energy transfer-processes in the subsurface region and at the top of the surface [30,37]. A laser-desorption experiment involves a mass analysis of desorbed particles and measurement of their velocity distributions. Our measurements were performed in an all-metal UHV system (base pressure < 10^{-7} Pa) and time-of-flight mass spectrometry was used for desorption registration. Details of the apparatus is published elsewhere [38,39]. For desorption excitation, 15 ns pulses from a neodymium laser were used. The radiation power density P on the surface was ranged $0.1 < P < 20$ MW/cm^2. A laser beam was focused on a ZnO(0001) (zink terminated) surface to a spot size of ~ 10^{-3} cm^2.

Adsorption-desorption investigations of carbon oxides on clean and sub-monolayer-copper-covered Zn(0001) surfaces are of present particular interest due to their importance in heterogeneous catalysis [6,29,40,41].

A transparent colourless zinc oxide crystal was used for these investigations. The procedure of surface preparation consisted of preliminary diamond polishing to 1 μm followed by in situ thermal annealing in UHV before measurements. A sharp LEED picture of the Zn(0001) surface was obtained after annealing at 500°C, yielding pronounced singlets with sixfold symmetry, which have been referred to as a 1x1 pattern in accordance with previous results [27,42]. With incident electron-beam-energy variation it is possible to trace the splitting of each of the LEED spots into a sextet of sixfold symmetry. The sample was exposed to CO (purity 99.997 %) at a pressure of 100 Pa for 4 hours at room temperature, followed by 10 min of 100 Pa CO at 600 K. The CO adsorption was followed by a weakening of the LEED spots, that may be related to a disordered CO adsorption. An alternative surface was prepared in the same manner, only before the CO adsorption about 0.3 monolayer of Cu was deposited on the clean ZnO(0001) 1x1 surface. The procedure of the CO adsorption was the same for both cases.

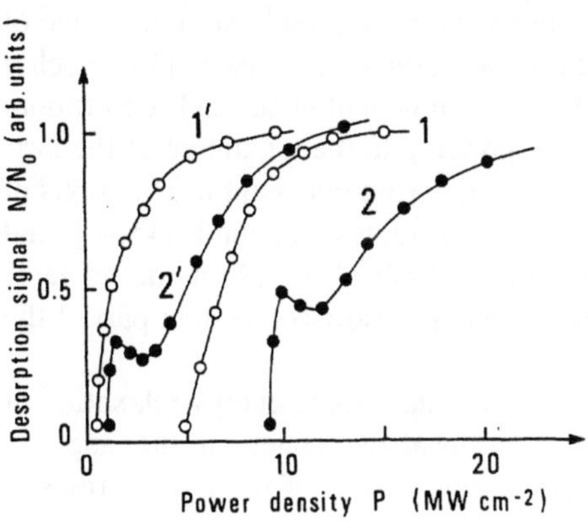

Fig. 3: Dependence on laser power density P of the LID signal N, normalized to background level N_0, for CO and CO_2 molecules desorbed (1,2) from clean and (1',2') from the submonolayer Cu-covered ZnO(0001) surfaces.

4.1 LID from Clean ZnO(0001) Surface

In the laser-induced desorption flow, CO and CO_2 molecules were registered as the main components. With increasing laser pulse-intensity there are threshold values at which desorption starts. The dependence of CO and CO_2 desorption signals on the laser power density P is shown in Fig. 3. We observe threshold values of P ~ 5 MW/cm^2 for CO and ~ 9.5 MW/cm^2 for CO_2, curves 1 and 2 in Fig. 3.

The TOF distributions of the desorbed molecules yield the most interesting information. For CO molecules a uniform distribution was obtained in the investigated range of laser fluency. A typical shape of the CO distribution is represented in Fig. 4 (curve 1), which contains only one maximum, located at about 340 μs. The behaviour of the TOF distributions of the CO_2 molecules turns out to be more complicated. At a laser fluency near to the threshold value we see a rather fast component of the desorbed molecules (Fig. 4, curve 2). The very sharp rise in curve 2 of Fig. 3 near the threshold is connected with the increase in intensity of this fast component. With increasing laser fluency the intensity of the fast component decreases and a slow component appears in the TOF distribution, and for P = 12-15 MW/cm^2 the intensities of the two components become equal, as it is represented in Fig. 4 (curve 3). The observed two-component TOF distribution may be associated at least with two different physical mechanisms in the molecular CO_2 desorption from the ZnO surface under light-quanta action. The CO_2 desorption observed in our experiment is a consequence of a catalytic CO oxidation at the ZnO surface since the surfaces under investigation were exposed to CO adsorption only. As was previously pointed out [43,44], CO molecules adsorb preferably on Zn^{2+} sites. We suggest that the CO interacts with lattice oxygen in the ZnO(0001) surface layer yielding chemisorbed CO, which then forms a surface $ZnCO_2^-$ complex. Laser quanta ($h\nu$ = 1.17 eV) absorption occurring on defect and impurity complexes located in the ZnO band gap is followed by their electronic excitation. The capture of a photoexcited hole to the surface $ZnCO_2^-$ complex is followed by a recharging of the adsorbed particle. At the same time, energy of recombination is transferred to the molecule to stimulate a desorption process. This photodesorption mechanism is quite similar to one of justifying relatively high kinetic energies of the desorbed molecules [30]. Such a mechanism explains an appearance of the fast component of the CO_2 desorption. In our observations the kinetic energy corresponding to the maximum of the fast-component is equal to 0.12 eV (Fig. 4, curve 3). Two-component distributions of laser ejected particles were previously observed for ionic compounds [45,46], and for laser-induced desorption of gas molecules [30,47]. A simple model of laser thermodesorption [30,48] cannot explain the delayed slow-component part of the distribution.

These may be understood in terms of the peculiarities of energy relaxation of laser-excited local complexes in ZnO. Well-localized states in the gap are characterized by long-living electronic excitations. Decay of these excitations is followed by the creation of energetic phonons which may migrate to the surface and there captured by adsorbed complexes. The absorption of phonons by the surface adsorbed complex may stimulate its vibrational excitation followed by its decay and by desorption of gas molecules [49]. This event may be referred to as a local heating of an adsorbed complex. Thus, a delay in the TOF distribution may be associated with the time of relaxation of the electronically excited states accompanied by the phonon creation, with the consecutive energy transferred to the adsorbed complexes.

4.2 LID from a Copper-Covered ZnO(0001) Surface

The desorption characteristics are drastically changed in the case of desorption from a ZnO(0001) surface covered with ~ 0.3 monolayer of copper. The CO and CO_2 desorption thresholds decrease to values as low as 0.5 MW/cm^2 and 1 MW/cm^2, respectively (curves 1' and 2' in Fig. 4). The intensities of desorption increase by about one order of magnitude. The maxima of slow components in the TOF distributions are shifted towards smaller time values reaching 135 μs and 180 μs for CO and CO_2, respectively. The location of the maximum of the fast component in the CO_2-distribution remains unchanged (curve 3' in Fig. 4).

The observed variation for the LID characteristics directly point to Cu-induced perturbation of physical and chemical properties of the surface. An increase of the CO_2 desorption flow by about an order of magnitude is in accordance with results of recent investigations [2], in which it was pointed out

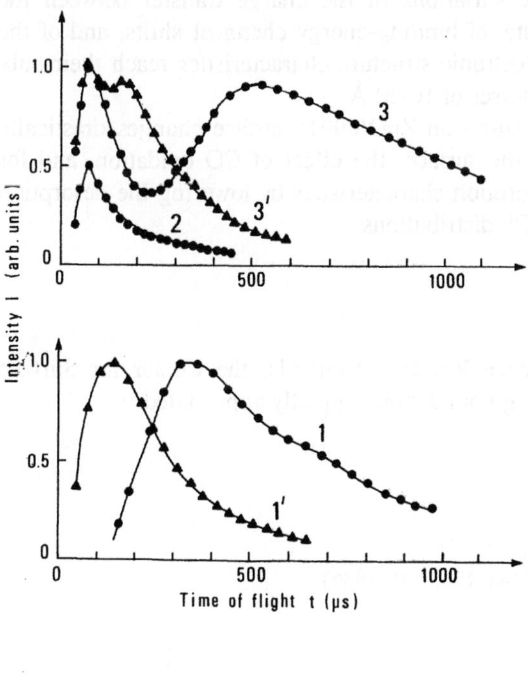

Fig. 4: Normalized time-of-flight distributions of CO (1,1') and CO_2 (2,3 and 3') molecules from a clean ZnO(0001) surface (1,2) P = 10 MW/cm^2; (3) P = 14 MW/cm^2, and from a Cu-covered surface (1') P = 2 MW/cm^2; (3') P = 3.5 MW/cm^2.

that deposited Cu atoms and Cu_xO complexes on a ZnO surface can act themselves as new adsorption and reaction sites. At the same time new active defect sites are created on the ZnO surface around adsorbed copper species. Both factors lead to an increase of adsorption capacity and catalytic activity of the investigated surface. The decrease of the desorption threshold and the slow-component delay may be associated with an increase in an optical absorption and with a change in the electronic properties in the subsurface region of ZnO due to an interaction with the Cu atoms. Adsorbed atoms, or clusters, may be considered as new centres of optical absorption. Stimulated by copper-oxygen interaction, new defects in the ZnO subsurface region also promote increasing light absorption. Action of the Cu atoms as donors [50] gives rise to a higher surface conductivity and to an acceleration of excited-state relaxation.

5. Conclusion

For submonolayer coverages the deposited copper grows as two-dimensional monolayer islands which may be both uniformly and not uniformly dispersed over the surface. Uniformly deposited islands are characterized by superstructure lattices whose lattice-constants for the first layers may be slightly different than the values for bulk copper. The Cu atoms are placed most readily in the a-top position of substrate metal sites. The following growth may be connected with three-dimensional cluster formation. In some cases the copper clusters grow directly in a preferred epitaxial orientation, in others after thermal annealing.

According to the type of a substrate and to the strength of adsorbate-adsorbent interaction there are variations in the charge transfer between the deposited atom and the substrate, of binding-energy chemical shifts, and of the electron energy loss spectra. Electronic structure characteristics reach their bulk values at mean deposited thicknesses of 10-20 Å.

Submonolayer copper deposition on ZnO(0001) surface changes drastically the CO-adsorption capacity of the surface, the effect of CO oxidation, and for laser-induced CO and CO_2 desorption characteristics by lowering the desorption thresholds and distorting the TOF distributions.

Acknowledgements

The support from Danish Science Research Councils, the Center for Surface Reactivity and from the Carlsberg Foundation is greatly appreciated.

References

[1] K. Klier: Adv. Catal. **31**, 243 (1982)
[2] L. Chan, G.L. Griffin: Surf. Sci. **173**, 160 (1986)

[3] J.W. He, P.J. Møller. Chem. Phys. Lett. **129**, 13 (1986)
[4] G.C. Chinchen, M.S. Spencer, K.C. Wangh, D.A. Whan: J. Chem. Soc. Faraday Trans. **83**, 2193 (1987)
[5] J.B. Friedrich, D.J. Young, M.S. Wainwright: J. Catal. **80**, 14 (1983)
[6] S.S. Fu, G.A. Somorjai: Appl. Surf. Sci. **48/49**, 93 (1991)
[7] C.T. Campbell, K.A. Caube, J.M. White: Surf. Sci. **182**, 458 (1987)
[8] P.J. Møller, J.-W. He: Nucl. Instrum. Meth. Phys. Res. B **17**, 137 (1986)
[9] Q. Guo, P.J. Møller: Vacuum **41**, 1114 (1990), Surf. Sci. **244**, 228 (1991)
[10] M.C. Wu, P.J. Møller: Surf. Sci. **235**, 228 (1990)
[11] K.L. Siefering, G.L. Griffin: Surf. Sci. **207**, 525 (1989)
[12] J.-W. He, P.J. Møller: Surf. Sci. **178**, 934 (1986), ibid. **180**, 411 (1987)
[13] P.J. Møller, M.-C. Wu: Surf. Sci. **224**, 265 (1989)
[14] J.E.T. Andersen, P.J. Møller: Thin Solid Films **186**, 137 (1990)
[15] P.J. Møller, Q. Guo: Thin Solid films **201**, 267 (1991)
[16] J.E.T. Andersen, P.J. Møller: Surf.Sci. **258**, 247 (1991)
[17] J.E.T. Andersen, P.J. Møller: J. Vac. Sci. Technol. **A10**, 497 (1992)
[18] Q. Guo, P.J. Møller: Acta Phys. Pol. **81**, 647 (1992)
[19] U. Diebold, J.-M. Pan, T.E. Madey, to be published
[20] I. Alstrup, P.J. Møller: Appl. Surf. Sci. **33/34**, 143 (1988)
[21] M.C. Wu, P.J. Møller: Phys. Rev. **B40**, 6063 (1989)
[22] M.C. Wu, P.J. Møller: Surf. Sci. **224**, 250 (1989), ibid. **235**, 228 (1990)
[23] P.J. Møller, J. Nerlov, to be published
[24] F. Boccuzzi, G. Ghiotti, A. Chiorino: Surf. Sci. **156**, 933 (1985)
[25] F. Boccuzzi, G. Ghiotti, A. Chiorino: Surf. Sci. **162**, 361 (1985)
[26] P. Esser, W. Göpel:. Surf. Sci. **97**, 309 (1980)
[27] W.H. Cheng, H.H. Kung: Surf. Sci. **122**, 21 (1988)
[28] J. Shapira, S.M. Cox, D. Lichtman: Surf. Sci. **54**, 43 (1976)
[29] P.J. Møller, S.A. Komolov, E.F. Lazneva: Surf. Sci. Lett. (1993) in press
[30] E.F. Lazneva: Radiation Effects, Defects in Solids **115**, 257 (1991)
[31] P.H. Lu, Y.L. Li, Q.K. Zheng, Q.Z. Qin: J. Phys. Chem. **96**, 4496 (1992)
[32] X.Y. Zhu, M. Wolf, J.M. White: J. Chem. Phys. **97**, 5868 (1992)
[33] C. Argile, G.E. Rhead: Surf. Sci. Reports **10**, 277 (1989)
[34] H.M. Neergaard, P.J. Møller: Proc. Joint Nordic Spring Meeting '92, Denmark 1992, p. 87
[35] P.J. Møller, H.N. Waltenburg, to be published
[36] D. Briggs, J.C. Riviere in: Practical Surface Analysis, eds. D. Briggs and M.P. Shah, John Wiley, London 1983, p. 119
[37] A. Namiki, S. Cho, K. Ichige: Jap. Appl. Phys. **26**, 39 (1988)
[38] P.J. Møller, E.F. Lazneva, S.A. Komolov, to be published
[39] E.F. Lazneva, A.M. Turiev: Instrum. & Exp. Tech. **27**, 926 (1984)
[40] S.V. Didziulis, K.D. Butcher, S.L. Cohen, E.I. Solomon: J. Am. Chem. Soc. **111**, 7110 (1989)
[41] J.P. Hindermann, M. Idress, A. Kienneman: Materials Chemistry and Physics **1D**, 513 (1988)

[42] S.C. Chang, P. Mark: Surf. Sci. **46**, 293 (1974)
[43] W. Hotan, W. Göpel, R. Haul: Surf. Sci. **83**, 162 (1979)
[44] S.F. Jen, A.B. Anderson: Surf. Sci. **223**, 119 (1989)
[45] A. Namiki, T. Kawai, K. Ichige: Surf. Sci. **166**, 129 (1986)
[46] R.C. Estler, J.E Anderson, E.C. Apel, N.S. Nogar: J. Vac. Sci. Technol. **B5**, 1519 (1987)
[47] E.F. Lazneva, A.M. Turiev: Poverkhnost **8**, 142 (1986)
[48] G. Wedler, H. Ruhman: Surf. Sci. **121**, 464 (1982)
[49] Z.W. Gortel, H.J. Kreuzer, P. Pierey, R. Teshime: Phys. Rev. **B27**, 5066 (1983)
[50] E. Mollwo: Solid State Commun. **13**, 1283 (1973)

Part IV

Adsorption on Oxide Surfaces: Theory

Quantum Chemical ab initio Calculations for the Adsorption of Small Molecules on NiO(100)

V. Staemmler

Lehrstuhl für Theoretische Chemie, Ruhr-Universität Bochum,
D-44780 Bochum, Germany

Abstract. We report on a series of quantum chemical ab initio calculations for the adsorption of small molecules (CO, NO, CH_3) on the ideal neutral NiO(100) surface. The adsorption site was represented by a NiO_5^{8-} cluster which was either embedded in a Madelung field or saturated by protons. The calculations were performed using standard ab initio techniques at different levels of accuracy (SCF, CASSCF, valence CI, multi-reference CEPA) and various basis sets (DZ, TZP, and extended sets). The results were always corrected for the basis set superposition error. In all cases studied the adsorption site is a surface Ni^{2+} ion in its $^3B_1$ ground state with a $3d^8$ configuration. Its low lying excited states (surface states) are quite distinct from the corresponding bulk states. CO, CH, and CH_3 are adsorbed linearly while NO is tilted by about 45° against the surface normal. The adsorption energies are rather small, ranging from about 0.1 eV (CO) to about 0.4 eV for the s-radical CH_3. The analysis of the molecular wavefunctions shows that the adsorption is a pure physisorption for CO and CH (electrostatic attraction and Pauli repulsion) whereas NO and CH_3 are bound by a weak chemisorption. Correlation effects play a marked role only in the bonding of CH_3 to the surface, van der Waals forces are almost negligibly small.

1. Introduction

While the adsorption of small molecules on metallic substrates has been studied in great detail both experimentally and theoretically for many years, oxidic systems and in particular transition metal oxides have attracted a wider interest only recently. The main reasons are that metallic substrates and particles are of larger importance for catalytic processes and that the preparation and characterization of well ordered oxidic surfaces or thin films is quite difficult experimentally.

Parallel to the recent experimental studies of adsorption processes on oxidic surfaces ab initio cluster calculations have been performed by several groups with the aim to understand the mechanism of the chemical bond between an oxide surface and adsorbed molecules and to provide reliable data for adsorption energies, geometries, vibration frequencies, electronic states etc. Most of these theoretical studies were concerned with the adsorption on neutral (=stoichiometric) low-index surfaces of perfect ionic crystals, such as H_2/MgO(100) [1], CO/MgO(100) [2-5], CO/NiO(100) [2,6], NO/NiO(100) [7-9], CH_3/NiO(100) [10]. But recently, it is also tried to extend such investigations to the adsorption on polar surfaces (e.g. CO/NiO(111) [11], CO/ZnO(0001) [12]) or on defect sites or steps [13-15] which cannot be so easily represented by small clusters.

In this contribution we summarize our recent ab initio cluster calculations for the adsorption of small closed-shell molecules (CO) and radicals (NO, CH, CH_3) on NiO(100).

2. Method of Calculation

Solid NiO has rocksalt structure with a Ni-O distance of 2.08 Å [16]. It is generally accepted to be highly ionic with charges close to the ideal values of $\pm$ 2e. The NiO(100) surface is not reconstructed and shows a slight contraction of the NiO distance in the top layer of about 2% normal to the surface [17]. Thus, the Ni^{2+} ions in the bulk have a perfect octahedral coordination, those at the surface have five O^{2-} neighbours.

In order to model the Ni^{2+} adsorption site at the NiO(100) surface we have used several different cluster environments [8]:

i) PC: A pure point charge model for the full semiinfinite crystal. Point charges of $\pm$ 2e are placed at the regular undistorted lattice points in order to generate the Madelung potential at the Ni^{2+} adsorption site.

ii) Ni^{2+}/MAD: One Ni^{2+} ion embedded in the Madelung field of the point charges.

iii) NiO_5^{8-}/MAD: A $Ni^{2+}(O^{2-})_5$ cluster embedded in the Madelung field.

iv) NiO_5^{8-}/ECP/MAD: The same cluster as in the NiO_5^{8-}/MAD model, but the next coordination shell of twelve Ni^{2+} ions is represented by effective core potentials in order to account for the Pauli repulsion between O^{2-} and the Ni^{2+} cores.

v) $Ni(OH_2)_3(OH)_2$: the NiO_5^{8-} cluster without the Madelung field, but with the unsaturated bonds and negative charges saturated by protons.

Of course, the crucial point in cluster calculations for bulk properties or adsorption processes is the dependence of the results on the size and the shape of the cluster. For metal clusters or surfaces, the cluster convergence is very slow since the valence electrons are generally delocalized over the whole crystal. But, by using special techniques of "bond preparation" Siegbahn and coworkers [18] could obtain rather satisfying results even for comparably small metal clusters.

The problem seems less difficult for ionic crystals like oxides, in particular for systems like MgO and NiO with a low degree of covalency. Since the adsorption site in both systems is the cation and neither the 2p-shell of Mg^{2+} nor the 3d-shell of Ni^{2+} show a significant amount of dispersion [7] a local description by a small cluster in the respective Madelung field will be a reasonable representation of the adsorption site. Actually, the five models given above show only minor differences for adsorption energies and geometries as long as the adsorption of CO and NO on the neutral surfaces NiO(100) [6-8] and MgO(100) [2-4] is concerned and the Madelung field itself, i.e. the effective charge at the ions, is not changed. For more covalent oxides, like ZnO, or for polar surfaces, like NiO(111), ZnO(0001), $Cr_2O_3(111)$, the design of reasonable clusters and the choice of effective charges is much more problematic as discussed by Pettersson [9] and by Siegbahn et al. [19] in the present volume.

All our ab initio calculations have been performed with the Bochum open-shell CEPA program package consisting of a restricted open-shell SCF (ROHF) program, a CASSCF part [20], a conventional graphical CI part [21], a single-reference [22] and a multi-reference CEPA [23] program. Gaussian basis sets of various sizes (DZ, TZP, extended sets) have been used starting from the basis sets of Roos et al. [24] and Wachters [25] for Ni and from Huzinaga sets [26,27] for H, C, O. For obtaining reliable results for adsorption energies and geometries it was necessary to correct the basis set superposition error (BSSE) by the Boys-Bernardi [28] counterpoise method.

3. Characterization of the Surface States

Ni^{2+} has the electronic configuration $3d^8$ with a ^3F ground state, the lowest excited electronic states are ^1D and ^3P at 1.74 eV and 2.07 eV, respectively [29]. In the octahedral environment the ^3F state of a Ni^{2+} ion in the bulk is split into a $^3A_{2g}$ ground state with the occupation $t_{2g}^6e_g^2$, a first excited $^3T_{2g}$ state with $t_{2g}^5e_g^3$ and a second excited $^3T_{1g}$ state which is a mixture of the configurations $t_{2g}^5e_g^3$ and $t_{2g}^4e_g^4$. In the reduced symmetry (C_{4v}) at the surface, the $^3T_{2g}$ state is further split into a ^3E and a $^3B_2$ component, $^3T_{1g}$ is split into $^3A_2$ and ^3E.

The electronic spectrum of NiO has been measured several times with different experimental techniques ([30] and references therein). Below the optical band gap at about 4 eV there are several weak bands which are attributed to local d-d transitions in the Ni^{2+} ions, the lowest ones of them are observed at about 1.10 eV, 1.75 eV and 1.95 eV. With surface sensitive techniques (e.g. electron energy loss spectroscopy (EELS) with primary electron energies of 13-20 eV) four electronic transitions at energies of 0.57 eV, 1.10 eV, 1.62 eV and 1.87 eV [30,31] are observed for the clean NiO(100) surface. The peaks at 0.57 eV and 1.62 eV are surface peaks which is proven experimentally by the observation that they are shifted to higher energies if the Ni^{2+} ions at the surface are covered by NO molecules [30].

The definite assignment of the low-lying surface and bulk states of NiO(100) was possible with the help of ab initio calculations. Table 1 contains our MC-CEPA results (multi-configuration coupled electron pair approximation [23], calculations at lower levels of sophistication yielded similar results, but are not

Table 1: Lowest d-d excitation energies for Ni^{2+} ions in bulk NiO (B) and at the NiO(100) surface (S) (in eV).

State		MC-CEPA[a]	MC-CEPA[a]	FOCI[b]	EELS[c]	EELS[c]
B	S	S	B	B	clean[d]	NO[d]
$^3A_{2g}$	$^3B_1$	0.	0.	0.	0.	0.
$^3T_{2g}$	^3E	0.62	1.00	1.04	0.57	0.90
	$^3B_2$	0.98			1.10	1.10
$^3T_{1g}$	$^3A_2$	1.21	1.81	1.77	1.62	1.75
	^3E	1.38			1.87	1.87

a Ab initio calculation, multi-configuration CEPA [30].

b First-order CI [32].

c Electron energy loss spectroscopy [30].

d Clean and NO covered NiO(100) surface.

documented here) in comparison with previous ab initio results of Janssen and Nieuwpoort [32] for NiO_6^{10-} and the experimental EELS data [30]. Apparently, the lowest transition at about 0.6 eV corresponds to a surface state, namely the ^3E component of $^3T_{2g}$ while the $^3B_2$ component of $^3T_{2g}$ remains unshifted at about 1.1 eV. The interpretation in terms of the 3d AOs is straightforward: The transition $^3B_1 \varnothing {}^3B_2$ corresponds to the excitation $\text{ɔa}_{xy} \rightarrow \text{ɔa}_{x^2-y^2}$ which is not affected if one O^{2-} ion in the z direction (=surface normal) is missing, whereas the transition $^3B_1 \varnothing {}^3E$ corresponds to ɔa_{xz}, $\text{ɔa}_{yz} \rightarrow \text{ɔa}_{z^2}$, and the ɔa_{z^2} AO of Ni^{2+} has a lower energy at the surface than in the bulk. Similarly, the $^3T_{1g}$ state is split into $^3A_{2g}$ and ^3E and both components have lower energies for surface than for bulk Ni^{2+} ions.

The Mulliken population analysis attributes charges of about ± 1.75 e to the Ni^{2+} and O^{2-} ions, without a significant difference between bulk and suface ions. This means that there is a certain, but small covalent contribution to the Ni-O bond and a Madelung field with effective charges of ± 2.0 e is probably slightly too strong [9]. In any case, one can be sure that the ground state of Ni^{2+} ion at the surface is a $^3B_1$ state with the electronic configuration $a_{xy}^{\uparrow\downarrow} a_{yz}^{\uparrow\downarrow} a_{xz}^{\uparrow\downarrow} a_{z2}^{\uparrow} a_{x2-y2}^{\uparrow}$ and the next excited state is about 0.6 eV higher in energy.

Surprisingly, NiO(111) shows nearly the same EELS spectrum as NiO(100) [33], though NiO(111) is a Madelung unstable surface, its geometrical and electronic structure are not yet fully understood. Preliminary calculations [11] have revealed that due to the reduced ligand field (three O^{2-} neighbours instead of six in the bulk) all d-d excitation energies are reduced by a factor of 2 from their values in the bulk, i.e. to about 0.6 eV and 0.9 eV instead of 1.1 eV and 1.8 eV, such that the transitions occur at nearly the same excitation energies as for NiO(100). There are indications that the NiO(111) surface can be stabilized by a certain amount of charge transfer from the first O^{2-} layer to the topmost Ni^{2+} layer. This needs the participation of the 4s and 4p-AOs on Ni while the 3d-AOs and the d-d transitions are not affected [11].

4. Adsorption of CO, NO, and CH_3 on NiO(100)

Most of the ab initio calculations presented in this chapter have been performed at the ROHF level with a quite flexible TZP basis set and with inclusion of the counterpoise correction for the BSSE. We will always refer to this level of accuracy unless otherwise indicated; the results of calculations with the other basis sets mentioned before do not differ much from this standard if the BSSE which is strongly basis set dependent is corrected. Furthermore, all calculations for the adsorption of CO and NO were performed for fixed C-O and N-O distances of 1.11 Å and 1.15 Å, respectively.

4.1 CO/NiO(100)

Figure 1 shows potential curves for the linear adsorption of CO and OC (and NO) on top of the Ni^{2+} ion on the NiO(100) surface in the $^3B_1$ ground state. The saturated $Ni(OH_2)_3(OH)_2$ cluster has been used in this calculation, but the results for the model NiO_5^{8-}/MAD are virtually identical. Both curves exhibit shallow minima with adsorption energies in the order of 0.08 eV for the adsorption of CO and 0.04 eV for that of OC at equilibrium distances of 5.0 - 5.5 a_o.

The analysis of the SCF wavefunction shows that the bonding between the surface and CO or OC is a pure physisorption [6]: The attractive part is given by the electrostatic energy of the permanent multipole moments of CO (mainly its quadrupole moment since the dipole moment of CO is very small) in the electric field above the surface, the inductive forces play only a minor role. At smaller distances the electrostatic attraction is compensated by the Pauli repulsion. Figure 2 contains contour plots of the orbitals involved and shows that the spatial extent of the 2p-orbitals at the O^{2-} ions in the Madelung field is not much larger

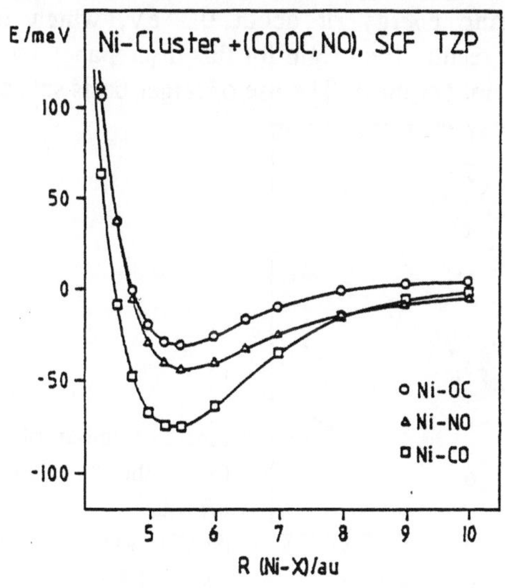

Fig. 1: Linear Adsorption of CO, OC, NO on NiO(100). Saturated $Ni(OH)_2)_3(OH)_2$ cluster.

than the extent of $3d_{z^2}$ at Ni^{2+}. Clearly, the repulsion between the 5σ-orbital of CO pointing towards the surface and the nearly fully occupied 3d shell of Ni^{2+} is the dominating contribution to the total Pauli repulsion.

In most discussions of the bonding mechanism between metal surfaces or clusters and CO the Blyholder model [34] is invoked: The 5σ-orbital of CO forms a σ-donor bond with a σ-type d-AO or a spd-hybrid at the metal atom which gives rise to a charge transfer from CO to the metal, while the π-type d-AOs of the metal form π-bonds with the low-lying unoccupied π^* orbitals of CO (π-backbonding, backdonation). This model is definitely not applicable to the bond between the NiO(100) surface and CO:

i) The 4s and 4p-AOs of Ni^{2+} at the surface are too high in energy and do not participate in the formation of hybrids directed towards the approaching CO.

ii) There is virtually no charge transfer between Ni^{2+} and CO, neither in the σ-orbitals nor in the π-orbitals.

iii) There is no indication of an increase in the CO bond length or a considerable decrease in the CO vibration frequency upon adsorption, which should be a consequence of the transfer of electrons into the antibonding π^* orbital of CO and which is indeed observed for the bonding of CO to metals.

Pacchioni and Bagus [2-4] arrive at the same conclusions concerning the bonding mechanism in their CSOV (constrained space orbital variation) analysis of CO/MgO(100).

For CO/NiO(100) as well as for CO/MgO(100) [2-5] the linear adsorption is the most favorable one. Both tilting of CO against the surface normal and translations parallel to the surface lead to an increase in the total energy.

The calculated binding energy for CO/NiO(100) of about 0.1 eV (ROHF, TZP basis) confirms experimental observations ([6] and references therein) that the bonding between CO and NiO(100) is rather weak, much weaker than between Ni metal and CO with a binding energy of about 1.3 eV. Recent TDS measurements [35] yielded a binding energy of about 0.3 eV, which is considerably larger than our ab initio result. The origin for this discrepancy (see also section 4.2 for NO/NiO(100)) is not yet clear: The use of larger basis sets at

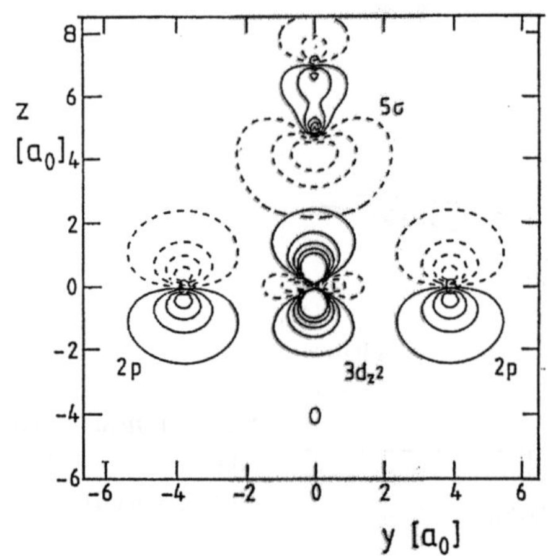

Fig. 2: Contour plots for the valence orbitals of $Ni(OH_2)_3(OH)_2$ at $R_{Ni\text{-}C} = 5.0\ a_0$.

the ROHF level did not lead to a stronger bond. Inclusion of electron correlation effects [6] by means of CASSCF and MC-CEPA calculations also did not change the potential curves markedly [6]. Whether the size of the cluster is not yet sufficient or the charges for the Madelung field have to be modified is still a matter of discussion [9]. However, the correction for the BSSE is absolutely necessary since this error is in the order of 0.1-0.2 eV in the bonding region, i.e. in the same order of magnitude as the binding energy itself.

4.2 NO/NiO(100)

The electronic structure of NO differs from that of CO by the extra unpaired electron that occupies the antibonding 2π orbital. Thus, the ground state of the system Ni^{2+}-NO at linear geometries is described by the electronic configuration $t_{2g}^6 e_g^2 (^3B_1) 2\pi^1$ which gives rise to a ^4E and a ^2E state. These two states are separated by a two-center exchange integral which is quite small at large distances. Therefore, the energy difference between the two states is small, ^4E being the ground state. All three singly occupied orbitals $a_{x^2-y^2}, a_{z^2}, 2\pi$ have different symmetry and cannot mix at linear adsorption geometries.

In Fig. 3 we have summarized our ROHF results for the linear adsorption of NO on Ni^{2+} in the ^4E ground state using the five different cluster models as described in section 2. Asymptotically, the curves are quite similar but at smaller separations large differences are observed [8]:

i) In the point charge model (PC), the Pauli repulsion between the surface and NO is missing, therefore this curve is only attractive and shows no minimum.

ii) If the Ni^{2+} ion at the surface is included (model Ni^{2+}/MAD), the Pauli repulsion starts at about 5.0 a_o and gives rise to a minimum at a Ni-N distance of about 4.5 a_o without changing the asymptotic behavior.

iii) Upon increasing the cluster size to NiO_5^{8-}/MAD the minimum becomes

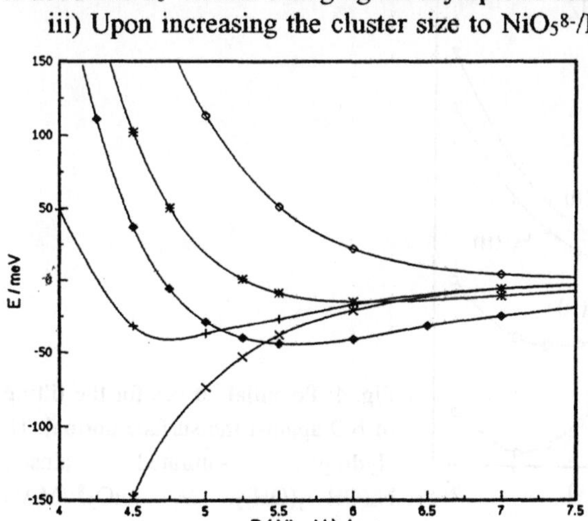

Fig. 3: Potential curves for the linear adsorption of NO to NiO(100). ROHF, TZP basis, ^4E ground state. xxx PC model, +++ Ni^{2+}/MAD, *** NiO_5^{8-}/MAD, ΓΓΓ NiO_5^{8-}/ECP/MAD, υυυ $Ni(OH_2)_3(OH)_2$.

shallower and is shifted to larger separations since now also the Pauli repulsion between NO and the O^{2-} ions is operative.

iv) If we further improve the cluster by representing the next Ni^{2+} coordination shell by effective core potentials (NiO_5^{8-}/ECP/MAD) the minimum even disappears and the potential is purely repulsive for all Ni-N distances.

v) The $Ni(OH_2)_3(OH)_2$ model yields a minimum with a well depth of about 0.05 eV at a distance of 5.5 a_o. This is rather similar to the corresponding minimum for the adsorption of CO or OC (compare Fig. 1).

vi) (Not documented here.) The adsorption of ON – i.e. O end of NO towards the surface – shows a similar behavior as the adsorption of NO, with slightly shallower minima [8].

Despite of these apparent differences between the various cluster models, they agree in the prediction that the binding energy is rather small - only in the order of 0.05 eV – and the equilibrium distance is large (5.0-5.5 a_o). The interpretation of the bonding mechanism for the linear adsorption of NO on NiO(100) is virtually the same as for CO/NiO(100).

The situation is changed considerably if NO is tilted against the surface normal. Now the singly occupied orbitals have no longer different symmetries, but belong to either a' ($3d_{z^2}$, $3d_{x^2-y^2}$, one component of 2π) or a'' (the other component of 2π). Therefore, a bonding linear combination of the a' orbitals can be formed and the $^2A'$ state in which this orbital is doubly occupied will be stabilized while both components of ^4E which need three singly occupied orbitals as well as the $^2A''$ state are not affected [7,8].

Figure 4 shows this behavior for the A' components of ^4E and ^2E: The $^4A'$ state is destabilized, the $^2A'$ state stabilized, with a minimum at a tilt angle φ of about 45°. Of course, since configurations with the bonding a' orbital doubly occupied are mixed into the wavefunctions with weighting factors which

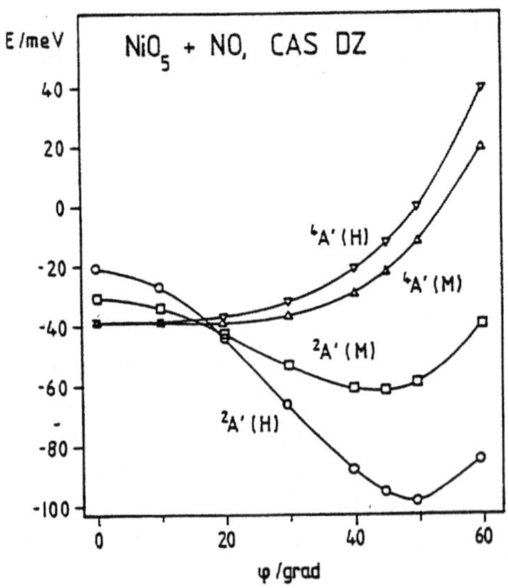

Fig. 4: Potential curves for the tilting of NO against the surface normal. H: Hydrogen saturated cluster $Ni(OH_2)_3(OH)_2$, M: NiO_5^{8-}/MAD model.

increase with φ, this bonding situation cannot be described at the ROHF level, but needs at least a CASSCF wavefunction.

We have performed similar calculations as those shown in Fig. 4 not only for the NiO_5^{8-}/MAD and $Ni(OH_2)_3(OH)_2$ model clusters, but also for the other models, different basis sets and with inclusion of dynamic correlation effects at the MC-CEPA level. The qualitative behavior does not change, the tilt angle is always between 40° and 60°, only the binding energy is slightly increased if correlation effects are taken care of. Our best value for binding energy is 0.10±0.05 eV [8], i.e. not much larger than for CO/NiO(100).

The experimental investigations of the system NO/NiO(100), both for thin NiO(100) films on Ni(100) and for NiO(100) single crystal surfaces obtained by cleaving NiO single crystals in vacuo, show that NO is indeed adsorbed tilted with a tilt angle of 45°. However, the experimental binding energy, as determined by TDS experiments, is about 0.5 eV [35]. Again, there is good reason to assume that the ionicity of the NiO(100) surface is not correctly described by our clusters and that a weaker Madelung field can lead to a larger binding energy [9]. However, the mechanism responsible for the tilt is not changed. Since a bonding a' orbital is formed and occupied in the tilted NO/NiO(100) adsorption one can call this bonding situation already a weak chemisorption.

4.3 CH_3/NiO(100)

While CO as a closed-shell molecule and NO as a π−radical with the singly occupied orbital parallel to the surface – at least for linear approaches – do not form genuine chemical bonds with NiO(100), CH_3 has a singly occupied $2p_z$-AO which points towards the surface and behaves like a σ-orbital in the planar approach of CH_3. Therefore, bonding and antibonding orbitals can be formed and CH_3 will be chemically bound with a binding energy larger than in the two other cases.

Recent ab initio calculations have confirmed this expectation [10]. For all Ni-C distances, the lowest $^2A'$ state is the ground state, the $^4A'$ state with nearly the same electronic configuration is slightly higher in energy.

Three effects not present for CO/NiO(100) or NO/NiO(100) enhance the binding energy of CH_3/NiO(100):

i) The CH bonds of CH_3 can bend away from the surface. This reduces the Pauli repulsion between the CH bond orbitals and the fully occupied O^{2-} 2p orbitals and creates a dipole moment in CH_3. Both effects contribute to a larger binding energy.

ii) The Ni^{2+} ion can be slightly pulled out of the surface.

iii) Dynamic correlation effects have a rather large contribution to the binding energy of CH_3/NiO(100). This is to be expected since we have the typical situation of a chemical bond with two subsystems each of them with unpaired electrons.

Our first preliminary results [10] for the $^2A'$ ground state of $CH_3/NiO(100)$ show that the bonding of CH_3 to this surface is indeed a weak chemisorption and considerably stronger than that of CO or NO. The Ni-C-H angle is slightly opened up to about 100° and Ni^{2+} is pulled by 0.16 a_o out of the surface. The calculated CASSCF binding energy amounts to 0.22 eV; we guess that dynamic correlation effects will enhance this value to 0.4-0.5 eV.

Experimental studies for the adsorption of CH_3 on NiO(100) have not yet been performed, since isolated CH_3 radicals are more difficult to prepare than e.g. CO. It is also not sure whether CH_3 adsorbes as CH_3 radical at NiO(100), since other processes, e.g. proton abstraction and the formation of OH species at the surface are also possible.

5. Conclusions

The present quantum chemical ab initio calculations for the adsorption of small molecules and radicals on NiO(100) have led to a detailed understanding of the surface states and the bonding mechanism: The Blyholder model with a σ-donor bond and π-backdonation is not applicable, but electrostatic forces and Pauli repulsion dominate. The present cluster models limit the accuracy of the calculated binding energies to ± 0.3 eV.

Acknowledgements

The author thanks R. Jaquet (Siegen), J. Wasilewski (Torun), U. Meier (CONVEX), and R. Fink (Basel) who participated in the development of the ab initio programs used in this study and M. Pöhlchen (TRIPOS), A. Freitag, and J. Freitag (Bochum) who performed most of the calculations. Without the many stimulating and helpful discussions with H.-J. Freund and H. Kuhlenbeck (Bochum) these investigations had never been possible. Financial support by the Deutsche Forschungsgemeinschaft, the Ministerium für Wissenschaft und Forschung des Landes Nordrhein-Westfalen and the Fonds der chemischen Industrie is gratefully acknowledged.

References

[1] H. Kobayashi, M. Yamaguchi, T. Ito: J. Phys. Chem. **94**, 7206 (1990)
[2] G. Pacchioni, G. Cogliandro, P.S. Bagus: Surf. Sci. **255**, 344 (1991)
[3] G. Pacchioni, G. Cogliandro, P.S. Bagus: Int. J. Quantum Chem. **42**, 1115 (1992)

[4] G. Pacchioni, P.S. Bagus: in: G. Pacchioni, P.S. Bagus and F. Parmigiani (eds), Cluster models for surface and bulk phenomena, NATO ASI series, Plenum Press, New York 1992
[5] K. M. Neyman, N. Rösch: Chem. Phys. **168**, 267 (1992); Ber. Bunsenges. Phys. Chem. **96**, 1711 (1992); J. Mol. Struct. **293**, 303 (1993)
[6] M. Pöhlchen, V. Staemmler: J. Chem. Phys. **97**, 2583 (1992)
[7] H. Kuhlenbeck, G. Odörfer, R. Jaeger, G. Illing, M. Menges, Th. Mull, H.-J. Freund, M. Pöhlchen, V. Staemmler, S. Witzel, C. Scharfschwerdt, K. Wennemann, T. Liedtke, M. Neumann: Phys. Rev. **B 43**, 1969 (1991)
[8] M. Pöhlchen, V. Staemmler, J. Wasilewski: to be published
[9] L.G.M. Pettersson: to be published
[10] J. Freitag, V. Staemmler: to be published
[11] A. Freitag, V. Staemmler: to be published
[12] N.U. Zhanpeisov, A. Freitag, V. Staemmler: to be published
[13] G. Pacchioni, T. Minerva, P.S. Bagus: Surf. Sci. **275**, 450 (1992)
[14] G. Pacchioni: Surf. Sci. **281**, 207 (1993)
[15] G. Pacchioni, P.S. Bagus: this volume
[16] R.W.G. Wyckoff: Crystal structures, 2nd edition, Vol. I, Interscience, New York 1965, p.89
[17] M.A. Van Hove, S.-W. Wang, D.F. Ogletree, G.A. Somorjai: Adv. Quantum Chem. **20**, 1 (1989)
[18] P.E.M. Siegbahn, U. Wahlgren: in: E. Shustorovich (ed.), Metal-surface reaction energetics, VCH Publishers, Weinheim 1991
[19] P.E.M. Siegbahn, U. Wahlgren, P. Boussard: this volume
[20] U. Meier, V. Staemmler: Theoret. Chim. Acta **76**, 95 (1989)
[21] J. Wasilewski: Int. J. Quantum Chem. **36**, 503 (1989)
[22] V. Staemmler, R. Jaquet: Theoret. Chim. Acta **59**, 487 (1981)
[23] R. Fink, V. Staemmler: to be published
[24] B. Roos, A. Veillard, G. Vinot: Theoret. Chim. Acta **20**, 1 (1971)
[25] A.J.H. Wachters: J. Chem. Phys. **52**, 1033 (1970)
[26] S. Huzinaga: J. Chem. Phys. **42**, 1293 (1965)
[27] S. Huzinaga: Approximate atomic functions, I. Preprint. University of Alberta, Canada 1971
[28] S.F. Boys, F. Bernardi: Mol. Phys. **19**, 553 (1970)
[29] C. Corliss, J. Sugar: J. Phys. Chem. Ref. Data **10**, 197 (1981)
[30] A. Freitag, V. Staemmler, D. Cappus, C.A. Ventrice Jr., K. Al-Shamery, H. Kuhlenbeck, H.-J. Freund: to be published
[31] A. Gorschlüter, H. Merz: J. Phys.: Cond. Matter: in press
[32] G.J.M. Janssen, W.C. Nieuwpoort: Int. J. Quantum Chem. Symp. **22**, 679 (1988)
[33] H.-J. Freund, H. Kuhlenbeck, M. Neumann: this volume
[34] G. Blyholder: J. Phys. Chem. **68**, 2772 (1964)
[35] H.-J. Freund: private communication

Adsorption and Reaction of CO and CO_2 at the Surface and Defect Sites of MgO and NiO

G. Pacchioni[a] *and P. S. Bagus*[b]

[a] Dipartimento di Chimica Inorganica, Metallorganica e Analitica, Universita' di Milano, via Venezian 21, 20133 Milano, Italy
[b] IBM Research, Almaden Research Center, 650 Harry Road, San Jose, California 95120-6099, USA

Abstract. *Ab initio* cluster model calculations on the interaction of CO and CO_2 with NiO and MgO surface and defect sites are reviewed. NiO behaves very similarly as the non-transition metal oxide MgO towards CO chemisorption because the bonding of CO to these surfaces is dominated by electrostatic terms and involves a very modest σ donation or, in the NiO case, π back donation. The role of the surface defects in determining the reactivity of oxides toward CO and CO_2 chemisorption is shown.

1. Introduction

The surface chemistry of metal oxides is of great importance in many catalytic processes [1,2]. Several experimental techniques have been developed to characterize the adsorbed species at the surface of a metal oxide. One of the most widely used techniques is the infrared, IR, spectroscopy of the surface species [3]. This technique has the advantage of being nondestructive and it can be employed to study powders and polycrystalline samples under normal conditions, e.g. in air, in solution, etc. Such samples are closely related to the topography of an actual catalyst. Only recently it has been possible to study adsorption on well defined oxide single crystal surfaces under ultra-high vacuum [4,5].

In recent years, we have investigated theoretically some fundamental aspects of the adsorption and reaction of small molecules at oxide surfaces by means of *ab initio* cluster model calculations [6-10]. Quite surprisingly, in the last 10-15 years while a great deal of attention has been devoted to the cluster modelling of metal and semiconductor surfaces [11], the number of theoretical studies on ionic surfaces is rather small [12-16]. The very rich behavior of oxide surfaces is not yet fully understood from a microscopical point of view; in addition, the technical and methodological aspects of the treatment of these surfaces have not yet been fully explored. Thus, two completely different classes of questions need to be answered by theory.

Springer Series in Surface Sciences, Vol. 33 Eds.: E. Umbach · H.-J. Freund
Adsorption on Ordered Surfaces of Ionic Solids and Thin Films

The ionic character of oxides cannot always be inferred by assuming O^{2-} anions and different degrees of ionicity are likely to be involved. The degree of ionicity in the bonding may be different for different metals and it may also be different at different sites in the oxide; for example, the M-O bond at a surface site may be quite different from the bond at a bulk site. There are open questions related to the theoretical treatment of highly (but not entirely) ionic systems. The cluster model approach for surface problems [11] is based on the fundamental assumption that the interaction between the adsorbate and the surface is dominated by local chemical bonding involving only the nearest neighbor surface atoms. Some limitations of the cluster model are related to the treatment of the rest of the crystal (embedding). There is a very similar need for embedding when clusters are used to model bulk properties of crystals. In particular, embedding is critical in the cluster description of ionic crystals because the Madelung potential represents an essential ingredient for the stabilization of the ions.

We have concentrated our attention on the interaction of two gas phase molecules, CO and CO_2, with the surface of highly ionic oxides like MgO and NiO. CO is a prototype molecule which has been widely used as probe to test the acid-basic character of the surface [3]. There is still confusion about the real mechanism of interaction of CO with the surface of an ionic material; it is often believed that the bond occurs because of the donation of charge from the CO 5σ MO to the surface cations according to the "classical" Blyholder model [17]. We present a quite different view that the interaction between CO and the ionic surface is largely electrostatic [6-9]; this view is fully consistent with the large body of experimental data. A dominantly electrostatic interaction also holds for the case of CO_2 [10]. In this latter case, however, the interaction can occur also at the oxide O^{2-} sites with formation of stable surface carbonates.

There are at least two important consequences of the electrostatic nature of the interaction of polar molecules with oxide surfaces. If two oxides have a highly ionic character and a similar crystal lattice, their detailed electronic structure becomes less relevant for the understanding of the adsorption processes. This is the case of the adsorption of CO at MgO(100) and NiO(100) surfaces which behave in a very similar way despite the presence of the d electrons on NiO. The second consequence is that low coordinated defect sites may exhibit a very different reactivity because the electric field at a surface site is different from that at corner or edge sites. Both these aspects will be briefly examined in this paper.

2. Cluster Models and Wavefunctions

We have performed all electron cluster calculations on the interaction of CO and CO_2 with both cationic, Mg^{2+}, and anionic, O^{2-}, sites of the MgO(100) and NiO(100) surfaces. We have considered regular, five-coordinated surface sites, Mg_{5c}^{2+}, Ni_{5c}^{2+} and O_{5c}^{2-}, and defect, four-coordinated step sites, Mg_{4c}^{2+} and O_{4c}^{2-}. Three-coordinated corner sites, Mg_{3c}^{2+}, have been considered only for CO adsorption. $[NiO_5]^{8-}$, $[MgO_5]^{8-}$, $[MgO_4]^{6-}$, $[MgO_3]^{4-}$, see Figs. 1 and 2, are the

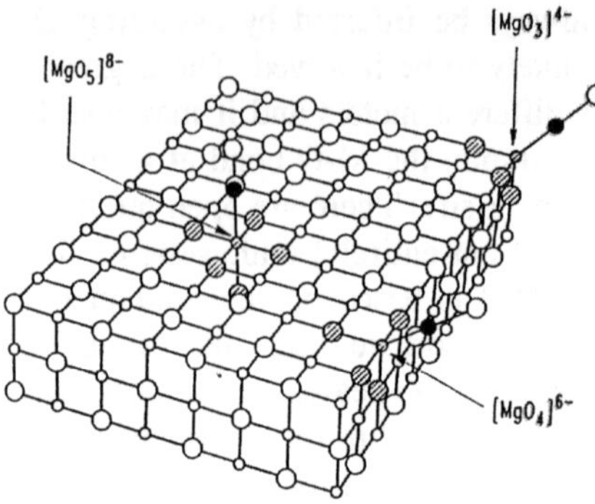

Fig. 1: Schematic representation of CO at surface, edge and corner sites of MgO (100); the different equilibrium CO orientations are shown. The atoms explicitly included in the cluster for the different sites are shown with hash marks. Similar clusters were used to simulate the NiO(100) surface.

clusters representing surface, step, and corner Ni^{2+} or Mg^{2+} sites. $[OMg_5]^{8+}$ and $[OMg_4]^{6+}$, Fig. 2, are the two clusters representing the surface and step O^{2-} chemisorption sites, respectively. In all these clusters only the atom where the interaction takes place and the first nearest neighbors have been included. The rest of the ionic crystal has been simulated by a very large set of point charges, PC = ±2, placed at the lattice positions. This array of PCs is large enough to ensure that the Madelung potential in the chemisorption region is reproduced with reasonable accuracy [9].

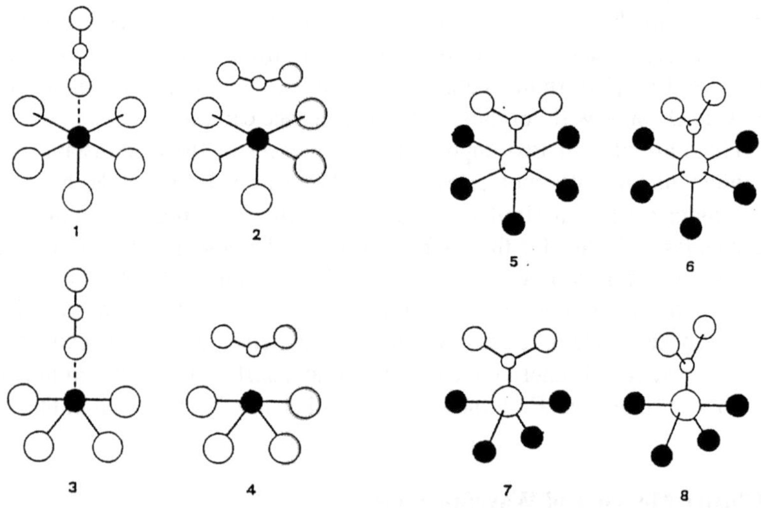

Fig. 2: Cluster models of CO_2 interacting with (a) a Mg_{5c}^{2+} surface site, 1 and 2, (b) a four-coordinated Mg_{4c}^{2+} step site, 3 and 4, (c) a O_{5c}^{2-} surface site, 5 and 6, and (d) a O_{4c}^{2-} step site, 7 and 8. In 1 and 3 CO_2 is physisorbed; in 7 CO_2 is chemisorbed (surface carbonate); the other structures are unstable toward dissociation.

The entire system, cluster + PCs, is neutral. Of course, this Madelung potential is appropriate for perfectly ionic MgO and NiO crystals. However, tests have shown that the cluster results do not depend strongly on this particular choice of Madelung potential; even when the point charges are reduced by large factors, there are only relatively small changes in the ionic character of the metal-oxygen bond.

Hartree-Fock SCF wavefunctions have been determined within the MO-LCAO approximation. Details about the basis sets used can be found in refs. [6-10]. Relaxation of the Mg^{2+} or O^{2-} chemisorption sites as well as the equilibrium geometry of the adsorbed molecules were determined by performing a geometry optimization for the SCF wavefunctions by means of analytical gradient techniques. Correlation effects have been introduced in selected cases by performing single-double configuration interaction, SD-CI, or multi-configuration SCF, MC-SCF, calculations at the SCF optimal geometry. For further details see [6-10].

3. Adsorption of CO at the MgO(100) and NiO(100) Surfaces

It is usually assumed that CO bounds to an ionic surface at a cation site and the traditional explanation of the observed shifts in the CO vibrational frequency ω is based on σ donation and π back-donation as the key bonding mechanisms [17]. It is argued that the positive shift of ω when CO is adsorbed on an oxide surface is due to the formation of a σ-dative bond from the 5σ lone pair of CO to the empty states of the metal cations [18,19]. The ω_e for gas-phase CO^+ ($5\sigma^1$) is 44 cm^{-1} higher than that of CO [20]; thus, the positive ω shift for adsorbed CO has been attributed to the chemical interaction of the CO molecule with the ionic surface, in particular to the decrease of the population of the CO 5σ MO [18,19]. It has been concluded that a correlation between CO ω shift and cation formal charge is possible only for cations with d^0 configuration, like Mg^{2+}, and problematic for d^n cations, like Ni^{2+} [18]. This is because the occurrence of $d\pi$-CO($2\pi^*$) back-bonding is a mechanism which leads to frequency shifts of opposite sign to those induced by the σ donation. Unfortunately, there is no independent justification, in terms of other bonding properties, for these assumptions.

We have recently formulated a completely different analysis for the nature of the bonding and the origin of the shifts in the CO stretching frequency [6-9]. This analysis provides a consistent and unified view of the shifts for both simple, sp, and transition metal ionic systems; it is based on the decomposition of the interaction energy E_{int} and of the vibrational frequency ω as sum of different contributions arising from electrostatic, polarization and dative bonding. It is worth noting that this view agrees with that obtained from Hartree-Fock band structure calculations [12,13] and has been confirmed by recent local density functional [21] and *ab initio* studies which include an extensive treatment of correlation effects [22,23].

3.1 Analysis of the Interaction Energy

The decomposition of the interaction energy between CO and the various sites of NiO and MgO is done according to the Constrained Space Orbital Variation, CSOV, method [24,25]. The results are for unrelaxed substrates and for a CO geometry close to equilibrium. At each step of the CSOV, the interaction energy is defined as E_{int}(CSOV step n) = E(CO) + E(MO) - E(MO/CO; CSOV step n) (M = Ni or Mg). The change in E_{int} between CSOV step n and the proceeding step n-1 represents the energetic importance of the new variational freedom allowed at step n, see Table 1.

The first step of the CSOV consists in simulating the ionic surface by an array of PCs. The M^{2+} cation and the O^{2-} anions are represented by PC = +2 and

Table 1: Contributions, from SCF wavefunctions, from various bonding mechanisms to the interaction energy, E_{int}, in eV, for CO chemisorbed on surface, edge and corner sites of the NiO and MgO surfaces.

cluster	NiO_5-CO	MgO_5-CO	MgO_4-CO	MgO_3-CO
site	surface	surface	edge	corner
z_e(M-C), in bohr	4.90[a]	4.90	4.60	4.50
Electric field at M^{2+}, a.u.	---	0.019	0.033	0.062
Bonding contributions, in eV				
1. Electrostatics	+0.24	+0.24	+0.42	+0.48
2. Pauli repulsion	-0.27	-0.21	-0.34	-0.28
3. MO polarization	+0.02	+0.02	+0.01	0.00
4. MO→CO donation	+0.12	+0.06	+0.04	+0.04
5. CO polarization	+0.04	+0.06	+0.12	+0.27
6. CO→MO donation	+0.07	+0.06	+0.07	+0.07
total D_e (SCF)	+0.24	+0.23	+0.33	+0.59
total D_e (SD-CI)	---	+0.32	+0.49	+0.83
total D_e (experimental)	≅ 0.30 [28]	≅ 0.25 [29]	---	---
$\omega_e/\Delta\omega$, cm^{-1} (SCF)[b]	2298/+29	2301/+31	2325/+55	2367/+97
$\omega_e/\Delta\omega$, cm^{-1} (SD-CI)[c]	---	2110/+43	2122/+55	2140/+73
$\omega_e/\Delta\omega$, cm^{-1} (exp.)[d]	≅ 2153/+10	≅ 2156/+13	≅ 2164/+21	≅ 2203/+60

a) Non-optimal distance; the optimal distance is 4.70 bohr [6]. b) $\Delta\omega$ is computed with respect to free CO ω_e(SCF) = 2270 cm^{-1}. c) $\Delta\omega$ is computed with respect to free CO ω_e (CI) = 2067 cm^{-1}. d) From [19], [26] and [27]; $\Delta\omega$ is computed with respect to free CO ω_e(exp.) = 2143 cm^{-1} [20].

PC = -2, respectively. At this first step, the charge density of the CO molecule is fixed as it is for the SCF wavefunction of the isolated molecule for the appropriate C-O bond distance. The value of E_{int} at this step provides a measure of the importance of the pure electrostatic interaction. At step 1 E_{int} is the same for MgO and NiO surface sites because it is assumed that both are perfect ionic crystals. At step 1 E_{int} is considerably larger for corner and edge sites than for a surface site, Table 1, because in the former cases the CO is adsorbed at a low-coordinated cation where the local electric field is stronger. In the second step of the CSOV we replace the few PCs defining the local chemisorption site with real M^{2+} and O^{2-} ions, but we do not allow the two charge densities, cluster and CO, to relax. The charge density for $[MO_x]$ is fixed as that for the SCF wavefunction for the isolated cluster and the CO charge density is kept at that for the isolated molecule. This step measures the extent of Pauli repulsion originating from the interpenetration of electronic charge at equilibrium positions, see Table 1. This energy contribution, which is purely repulsive, depends on the surface-CO distance but also on the number of O^{2-} ions on the surface. In $[MO_5]^{8-}$/CO there are four spatially extended O^{2-} anions which prevent the CO molecule from approaching too close to the surface; in $[MO_4]^{6-}$/CO there are only two O^{2-} ions on the first layer of the cluster, while in a corner site the O^{2-} ions are in the second layer, see Fig. 1. The Pauli repulsion is smallest for the surface and largest for the edge, Table 1, because it results from the balance of the two effects, i.e. the distance from the surface and the number of O^{2-} ions which are nearest neighbors of CO.

The substrate polarization, step 3, measured by allowing the $[MO_x]$ cluster orbitals to vary, is negligible for both MgO and NiO and remains small on the different sites. Also the oxide to CO charge transfer, measured at CSOV step 4, is small, Table 1; furthermore, the charge transfer is exaggerated by the occurrence of the basis set superposition error, BSSE (*vide infra*). While it is not surprising that the d^0 Mg^{2+} cation does not donate charge to the CO molecule, it is important to note that the d^8 Ni^{2+} cations also do not back-donate electrons to CO. Actually, the $d\pi$-$2\pi^*$ interaction in NiO-CO, 0.10 eV, is small but not negligible and it may indicate the occurrence of a weak back donation. In CSOV step 5, the CO molecule is free to polarize but the $[MO_x]$ charge density is frozen as determined at the previous CSOV step. The CO polarization can be viewed as arising from two origins: the Pauli repulsion and the electrostatic potential at the bonding site due to the M^{2+} and O^{2-} ions. The CO polarization, which is virtually the same for NiO and MgO, is considerably larger for a corner site than for a regular surface site; the edge site is intermediate. The dominant effects determining the magnitude of the CO polarization are the electric field and the bond distance associated with the different adsorption sites. Of course, the two effects are coupled because the larger electrostatic attraction leads to a shorter bond distance. The charge donation from the CO molecule to the surface, see CSOV step 6 in Table 1, is small, and again overestimated by the BSSE [6-9]. Differently from what is often believed, *there is little σ-donation from CO to* *the ionic surface*; this is true for both surfaces, NiO and MgO, and even for low-coordinated sites [6-9].

The previous analysis shows that the bonding of CO to NiO is very similar to that of MgO. In both oxides the interaction does not involve a significant charge transfer, neither from the Ni^{2+} $d\pi$ orbitals to the CO $2\pi^*$ MO nor from the CO 5σ carbon lone pair to the empty levels on the metal cations. The different binding energies computed for CO on the three sites of MgO arise principally from the very different local electric field at the surface cation. For a five-coordinated surface site this electric field is small, the initial electrostatic attraction is partially offset by the Pauli repulsion and the weak bonding is mainly due to the electrostatic term plus the CO polarization; for a four- and a three-coordinated site, e.g., an edge or a corner site, the electric field becomes larger hence both the electrostatic attraction and the CO polarization increase. As a consequence, the chemisorption energy of a CO molecule on a corner site is almost three times larger than on a five-coordinated site; this is found for both SCF and correlated CI wavefunctions, see Table 1.

3.2 Vibrational Shifts of Adsorbed CO

Experimentally it has been observed that when CO is adsorbed on MgO or NiO surfaces the vibrational shift is toward higher frequencies and of similar magnitude, ≈ 10 cm^{-1} [10]. This is perfectly consistent with our results which indicate a comparable shift for the two systems, Table 1. The main line in the IR spectrum of CO on MgO microcrystals and on MgO powders, at 2156 cm^{-1}, is accompanied by a low-intensity feature at 2164 cm^{-1} [26,27] which has been attributed to CO molecules adsorbed on the edges of the crystallite; the low intensity is consistent with the fact that the edge sites do not exceed the 5% of the total number of surface sites on the crystallite, the rest being regular, non-defect, surface sites. When CO is adsorbed on polycrystalline MgO, which has a high concentration of defect sites, a third band is observed at 2203 cm^{-1} [26]. This band has been assigned to CO adsorbed on low-coordinated Mg_{3c}^{2+} sites; an assignment which is fully supported by the present calculations, see Table 1. The SCF ω_e for an edge Mg_{4c}^{2+} site is 24 cm^{-1} larger than for the regular site, compared to the experimental assignment of the frequency shift being $\approx +10$ cm^{-1}. For a corner, Mg_{3c}^{2+}, site, the SCF ω_e is 66 cm^{-1} larger than for the regular, Mg_{5c}^{2+}, site while the experimental ω_e is ≈ 50 cm^{-1} larger than for the regular site. A similar trend is found when ω_e(C-O) is computed with CI wavefunctions [8].

The ω shift for CO adsorbed on NiO and MgO is basically due to two mechanisms: (1) the interaction between the non uniform electric field at the surface and the CO dipole, more precisely the CO dynamic dipole, and (2) the "wall effect"; the wall effect occurs because the molecule stretches in the presence of the "rigid" surface. Both mechanisms act to raise ω_e [6-9]. For a coordinatively unsaturated corner site the field-dipole interaction is dominant with respect to the wall effect; for a five-coordinated surface site the field-dipole term is small and the shift is largely due to the wall effect [8]. This is different from the interpretation which attributes the shift to the formation of a surface σ-complex where charge flows from the CO 5σ orbital to the substrate [18,19].

3.3 Dependence of D_e on Correlation, BSSE, and Cluster Size

The computed adsorption energy for CO on MgO and NiO, ≈0.25 eV, agrees well with the experimental values [28,29]. However, this agreement is, in part, fortuitous because at least three effects which may affect the D_e have not been taken into account so far: (i) correlation effects, (ii) basis set superposition error (BSSE) [30], and (iii) cluster size dependence. In this paragraph we discuss briefly these effects. More details can be found elsewhere [6-9,31].

(i) Correlation effects: Correlation effects, which are usually essential for the accurate determination of the stability of a molecular complex, are less relevant here where the bonding is largely electrostatic [9]; however, SD-CI calculations have shown a shortening of the Mg-CO distance and a modest strengthening of the D_e [31].

(ii) BSSE: The BSSE leads to an overestimate of the D_e; the D_e value corrected by the standard counterpoise procedure [30] for the [MgO_5] cluster is 0.17 eV; the BSSE for this cluster is thus of the order of 0.1 eV.

(iii) Cluster size: In the [$Mg_{13}O_{13}$] cluster the central [MgO_5] unit, treated with a triple-zeta (TZ) basis set, is surrounded by 12 Mg and 8 O neighbors treated either with double-zeta (DZ) or with single-zeta (SZ) type bases; we use the notation TZ/DZ to indicate that two different basis were employed to treat the central and the peripheral atoms. On [$Mg_{13}O_{13}$] TZ/DZ cluster the computed D_e, corrected by the BSSE, is 0.08 eV only. This is because the replacement of the PCs with real ions decreases the electrostatic potential in the chemisorption region, see Table 2. Also the ω shift is reduced with respect to the [MgO_5] cluster (but this is due also to the occurrence of the BSSE [31]). Using a SZ basis on the neighboring atoms of the [MgO_5] unit a larger cluster, [$Mg_{21}O_{21}$] TZ/SZ, is constructed; significant comparisons can be done with the [$Mg_{13}O_{13}$] TZ/SZ cluster, Table 2. Of course, with a SZ basis the BSSE increases, Table 2. It is apparent, however, that going from the medium-size to the large-size cluster there are basically no changes in the chemisorption properties, indicating a substantial convergence of the results.

To summarize, taking these corrections into account it is possible to estimate that CO is physisorbed at Mg^{2+} surface sites of MgO with a rather low D_e, of the order of ≈0.1 eV.

4. Adsorption of CO_2 at the MgO Surface

Differently from CO which bonds on metal as well as on oxide surfaces with the C atom down and the molecular axis parallel or slightly tilted from the surface

Table 2. Size dependence of the chemisorption properties

cluster[a]	MgO_5-CO	$Mg_{13}O_{13}$-CO		$Mg_{21}O_{21}$-CO
basis set	TZ	TZ/DZ	TZ/SZ	TZ/SZ
r_e(Mg-C), bohr	4.915	---	4.906	---
r_e(C-O), bohr	2.119	2.121	2.123	2.123
$\Delta\omega$, cm^{-1} [b]	+31	+17	+7	+4
D_e, eV	0.24	0.15	0.40	0.53
D_e, corr. by BSSE, eV	0.17	0.08	0.07	0.08
electrostatic z = 2	0.409	0.369	0.337	0.327
potential, a.u., z = 5	0.050	0.034	0.010	0.000
at various points z = 8	0.021	0.017	0.000	-0.008

[a] All clusters are surrounded by a large set of point charges to reproduce the Madelung potential. [b] With respect to free CO.

normal, CO_2 can bound in several modes at the surface of an oxide. At least five different bonding modes have been found for CO_2 coordinated to metal atoms or in metal complexes: side-on, C-coordination, chelating bent, end-on bent, and end-on linear [10]. All these bonding modes are possible candidates for the bonding of CO_2 with the cationic sites of an ionic crystal. In addition, CO_2 can bound with the oxide O^{2-} sites of an oxide surface and form surface carbonates.

Various orientations of the CO_2 molecule on the $[MgO_5]^{8-}$ and $[MgO_4]^{6-}$ cluster models of surface and step sites, respectively, of the MgO surface have been considered, see Fig. 2; however, only the end-on linear bonding mode is stable, with a D_e of about 0.2 eV which increases slightly when the adsorption occurs at a step site. Actually, the CO_2 may act also as a "bidentate" ligand and interact with the O-ends with two adjacent Mg^{2+} sites assuming a flat orientation. This bonding mode is almost isoenergetic with the perpendicular one and the most favored bonding mode at zero coverage, perpendicular or flat, cannot be easily predicted at this level of treatment [10]. All the other orientations considered, side-on and C-coordination in particular, are unfavorable and spontaneously dissociate into CO_2 and MgO fragments. The electronic structure of CO_2 adsorbed at cationic sites is relatively unperturbed, as shown for instance by the very small changes in the C-O equilibrium bond distances with respect to the free CO_2 molecule [10].

The interaction of CO_2 with the oxide, O^{2-}, sites of MgO can lead to the formation of a surface carbonate. This can be schematically described as a two-electron, two orbitals HOMO-LUMO interaction involving charge transfer from a filled sp_z hybrid orbital of O^{2-} to the empty $2\pi_u$* MO of linear CO_2. This charge transfer induces the bending of the CO_2 molecule. The same mechanism leads to the formation of C-coordinated CO_2 on transition metal surfaces [32], atoms [33], or complexes [34]. The formation of carbonates at the surface of MgO has been investigated with the help of the $[OMg_5]^{8+}/CO_2$ and $[OMg_4]^{6+}/CO_2$ cluster models, see Fig. 2.

At a five-coordinated O^{2-} surface site, the formation of a carbonate is an endothermic process; the surface carbonate, in fact, is 0.4-0.6 eV less stable than the separated units, $[OMg_5]^{8+}$ and CO_2, depending on the orientation of the molecule, see structures **5** and **6** and Table 2. At a distance of about 1.5 Å from the surface, we found a local minimum on the potential energy surface; a very small energy barrier separates this minimum from the dissociation limit [10]. The small energy difference between structures **5** and **6** suggests that the electrostatic attraction between the surface Mg^{2+} cations and the oxygen atoms of CO_2, which become negatively charged because of the partial charge transfer from the surface, plays a non-negligible role in the stabilization of the surface carbonate. In fact, in **5** the shortest distance between a Mg^{2+} ion and the O-end of CO_2 is about 2.5 Å while it is 2 Å in **6**. This effect has a dramatic influence on the reactivity of the step O_{4c}^{2-} site, see Table 2. The surface carbonate **7** is bound by 1.60 eV at the SCF level. Structure **8**, on the contrary, is 0.5 eV less stable than the separated fragments. The two structures, **7** and **8**, differ only by the fact that the CO_2 molecule is rotated by 90°, see Fig. 2. In **7** the shortest distance between Mg^{2+} and the O-end of CO_2 is about 2 Å, while it is about 3 Å in **8**. Clearly, local electrostatic effects largely contribute to stabilize a given conformation. The different stability of the two forms is also reflected in the equilibrium geometry and in particular in the O-CO_2 distance which is substantially shorter for the bound complex **7**, see Table 3.

Thus, the same mechanism leads to the formation of a carbonate at a surface and at a defect site: as the CO_2 molecule approaches the O^{2-} site, charge starts to flow into the $2\pi_u$* orbital and the molecule begins to bend. However, the energetics of the two processes is completely different at surface or step sites. This is due to the distinct nature of five- and four-coordinated surface O^{2-} ions. The O^{2-} ion is unstable in the gas-phase, where it dissociates into O^- + e^-, but is stabilized in ionic crystals by the Madelung potential; the Madelung potential for

Table 3. Optimal geometrical parameters and binding energy from SCF wavefunctions for CO_2 bound at surface and step O^{2-} sites of the MgO(100) surface (see Fig. 2).

Site and structure		r_e, Å O_{nc}-C	C-O	α_e O-C-O	D_e, eV
Surface site $[O_{5c}\text{-}CO_2]^{2-}$	(5)	1.525	1.202	140°	-0.62
	(6)	1.506	1.206	138°	-0.40
Step site $[O_{4c}\text{-}CO_2]^{2-}$	(7)	1.398	1.224	134°	1.60
	(8)	1.481	1.211	136°	-0.52

a surface O_{5c}^{2-} site is considerably larger than for a step O_{4c}^{2-} site [9,16]. As a consequence, the electron cloud of a step O_{4c}^{2-} site is more spatially diffuse and can overlap with the CO_2 empty orbitals at much larger distances from the surface. The interaction is of HOMO-LUMO type, and the position of the HOMO in the two sites is also very important: the charge transfer will occur at a lower energy cost if the "donor" orbital is at higher energy. Indeed, the a_1 HOMO of an O_{4c}^{2-} site is about 1.9 eV higher in energy than the corresponding level for a O_{5c}^{2-} site. This is roughly the difference in stability between the two surface carbonates **5** and **7**, see Table 3.

Acknowledgement

The support of NATO through the Collaborative Research Grant N. 900031 is gratefully acknowledged.

References

[1] H.H. Kung: Transition Metal Oxides: Surface Chemistry and Catalysis, Elsevier, Amsterdam 1989
[2] V.F. Kiselev, O.V. Krilov: Adsorption and Catalysis on Transition Metals and their Oxides, Springer, Heidelberg 1989
[3] A.A. Davydov: Infrared Spectroscopy of Adsorbed Species on the Surface of Transition Metal Oxides, John Wiley, New York 1990
[4] H. Kuhlenbeck, G. Odörfer, R. Jaeger, G. Illing, M. Menges, Th. Mull, H.-J. Freund, M. Pöhlchen, V. Staemmler, S. Witzel, C. Scharfschwerdt, K. Wennemann, T. Liedtke, M. Neumann: Phys. Rev. **B 43**, 1969 (1991)
[5] J.W. He, C.A. Estrada, J.S. Corneille, M.-C. Wu, D.W. Goodman: Surf. Sci. **261**, 164 (1992)
[6] G. Pacchioni, G. Cogliandro: P.S. Bagus, Surf. Sci. **255**, 344 (1991)
[7] G. Pacchioni, P.S. Bagus: in ref. 11, p. 305
[8] G. Pacchioni, T. Minerva, P.S. Bagus: Surf. Sci. **450**, 275 (1992)
[9] G. Pacchioni, G. Cogliandro, P.S. Bagus: Int. J. Quant. Chem. **42**, 1115 (1992)
[10] G. Pacchioni: Surf. Sci. **281**, 207 (1993)
[11] G. Pacchioni, P.S. Bagus, F. Parmigiani (Eds.): Cluster Models for Surface and Bulk Phenomena, NATO ASI Series, Vol. 283, Plenum Press, New York 1992, and references therein
[12] C. Pisani, R. Dovesi, R. Nada, S. Tamiro: Surf. Sci. **216**, 489 (1989)
[13] R. Dovesi, R. Orlando, F. Ricca, C. Roetti: Surf. Sci. **187**, 267 (1987)
[14] E. A. Colbourn, W.C. Mackrodt: Surf. Sci. **143**, 391 (1984)
[15] J. Sauer: Chem. Rev. **89**, 199 (1989)
[16] E.A. Colbourn: Surf. Sci. Rep. **15**, 281 (1992)
[17] G. Blyholder: J. Phys. Chem. **68**, 2772 (1964)
[18] M. I. Zaki, H. Knözinger: J. Catal. **119**, 311 (1989)

[19] E. Escalona-Platero, S. Coluccia, A. Zecchina: Surf. Sci. **117**, 465 (1986)
[20] G. Herzberg: Molecular Spectra and Molecular Structure, Vol. 1, Van Nostrand, Princeton 1950
[21] K. Neyman, N. Rösch: Chem. Phys. **168**, 267 (1992)
[22] M. Pöhlchen, V. Staemmler: J. Chem. Phys. **97**, 2583 (1992)
[23] M.A. Nygren, L.G.M. Pettersson, Z. Barandiaran, L. Seijo: (to be published)
[24] P.S. Bagus, K. Hermann, C.W. Bauschlicher: J. Chem. Phys. **81**, 1966 (1984)
[25] P.S. Bagus, K. Hermann, C.W. Bauschlicher: J. Chem. Phys. **80**, 4378 (1984)
[26] A. Zecchina, S. Coluccia, G. Spoto, D. Scarano, L. Marchese: J. Chem. Soc. Faraday Trans. **84**, 703 (1990)
[27] L. Marchese, S. Coluccia, G. Martra, A. Zecchina: Surf. Sci. **269/270**, 135 (1992)
[28] K. Klier: Collect. Czech. Chem. Commun. **28**, 2996 (1963)
[29] R. Henry, C. Chapon, C. Duriez: J. Chem. Phys. **95**, 700 (1991)
[30] S. Boys, F. Bernardi: Mol. Phys. **19**, 553 (1970)
[31] G. Pacchioni, P.S. Bagus: (to be published)
[32] H.-J. Freund, R.P. Messmer: Surf. Sci. **172**, 1 (1986)
[33] G.H. Jeung: Mol. Phys. **65**, 669 (1988)
[34] S. Sakaki, K. Kitaura, K. Morokuma: Inorg. Chem. **21**, 760 (1982)

Cluster Models of Zinc Oxide Including Ionic and Covalent Effects

P. Boussard, P.E.M. Siegbahn, and U. Wahlgren

Institute of Theoretical Physics, University of Stockholm, Vanadisvägen 9, S-11346 Stockholm, Sweden

Abstract. Two different cluster models mimicking bulk zinc oxide have been developed and tested. The goal has been to incorporate both the ionic and the covalent effects which are present in the bonding between zinc and oxygen. In the first model, termed the dipole model, the building units are neutral ZnO molecules. The binding within this unit is allowed to have covalent contributions, but the binding between different units is assumed to have only small covalent contributions. In the second model, termed the covalent model, the covalent character of the bonding between zinc and oxygen is emphasized. The building units are one-dimensional covalently bound chains of ZnO units. The dangling bonds in these chains are closed by adding hydrogen atoms and/or copper atoms. Both the dipole and the covalent model of zinc oxide give results for the band-gap which are at most in qualitative agreement with experimental measurements. For these models to be useful as quantitative tools in the future, further development is needed.

1. Introduction

Ab initio quantum chemical methodology is now in a stage where molecular systems containing at least 10 atoms can be routinely handled to high accuracy. This means that quite realistic models of gas-phase systems can be relatively easily set up. However, for processes occurring in the solid phase and on solid surfaces it is much more difficult to design realistic models. At least four different types of situations are of interest to model. First, there is the important area of metal surfaces. In this area significant progress has been made during the last decade both in the use and the understanding of the cluster model [1]. Clusters with 20-30 atoms can be routinely handled and can provide information which is of quantitative significance for processes occurring on real surfaces without using any direct embedding technique. Secondly, there are purely covalent systems like graphite and silicon surfaces. In this case the dangling bonds are so pronounced that an embedding has to be used. The common approach is to add hydrogen atoms to close the dangling bonds. The third area concerns the modelling of purely ionic systems. In this case the embedding

Springer Series in Surface Sciences, Vol. 33 Eds.: E. Umbach · H.-J. Freund
Adsorption on Ordered Surfaces of Ionic Solids and Thin Films

problem is quite trivial and can be achieved by placing the chosen cluster in a Madelung field. The fourth area where cluster models are of interest is for ionic crystals where covalent effects can not be neglected, and is probably the area where it is most difficult to find adequate models. One such example is the modelling of zinc oxide, which is the topic of the present contribution.

Purely ionic crystals in which all constituents have a closed-shell electronic structure have close-packed cubic structures. The zinc oxide crystal has a hexagonal wurtzite structure. The fact that ZnO has a hexagonal and not a cubic structure shows that covalency must play a role for this system. Solid ZnO can be visualized as consisting of separate ZnO units parallelly oriented along the crystal c-axis. The geometry of ZnO used in this study is taken from Wykoffs tables [2], where two different Zn-O distances are given, 2.04 Å and 1.80 Å. After the conference it was brought to our attention that this geometry is not entirely correct, and that all Zn-O distances in the crystal are essentially equal (approx. 1.98 Å [3]). We have redone some of the calculations using the new geometry and found only minor effects on the calculated excitation and ionization energies (0.2-0.4 eV). None of the conclusions in the paper are affected by this error.

In the cluster model a cluster of atoms, or aggregates of atoms, embedded in a suitable surrounding, is used to model bulk or surface properties of a solid. In the case of zinc oxide we have used ZnO units to construct the cluster.

In the first model, which we refer to as the dipole model, we make the assumption that covalent bonding primarily occurs between one zinc and one oxygen ion. We thus view the crystal as consisting of neutral, *albeit* polar, ZnO units rather than of separate Zn^{2+} and O^{2-} ions, implicitly assuming that long-range covalency effects, i.e. covalent bonding delocalized over the whole crystal, are of minor importance. The effects of the remaining crystal have been included in the form of a permanent electric field and projection operators mimicking orthogonality effects on lattice positions close to the cluster.

The fundamental difference between the dipole model and a standard ionic model is that the clusters are chosen as aggregates of neutral diatomics rather than as a charged cluster of ions. The standard ionic model has been used by several authors, e.g. Bagus and Wahlgren [4] and by Janssen and Nieuwpoort [5] who used a NiO_6^{10-} cluster as a model of nickel oxide. The type of cluster used in the dipole model allows for local covalent effects.

In the second model, the covalent model, we do not make the assumption that covalent bonds only occur between one zinc and one oxygen. Instead we view zinc oxide as a completely covalent crystal, like e.g. silicon, and in this picture each zinc (and each oxygen) will form two covalent bonds. In the covalent model we use an embedding scheme where we saturate the dangling bonds with suitable atoms.

2. Methods

Our largest cluster consists of seven ZnO units. All Zn atoms are described by means of an Effective Core Potential (ECP) where all electronic shells except for the 4s shell (and the 4p shell) are parametrized. This ECP is similar to the ECP's used previously in metal cluster calculations [6]. In addition all-electron calculations have been carried out on ZnO embedded in the crystal.

Several different basis sets have been employed. At the all-electron level a standard [14s, 11p, 5d] basis set [7], augmented with two diffuse p-functions and one diffuse d-function was used for Zn. This basis set was contracted to [5s, 4p, 3d] basis functions using the Rafenetti [8] contraction. In calculations on polarizabilities this basis set was extended with 2 extra diffuse d-functions and 3 f-functions. Three different basis sets were used on oxygen: a [3s, 3p] basis set used on the six outer ZnO units in $(ZnO)_7$, a [3s, 4p, 1d] basis set used on the oxygen in the central ZnO unit and an extended basis set including f-functions in the calculation of polarizabilities. All the oxygen basis sets were Rafenetti contracted from a [9s, 6p] primitive basis set [9].

In the largest cluster it was not possible to correlate all valence electrons. The valence correlation effects were instead estimated from model calculations where the molecular orbitals (MO's) of the six ZnO units surrounding the central one were frozen. These calculations were done as follows. In step 1 molecular orbitals were generated for one ZnO in the field of the crystal, with projection operators mimicking the overlap effects from ions surrounding the ZnO unit. In step 2 these orbitals were used on the central ZnO and were kept frozen in a calculation on $(ZnO)_7$, where all other orbitals were allowed to relax. In step 3 the resulting orbital set was used in a second calculation on $(ZnO)_7$, but this time the "ligand" orbitals were frozen from step 2 while the original MO's on the central ZnO unit were allowed to relax. A Size-consistent Modified Coupled Pair Functional (MCPF) calculation was then carried out in the orbital basis from step 3, correlating only the electrons associated with the central ZnO unit.

Core-valence correlation effects were estimated using the Core Polarization Potential (CPP) method suggested by Müller and Meyer [10] and modified for transition metals by Pettersson [11]. The CPP effects were calculated at the Self-Consistent Field (SCF) level and added to the final correlated results. The additivity of the valence correlation and CPP effects was verified for the case of one embedded ZnO molecule.

Since limited basis sets had to be used on all ZnO units except for the central one, and since the surrounding ZnO's were not correlated, we also had to correct for the lower polarizability of the surrounding ZnO units. This was done by a scaling procedure. The relaxation energy of the $(ZnO)_6$ surrounding, that is the energy decrease obtained when the orbitals were allowed to relax in $(ZnO)_7$ (relaxation from step 3 above), was assumed to represent the polarization energy of the outer six ZnO's in the cluster. The SCF polarizability of one ZnO in the basis set used was 6 a_0^3 while the polarizability for ZnO using large basis sets (including f and d-functions on Zn and O) was 16 a_0^3. The calculated polarization energy (see section 3.3 below) was thus scaled by a factor 2.7.

3. The Dipole Model of Zinc Oxide

In the dipole model a cluster of ZnO units is embedded in the remaining crystal. The effects from the crystal are assumed to be of two types: electrostatic effects from the dipoles in the remaining crystal and orthogonality effects from the ZnO units neighbouring the cluster. The dipole field is approximated by point charges at lattice positions, with charges chosen to reproduce the dipole moment of a ZnO in the crystal. The field was calculated using the standard Madelung procedure. The orthogonality effects of nearest neighbours were accounted for by projection operators in the lattice positions closest to the cluster.

In free ZnO there are two states which are approximately degenerate, a $^1\Sigma^+$ state which presumably is the ground state and a $^3\Pi$ state which Bauschlicher et al [12] find to be 0.03 eV below the singlet state. However, when the effect of the crystal surrounding is taken into account the ground state becomes a singlet with a closed shell structure. Assuming an ionicity of 2 (see the next section, Table 2) the excitation energy of one ZnO embedded in the crystal to the first excited triplet state becomes 5.3 eV (at the SCF level).

A cluster consisting of seven neutral ZnO units embedded in the crystal was used in the calculations of the ionization potential and the excitation energy.

3.1 The Ionicity of the Crystal

The charges at the lattice positions were determined by varying the ionicity of the crystal and comparing this ionicity with the calculated dipole moment of one ZnO unit embedded in the crystal. The dipole field in the cluster was assumed to be well described by point charges in the lattice positions which reproduce the calculated dipole moment. Self-consistency was obtained for an ionicity of 2.3. This value is somewhat high, it does not seem reasonable with an ionicity exceeding 2. Clearly our dipole approximation is rather approximative, since the distances between the nearest neighbours and the central unit is of the same order as the intermolecular distance in ZnO, and this may be the reason for the apparently overestimated ionicity. In order to check the calculated ionicity further another set of calculations, where the ionicity was correlated with the Mulliken charges of an embedded ZnO, was therefore also carried out. The self-consistent value of the atomic charges, as obtained from Mulliken's population analysis was 1.8.

Both the dipole moment and the gross atomic charges thus yield an ionicity close to two. We have thus chosen to use an ionicity of two in the model.

3.2 Factors Contributing to the Excitation Energy

In a limited cluster model of a mainly ionic crystal there are two fundamental factors which influence excitation and ionization energies. The first one is the electric field due to the surrounding crystal. The second and more complicated factor is the polarizability effect induced by an electronic ionization or excitation on the surrounding crystal.

In addition to these two fundamental factors there are a number of additional corrections which are introduced for technical reasons.

3.3 The Effects of the Polarizability of the Crystal

In a crystal possessing central symmetry the net electric field at the lattice positions is zero. In a polar crystal such as ZnO this is not the case, however. The standard expression for the polarization energy is:

$$E = -\frac{1}{2}\alpha \left|\underline{f}\right|^2 \tag{1}$$

where α is the polarizability. This expression shows that the effect of the polarizability on, say an ionization process, must be calculated using the complete field both for the ground state and the ionized state.

Alternatively, we can Taylor-expand the polarization energy around the field appropriate for the ground state. Let $\underline{f}$ be the field at a lattice site in the unperturbed crystal, and $\Delta\underline{f}$ the field induced by some change, for example an ionization process. In this case $\Delta\underline{f}$ would simply be the field of a +1 point charge at the position where the ionization takes place. The Taylor expansion of the energy in the perturbed field is (in obvious notations):

$$E = E_0 + \frac{\partial E}{\partial \underline{f}}\Delta\underline{f} + \frac{1}{2}\frac{\partial^2 E}{\partial \underline{f}^2}\left|\Delta\underline{f}\right|^2 + \ldots \tag{2}$$

The linear term contains the interaction between the ground-state dipole and $\Delta\underline{f}$, and this effect is, to the extent that the external field really represents the true ground-state dipole field, in principle already present in the calculated energy of the ionized state. If this is the case equation (1), with $\underline{f}$ taken as the change in the field due to the excitation or ionization, can be used to calculate the polarization contribution to the excitation/ionization energy.

Nieuwpoort et al. have used equation (1) to calculate the polarization contribution in a cluster calculation of the band gap in NiO. However, in NiO the electric field at the lattice positions is zero for symmetry reasons, and the polarization effect can thus be calculated directly from the change in the field. In zinc oxide, the field which is obtained directly from the ionic model, assuming +2 and -2 charges at the lattice positions, is 0.003 a_0^{-2}, a value which is larger than the field from a unit charge at the centre of the cluster at all extra-cluster lattice points. Although the effect linear in the field in principle is accounted for in the calculation on the ground state, the uncertainties are large since the model only approximatively mimics the dipole field from the crystal. However, the result obtained with equation (1) is, as will be seen below, of the right sign and of the right order of magnitude, which serves as an indication that the linear term describes the interaction with the permanent electric field reasonably well.

3.4 The Additional Corrections

Three approximations/corrections of this type were used. The CI effects were estimated from a calculation on a partially frozen cluster, core-valence correlation effects were calculated at the SCF level using the CPP method and a scaling procedure was used to correct for the limited polarizability of the cluster. These approximations were described in detail in section 2.

As mentioned in section 3, the electric field from the surrounding crystal strongly stabilizes the singlet closed shell structure of the individual ZnO's. The effect on the first intra-molecular excitation, which to a first approximation can be described as an oxygen to zinc charge transfer excitation, is to raise the excitation energy.

3.5 Results Obtained with the Dipole Model

The monomer. The calculations on the monomer were carried out mainly for test purposes. Three aspects of the system were investigated: The ionicity of the crystal as obtained from the dipole moment and from the gross populations of an embedded ZnO, the effect of the ionicity on the excitation energy and the polarizability of an embedded ZnO unit.

The computed dipole moment and gross population of ZnO as functions of the ionicity of the crystal are shown in Table 1. The polarity of the embedded ZnO unit obtained from the dipole moment is always larger, than the polarity obtained from the gross populations. In terms of the ionicity of the crystal the difference is about 0.5 charge units. This may partly reflect a weakness of the dipole approximation; clearly the field from point charges representing ZnO deviates from a dipole field at short distances.

The effect of the ionicity of the crystal on the excitation and ionization energies are shown in Table 2. The effect is quite dramatic, with an increase in the excitation energy from 2.9 to 5.3 eV for the excitation out of the oxygen $2p_z$ shell when the ionicity was increased from 1 to 2. It is noteworthy that the difference between this excitation energy and the $O_{2p_x,2p_y}$ to $Zn_{4s,4p}$ excitation

Table 1: Populations and dipole moments of one ZnO unit in the crystal for different ionicities. Dipole moment in a.u.

Ionicity	$\text{Charge}_{Mulliken}$	Dipole moment	Charge_{dipole}
1.6	1.75	6.91	2.02
1.7	1.79	7.04	2.07
1.8	1.82	7.16	2.11
1.9	1.85	7.29	2.15
2.0	1.88	7.40	2.18
2.1	1.91	7.52	2.22
2.2	1.93	7.63	2.25
2.3	1.96	7.75	2.28

decreases with the ionicity of the crystal. The excitation energies are approximately equal for an ionicity of 2. This is in fact in agreement with observations. There are two observed structures in the reflectance spectrum of ZnO [13,14]; at 3.30 eV and at 3.34 eV. This is exactly the splitting we obtain for an ionicity of 2. Although this agreement is partly fortuitous, the results indicate that the ionicity should be about 2.

The calculated polarizabilities of O^{2-}, Zn^{2+} and ZnO, in the Madelung field of the crystal, are shown in Table 3. It is noteworthy that the polarizability of ZnO is substantially larger than the sum of the polarizabilities of O^{2-} and Zn^{2+}. The covalency in ZnO thus has a substantial effect on the polarizability of ZnO. This result supports the dipole model *ansatz*.

The cluster. Four different calculations of the ionization potential and the excitation energy were carried out: (1) An SCF calculation with the MO's of the outer six ZnO units frozen as detailed in section 2, (2) an MCPF calculation on the same system, (3) an SCF calculation in the fully relaxed system and (4) a calculation using the Core Valence Polarization (CPP) operators, at the SCF level, on the relaxed system. From the results of these four calculations we can extract, at least approximately, the CI effect, the polarization effect and the CPP

Table 2: The excitation energy as a function of the ionicity of the crystal. SCF results for the charge transfer excitation O_{2p_z} and $O_{2p_x,2p_y}$ to $Zn_{4s,4p_z}$. The z-axis is along the molecular axis. All energies in eV.

Ionicity	$E_{(O_{2p_z}-Zn)}$	$E_{(O_{2p_x,2p_y}-Zn)}$
0.8	1.69	2.40
1.0	2.86	3.47
1.2	3.83	4.39
1.4	4.42	4.94
1.6	4.78	5.16
1.8	5.05	5.26
2.0	5.28	5.32
2.2	5.49	5.35
2.4	5.69	5.37

Table 3: Calculated polarizabilities, atomic units.

System	Method	Component[a]	SCF	MCPF
ZnO	AE	zz	14.22	16.38
ZnO	AE	xx,yy	11.97	13.47
ZnO	ECP	zz	11.06	-
O^{2-}	AE	zz	7.27	7.61
O^{2-}	AE	xx,yy	7.35	7.68
Zn^{2+}[b]	AE	xx,yy,zz	1.50	1.66

[a]Component of the polarizability tensor
[b]Without field

effect on the excitation energy and the ionization energy of the cluster. The results are shown in Table 4.

The CI effect, which is substantial, is to raise both the excitation and the ionization energies. This is not surprising. In the case of ionization there is one less electron to correlate in the final state, and in the excitation the excited electron is not close to any other valence electrons. The CI effects on the ionization potential and the excitation energy are therefore reasonably similar, 1.28 vs. 0.99 eV. According to the population analysis the excited orbital has a population on the central zinc of 0.77, which indicates that the electron is fairly localized. The excitation is thus best described as an oxygen 2p to zinc 4s,4p charge transfer excitation.

The CPP effect on the other hand decreases both the excitation and the ionization energies by about 0.6 eV. The relaxation energy, which reflects the polarization properties of the cluster, is 0.40 eV for the excitation and 0.75 eV for the ionization. A separate investigation was carried out to test the additivity of the relaxation of the individual "ligand" ZnO's. Calculations were done on one of the two types of ligand ZnO's (one closer and one further away from the oxygen centre in the central ZnO) in the field of the crystal but with a charge of -2 and -1 at the central oxygen site. Adding the obtained relaxation energies gave a sum of 0.63 eV, in reasonable agreement with 0.75 eV for the cluster.

The polarizability of the ZnO in the small basis set used on the "ligand" ZnO's is about 6 (an averaged value over the diagonal elements in the polarizability tensor). However, it is more complicated to use the polarizability of the "ligand" ZnO's directly, since it is not clear where to centre it. This problem is due to the inadequacy of the multipole expansion at these short internuclear separations. Using the polarizability of ZnO in the small basis set used for the

Table 4: Results for $(ZnO)_7$. Excitation and ionization energies. All energies in eV.

Method	Fro/rel	$E_{excitation}$	$E_{ionization}$	$E_{affinity}$
SCF	frozen	7.72	9.63	-
MCPF	frozen	8.71	10.91	-
SCF	relaxed	7.32	8.88	-1.63
SCF+CPP	relaxed	6.74	8.22	-

"ligand" ZnO's we can reproduce the relaxation energies obtained for the $(ZnO)_7$ system if the polarizabilities are centered 0.5 a.u. from the oxygen centre in the direction of Zn.

The electron affinity of the cluster is negative, -1.63 eV. This indicates that if the cluster accepts an electron into the conduction band it must be a delocalized state.

There are observed transitions at 3.30, 3.34, 7.00, 7.05, 8.95, 9.2, 12.6, 13.8-15.6 and 17.1-20.8 eV in the reflectance spectra of ZnO [14]. The first transitions, at 3.30 and 3.34 eV, have been assigned to an exciton excitation in ref. [14] and [15]. In our picture this would correspond to a charge transfer transition.

However, the numbers in Table 4 cannot be used directly to assign the peaks. This is evident from the numbers themselves: our first transition would occur at 8.3 eV (the relaxed SCF result corrected for CI and CPP effects). There are two corrections which have to be done. First, there is a correction due to the polarizability of the medium surrounding the cluster, and secondly, the relaxation/polarization effects from the cluster are underestimated by the calculation.

As detailed in a previous section, the correction to the underestimated near-field polarization can be done by scaling the relaxation energy of the nearest neighbouring ZnO units by a factor 2.7.

The polarization effect from the remaining crystal is somewhat more problematic, as discussed above. We have used the standard procedure, i.e. equation (1), where the field was either taken to be the field from a unit charge or a dipole at the centre of the cluster. The polarization energies induced by a unit charge and a dipole are 3.2 eV and 1.0 eV, respectively. The polarization energy associated with electron capture should be very small since the accepting orbital has band character.

It is not clear whether the structures at 3.30 and 3.34 eV are due to an exciton, i.e. a charge transfer transition, or an ionization/electron capture process, i.e. a transition to the conduction band. In the first case the polarization energy is probably somewhat larger than 1 eV, due to the diffuse character of the excited orbital in the cluster. In the case of the ionization/electron capture process the accepting orbital is delocalized (since the computed electron affinity is negative), and the polarization contribution to this state is probably very small. We do thus only include the polarization energy correction for the ionized state. The electron affinity is poorly described in our model, and in order to get a low estimate we put it to zero.

The ionization and excitation energies are thus estimated as

$$\Delta E_0 + \Delta E_{CPP} + 2.7\,\Delta E_{pol} + \Delta E_{pol}^{ext}$$

where ΔE_0 is the frozen MCPF ionization/excitation energy, ΔE_{CPP} is the CPP contribution and ΔE_{pol} is the contribution from the intra-cluster relaxation.

ΔE^{ext}_{pol}, which is the contribution from the extra-cluster polarization, is 3.2 eV for the ionization and 1.0 eV for the excitation.

Using these numbers we arrive at an estimated exciton energy of 6.06 eV and an estimated ionization/electron capture energy of 6.68 eV (the ionization energy calculated using the above formula + the electron affinity 1.63 eV). It is likely that the first number is overestimated. The polarizability effects on the charge-transfer excitation is most likely larger than 1 eV, since the excited orbital probably is rather delocalized. Using a value of 3.2 for the polarization energy we arrive at an excitation energy of 3.86 eV. The calculated excitation energy thus falls in the interval 3.9-6.1 eV, compared to the experimental result 3.3 eV.

Our results are consistent with an assignment of the first structures in the reflectance spectrum as an exciton excitation, i.e. to a localized charge-transfer transition. The numbers are poor, however. One reason may be an underestimated polarizability of the crystal. This is likely to be the case, for at least three reasons: 1) The estimate of the polarizability contribution from equation (1) is very uncertain. There may well be important corrections due to effects linear in the field (the linear term in equation (2)). 2) It is difficult to calculate the polarizability correctly even on the diatomic. 3) It is quite likely that even a weak tendency towards covalent binding between the ZnO units in the crystal would increase the polarizability of the crystal substantially.

4. The Covalent Model of Zinc Oxide

As mentioned in the introduction, there are essentially two possible ways to view the bonding in zinc oxide. The first viewpoint, described in the previous sections, is the more conventional one where ionic bonding and the electrostatic picture determines the electronic structure. In the second viewpoint, which will be described in this section, the covalency in the bonding between zinc and oxygen is emphasized.

The background of the covalent model of zinc oxide is the electronic structure of diatomic ZnO. The ground state of diatomic ZnO is $^1\Sigma_g^+$ which thus has the same spin as the building block of crystalline zinc oxide. However, the bond-distance of the diatomic is much shorter than for the crystal. At the correlated level the bond distance of ZnO is 1.56 Å as compared to the Zn-O distance in the crystal of 1.98 Å. The charge on zinc in diatomic ZnO is found to be only +0.54 at the correlated level. This charge stays nearly constant as the bond distance is increased from the diatomic to the crystalline value. It is clear that the charge will increase in the crystalline surrounding, but it is initially far from obvious that it will actually increase all the way to +2, as required in a totally ionic model.

Additional background to the covalent model of zinc oxide is provided by a study of the bonding of two units of ZnO. In the following, most calculations are only done at the SCF level. As described above, correlation effects can be estimated to increase the band-gap by approximately 0.4 eV. The geometric positions of the atoms in this model are taken directly from the crystal [2].

The resulting electronic structure in this two-unit model is interesting. There are three possibilities. First, the bonding could be totally ionic with a charge close to +2 for zinc. The second possibility is a total singlet state with some covalent bonding within each unit and only a dipole interaction between the different units. The third possibility is a covalent structure with single bonds between all the atoms, which will lead to a triplet structure. It turns out that it is the third possibility that is the ground state. At the SCF level the excitation energy to the second structure, without covalent bonding between the units, is as high as 3.8 eV.

The two open shells in the two-unit model, described above, are situated at the end atoms as expected. This leads to the first step in the build-up of the covalent cluster model of zinc oxide. It is clear that in the modelling of infinite zinc oxide, these two open shells have to be removed in some way since they should be regarded as edge artifacts. The simplest way to perform this embedding is to add two hydrogen atoms close to the end atoms of the two-unit model. This leads to a singlet structure of $(ZnO)_2H_2$, from which a first estimate of the bulk band-gap can be made. The result at the SCF-level is a small excitation energy of 0.6 eV, where the orbital from which the electron is excited is required to be in the plane of the $(ZnO)_2H_2$ molecule. If the oxygen orbital is allowed to be an out-of-plane orbital the excitation energy is even lower with a value of only 0.2 eV. Adding expected correlation effects of 0.4 eV to the in-plane band-gap leads to a value of 1.0 eV which is still much lower than the experimental band-gap of zinc oxide of 3.3 eV. However, it is interesting to note that the simplest ionic dipole model, with one ZnO embedded in a Madelung field of ± 2 charges, leads to a predicted band-gap of 5.7 eV (correlation effects of 0.4 eV included) which is further away from the correct value for infinite zinc oxide.

The two-unit model described above should be regarded as the simplest possible covalent model of zinc oxide, and further improvements of this model are definitely required. The next natural step in this modelling is to increase the one-dimensional chain of ZnO units. Two hydrogen atoms are still placed at the ends to avoid dangling bond effects. For a three-unit one-dimensional model of this kind the excitation energy increases by approximately 0.9 eV and for a five-unit model there is an additional increase of 0.6 eV. With the estimated correlation effects added, the five-unit one-dimensional model, $(ZnO)_5H_2$, thus has an in-plane excitation energy of 2.5 eV, not far away from the bulk value of 3.3 eV. The out-of-plane excitation energy is about 0.5 eV lower. The charges on zinc and oxygen are close to one for this model, rather than two as expected in a purely ionic model.

The next step in making the covalent model of zinc oxide more realistic is to make a two-dimensional model. A five-unit model with a single ZnO unit surrounded by four other units is a natural extension to two dimensions. One problem here is that it is not obvious how many hydrogen atoms should be added and where they should be placed in order to close the dangling bonds. The embedding model finally decided on here is to consider the two-dimensional model as built up of one-dimensional chains of ZnO units. In the five-unit two-dimensional model, three ZnO units would then form one one-dimensional chain which should be embedded by one hydrogen atom in each end. The remaining two ZnO units in this two-dimensional model would then form one one-dimensional chain each (including only one ZnO unit) which also should be ended by hydrogen atoms. The final two-dimensional model system chosen is thus $(ZnO)_5H_6$. Since this is a non-symmetrical orbital model of zinc oxide, it should be regarded as one resonance structure. Symmetry can as usual be regained in principle by including also the other non-orthogonal resonance structures in the calculations. These structures are not included in practice and the results of this model therefore relies on the fact that the interaction between different resonance structures is small. The resulting excitation energy using this two-dimensional model is 2.5 eV, where electron correlation effects have been estimated as above, and where the oxygen orbital is required to be in the plane of the cluster. If an electron is taken from an out-of-plane oxygen orbital the excitation energy decreases by 1.0 eV. Clearly, the in-plane excitation is a more realistic model of what actually happens in real zinc oxide, since the oxygen atom should feel the Pauli repulsion from neighbouring atoms. The charges on zinc and oxygen are somewhat larger than one as for the one-dimensional case. In summary, also this two-dimensional covalent model gives an excitation energy in reasonable agreement with experiment.

The three-dimensional covalent model is built on the same $(ZnO)_7$ cluster as described in the previous section for the dipole model. This cluster can be described as one central ZnO unit surrounded by six ZnO units arranged in two equilateral triangles keeping C_{3v} symmetry. Using the same picture of the electronic structure as for the two-dimensional model, the dangling bonds should be closed by ten hydrogen atoms placed at the ends of five one-dimensional chains of ZnO units. Since this three-dimensional model should be the most realistic covalent model of zinc oxide used here, some additional improvements of the model were made. First, rather than having hydrogen atoms to model zinc-ions, one-electron ECP copper atoms were used. Secondly, when this is done, it is then possible to place the five copper atoms and the five hydrogen atoms (modelling oxygens) at actual lattice positions without using unreasonable bond-distances anywhere. The calculated excitation energy using this model is 4.3 eV (including estimated correlation effects) which is now substantially larger than the experimental value of 3.3 eV. The charge distribution is still similar to what was obtained for both the one-dimensional and the two-dimensional model with zinc and oxygen charges somewhat larger than one. The charge distribution is also quite even without large oscillations. At this point it is clear that the limit of how far this simple covalent model can be taken is more or less reached. The conclusion must be that it is a model of some qualitative interest which shows that an alternative picture than the usual ionic one is not unreasonable. However, the model does not seem to be possible to use to obtain results of quantitative accuracy.

A few additional modifications of the three-dimensional model were finally tried. First, the embedding copper and hydrogen atoms were placed at positions where there should be less strain in the system. If the model would behave satisfactorily, this modification should not affect the results significantly. The actual result is a decrease of the excitation energy by as much as 0.6 eV, which shows that there are indeed problems with this model. The second improvement attempted with the covalent model was to embed it in a Madelung field. Since the charges on zinc and oxygen are close to one in the cluster calculations, point charges of ±1 were placed at lattice positions surrounding the cluster. However, as reasonable as it may seem, this model lead to very large artificial cluster boundary effects. How these problems should be solved were not clear at the start of this conference.

5. Conclusions

Results obtained with two cluster models, the dipole model and the covalent model, have been presented and discussed. In the dipole model, zinc oxide is considered as an aggregate of neutral ZnO units, allowing for one covalent bond per atom. In the covalent model zinc oxide is treated similarly to semiconductors like silicon, allowing for extended covalent bonding (formally two covalent bonds per atom in zinc oxide). A characteristic difference between the two models is that for the smallest cluster sizes the first excitation energy appears to be underestimated by the covalent model, while it is overestimated by the dipole model. It appears that a stabilizing external electric field is necessary also for the covalent model for the cluster sizes used in the present study. It remains unclear how large clusters which need to be used to get close to experiment, with or without an external electric field. An evident weakness of the dipole model is that the electric field from the surrounding medium gives a much too high, by 5-6 eV, transition energy. The decrease in the calculated excitation energy resulting from the polarizability of the surrounding medium is of the same order. It is not satisfactory with models which depend that strongly on approximative corrections. An interesting development would be to use the multipole method by Panas [16], in which the surrounding multipole field is more rigorously treated. Another possibility is to use a band structure approach [17].

References

[1] P.E.M. Siegbahn, U. Wahlgren in: "Cluster Modelling of Chemisorption Energetics, in Reaction Energetics on Metal Surfaces: Theory and Application", (edited by E. Shustorovich, VCH Publishers, New York 1990)

[2] R.W.G. Wyckoff in: "Crystal Structures", Vol. 1, John Wiley, New York 1965

[3] E.H. Kisi, M.E. Elcombe: Acta Cryst. **c45**, 1867 (1989)

[4] P.S. Bagus, U. Wahlgren: Mol. Phys. **33**, 641 (1977)

[5] G.J.M. Janssen, W.C. Nieuwpoort: Phys. Rev. **B 38**, 3449 (1988)

[6] I. Panas, P. Siegbahn, U. Wahlgren: Chem. Phys. **112**, 325 (1987)

[7] A.J.H. Wachters: J. Chem. Phys. **52**, 1033 (1970)

[8] R.C. Rafenetti: J. Chem. Phys. **58**, 4452 (1973)

[9] T.H. Dunning Jr.: J. Chem. Phys. **53**, 2823 (1970)

[10] W. Müller, J. Flesch, W. Meyer: J. Chem. Phys. **80**, 3297 (1984)

[11] L.G.M. Pettersson, H. Åkeby: J. Chem. Phys. **93**, 4954 (1990)

[12] C.W. Bauschlicher Jr., S.R. Langhoff: Chem. Phys. Lett **126**, 163 (1986)

[13] J.L. Freeouf: Phys. Rev. **B 7**, 3810 (1973)

[14] R. Klucker, H. Nelowski, Y.S. Park, M. Skibowski, T.S. Wagne: Phys. Stat. Sol. **45**, 265 (1971)

[15] W.Y. Liang, A.D. Yoffe: Phys. Rev. Letters **20**, 59 (1968)

[16] I. Panas: Chem. Phys. Letters **194**, 239 (1992), Chem. Phys. Letters **201**, 255 (1993)

[17] C. Pisani, R. Dovesi, C. Roetti: Hartree-Fock Ab Initio Treatment of Crystalline Systems, Lecture Notes in Chemistry 48, Springer, Berlin

Density Functional Investigations of Adsorption at Metal Oxide Surfaces

N. Rösch, K. M. Neyman, and U. Birkenheuer*

Lehrstuhl für Theoretische Chemie, TU München, D-85747 Garching, Germany
*Institute of Chemistry of Natural Organic Materials,
660049 Krasnoyarsk, Russia

Abstract. Density functional investigations of molecular adsorption at ionic metal oxide surfaces are reviewed. As a typical example, we have considered the adsorption of CO at a perfect and at a Ni-doped MgO(001) plane as well as at edge and corner sites of MgO. The geometry of the complexes, the binding mechanism and the spectroscopic manifestation of the surface species are analyzed. Special attention is payed to vibrational frequencies and intensities. A reduction of the Mg-O bond ionicity below the value of complete charge separation (±2 au) and its possible effect on values of computed parameters are also discussed.

1. Introduction

In the past decade a growing number of investigators have turned their attention to the surface chemistry and physics of ionic solids. These studies have demonstrated that the chemical bonds and the structure of adsorption complexes at insulating surfaces may differ significantly from those found at metals and semiconductors [1,2 and other articles in this volume]. Theoreticians responded quite strongly to these experimental activities. Their efforts to rationalize the adsorption and reaction phenomena on ionic materials using various quantum chemical models have been reviewed elsewhere [3,4 and other articles in this volume]. However, one of the most powerful tools of quantum chemistry, the first-principles density functional approach, has not been used until recently for the elucidation of this type of surface problems. Here, we will review the applications of the linear combination of Gaussian-type orbitals local density functional (LCGTO-LDF) cluster method [5-7] to adsorption complexes at ionic metal oxide surfaces. This method combines a rather accurate description of structural and vibrational spectroscopy parameters with computational efficiency which permits one to consider realistic cluster models, also containing heavy many-electron atoms [5,7]. In spite of some deficiencies of the LDF approach, e.g. the underestimation of band gaps and the insufficient localization of the valence electrons in narrow-gap oxides of 3d metals [8], one may hope for a

Springer Series in Surface Sciences, Vol. 33 Eds.: E. Umbach · H.-J. Freund
Adsorption on Ordered Surfaces of Ionic Solids and Thin Films

successful modelling of *ground-state features* of wide-gap insulators. Moreover, where necessary, the LDF results can be systematically improved with the help of self-interaction corrections [8]. As will be demonstrated, the LCGTO-LDF method is a very useful instrument for investigations of molecular interactions and processes at ionic surfaces.

Alkali-metal halides and oxides of the alkaline-earth metals of the rock salt structure exhibit a rather similar adsorption affinity to CO molecules which are commonly used as a probe for detecting Lewis acid sites with the help of infrared (IR) spectroscopy [9]. On the (001) plane, CO is found to adsorb only at low temperatures, on top of the cations and with its molecular axis normal to the surface [10-12]. (A different structure of the NaCl(001)/CO monolayer has been reported below 35 K [13].) For the adsorption on cations in this coordination, a blue shift $\Delta\omega$(C-O) of the C-O frequency by 10 - 20 cm^{-1} (relative to its value in the gas phase) is observed in the zero-coverage limit. The frequency gradually moves to smaller values when the coverage increases. The adsorption of CO on low-coordinated cations of sintered polycrystallites [9,11] or smoke [12] is usually accompanied by larger values of $\Delta\omega$(C-O). The vibrational polarizability of adsorbed CO, which is proportional to the IR intensity of the C-O mode, was determined on the surface of MgO and of alkali-metal halides to be half that of the free molecule [10]. The polarizability increases noticeably for the carbonyl complex with Ni^{2+} impurity at MgO(001) [11]. However, no direct measurement of the C-O molar absorption coefficient has been reported so far.

Here, the adsorption system MgO/CO is discussed as a representative for weakly bonded carbonyl complexes at ionic surfaces. We will review the results of the first attempts [14-17] to rationalize a large body of structural, binding and vibrational spectroscopy parameters (including IR intensities) of various MgO/CO adsorption species within the framework of LCGTO-LDF model cluster approach. The results are supplemented by LCGTO-LDF calculations of MgO slabs [18] based on the FILMS program package [19,20] to examine in detail the charge distribution and the Madelung field at the adsorption sites of a clean surface.

2. Computational Details

The LCGTO-LDF method [5-7] is employed in the Xα approximation ($\alpha = 0.7$). Spin-polarization has been taken into account for the open-shell models of the Ni impurity centers ($^3B_1$ state). A five-fold coordinated site, Mg_{5c}, on a perfect MgO(001) surface is represented by the clusters Mg_9O_9(9,9) and $Mg_{13}O_{13}$(21,5) of C_{4v} symmetry [14,15]. The designation (*n, m, ...*) indicates a cluster with *n* substrate atoms in the first layer, *m* atoms in the second layer etc. The cluster $Mg_{13}O_{13}$(21,5) offers five cationic adsorption sites for CO. A Ni^{2+} impurity is modeled by the cluster $NiMg_8O_9$(9,9). The clusters Mg_6O_6(3,6,3) (C_{2v}) and Mg_4O_4(1,3,3,1) (C_{3v}) are chosen to describe low-coordinated cations at edges (Mg_{4c}) and corners (Mg_{3c}) of microcrystallites, respectively [17]. The central position of these clusters is occupied by either a Mg or a Ni cation. All cluster models exhibit a stoichiometric constitution. They are embedded in electroneutral matrices of point charges (PC) to reproduce the Madelung field at the adsorption site. The PC arrays contain up to about 1700 charges of ±2 au. In some cases, the charge was reduced from its nominal value to ±1.5 au for comparison. Simplified substrate clusters consisting only of the PC arrays were used to evaluate electrostatic effects. The results are reported for unrelaxed substrate models based on a lattice parameter of 4.21 Å (for a validation of this approach see refs. [16,17]).

Two of the four conceivable MgO/CO vertical adsorption situations at on-top sites, namely the -CO and -OC binding to anions, are ruled out because of the repulsive character of the corresponding interactions [14,21]. The distances M-CO (M = Mg, Ni) and C-O were computed for the complexes where the CO molecule was assumed to be oriented upright along the main (z) axis of a corresponding point group C_{nv}, either C- or O-end down. First, the M-C (M-O) distance was optimized for a fixed C-O bond length and then the C-O distance was varied, keeping the CO center of mass fixed as calculated in the previous step. The normal modes of MgO/CO have been approximated by the MgO-CO and C-O internal modes. Absolute IR intensities have been calculated in the double-harmonic approach [22] according to which an IR intensity is proportional to the square of the dynamical dipole moment of the i-th normal mode, $\partial\mu/\partial r_i$. Basis sets for the LCGTO-LDF cluster models have been described elsewhere [6,7,14]. The same orbital basis sets were employed in the cluster and in the slab calculations [18] in order to minimize discrepancies between both methods.

3. Adsorption of CO Molecules at the Ideal and Ni-Doped MgO(001) Surface

3.1 Isolated CO at Perfect MgO(001)

We start our discussion with the case of an isolated CO on a perfect MgO(001) surface (Table 1, models I, II, V and VI). The following questions are relevant to this topic:

1. What is the structure of the surface complexes manifested in the IR spectra?
2. Is the adsorption bond purely electrostatic or does a charge transfer contribute as well?
3. Which interactions determine the adsorption-induced shift of the C-O frequency?
4. How and why does the IR intensity change of the C-O mode take place?
5. How rigorous is to embed the cluster into an array of point charges of ±2.0 au?

Both the adsorption energies and the frequencies ω(M-CO) indicate a rather weak interaction which is somewhat overestimated in the LDF cluster approach [14]. The computed results, low frequency < 200 cm^{-1} and small intensity of ≈ 0.5 km/mol, are in line with the fact that no detection of the corresponding vibrational peak has been reported. The classical C-down orientation of the carbonyl (model I) is favoured energetically. However, model VI with CO coordinated via oxygen also exhibits an energy minimum along the Mg-CO coordinate, possibly corresponding to a stable complex. In the case of C-end adsorption, a calculated blue frequency shift +54 cm^{-1} of the C-O mode with respect to the value for free CO is consistent with the experimental value of +14 cm^{-1} [11], but far too large. Model V for the O-end CO adsorption yields a red shift (Table 1). Thus, the alternative orientation of CO with the oxygen "downward" may be ruled out by this comparison. Nevertheless, it would be interesting to define experimental conditions under which such an unusual "iso-carbonyl" moiety may be detectable and to predict its properties (models VI-X). The Mg-CO distance has not yet been measured and the LDF data can only be compared to Hartree-Fock (HF) results [23] which are noticeably longer (by 0.1 to 0.3 Å) than those for the models I and V. This difference seems to reflect a general tendency of the HF level of theory to underestimate the weak M-CO bonding for oxide surfaces compared to the LDF method [14,17].

The Mg_{5c}-CO bonding, being mostly electrostatic [23], was shown to be accompanied by a small σ charge redistribution directed to the substrate [14,15]. The transferred charge is about 0.1 au. Interpreting the dynamical dipole moment $\partial\mu/\partial z$ as an effective charge, one can see from Table 1 that the Mg_{5c}-CO bond stretching causes a displacement of 0.12 au of the electronic density from CO to the surface. It is not easy to isolate the charge transfer contribution to the adsorption energy, but it can be illustrated by the effect on the overall C-O frequency shift. The value due to charge transfer alone, 17 cm^{-1} [14], is comparable both to the experimental shift of 14 cm^{-1} [11] and to that of Pauli repulsion component calculated by HF method [23]. The direct interaction of a CO molecule with the Madelung field of the surface (model II) generates a small red shift. One may therefore conclude that a substantial part of the overall shift $\Delta\omega$(C-O) = 54 cm^{-1} computed for the model I is associated with the charge transfer and thus is "chemically" induced.

The discrepancy between the LDF and the experimental value of $\Delta\omega$(C-O) warrants a further remark. While the direct influence of the surrounding point charges on the C-O frequency shift is small, their indirect effect through a modification of the substrate cluster may be quite important. This follows from a study where the environment of the cluster Mg_9O_9-CO was constructed from charges of ±1.5 au. In this model, the shift $\Delta\omega$(C-O) is reduced to 16 cm^{-1} (Table 1) which (fortuitously) agrees with the experimental value. A similar tendency to "improve" the computed shift for the models IV and V (Mg_{4c}-CO and Mg_{3c}-CO) with point charges of ±1.5 should also be mentioned (Section 4 and ref. [17]). This finding supports the assumption that field effects are overestimated in embedded cluster models which are based on the nominal ionicity Mg^{2+}-O^{2-}. Further arguments along the same line will be given in the next section and can be also found elsewhere [14,17].

The computed IR intensity of the C-O mode for adsorption on MgO(001) does not reveal a decrease relative to the value for the gas phase CO, as might be

Table 1: LCGTO-LDF results for cluster models of MgO/CO adsorption complexes and for the free CO molecule: equilibrium height z_e of the nearest adsorbate atom above the surface cationic center, C-O distance r_e, energy ω(M-CO) of the substrate-adsorbate stretching fundamental vibrational transition, shift $\Delta\omega$(C-O) of the C-O vibrational frequency (with respect to the value calculated for the CO molecule) and the corresponding absolute IR intensity I(C-O), adsorption energy D_e, induced dipole moment $\Delta\mu = \mu(M_nO_n\text{-CO}) - \mu(M_nO_n) - \mu(CO)$, derivatives of the dipole moment $\partial\mu/\partial z$ and $\partial\mu/\partial r$ along with the σ and π components $\partial\mu_\sigma/\partial r$ and $\partial\mu_\pi/\partial r$.

	Cluster	z_e Å	ω(M-CO)[a] cm^{-1}	$\partial\mu/\partial z$[b] au	r_e Å	$\Delta\omega$(C-O)[a] cm^{-1}	I(C-O) km/mol	$\partial\mu/\partial r$[b] au	$\partial\mu_\sigma/\partial r$[b] au	$\partial\mu_\pi/\partial r$[b] au	D_e eV	$\Delta\mu$ D
I	Mg_9O_9-CO	2.22	183 (158)	0.12	1.124	54 (16)	124	-0.93	0.50	-1.42	0.97	1.19
II	PC(001)-CO	2.22[c]	-	-	1.132	-8 (0)	40	-0.53	0.54	-1.07	0.62	0.64
III	$NiMg_8O_9$-CO	1.88	327	0.41	1.126	45	283	-1.41	0.72	-2.15	1.66	1.67
IV	Mg_6O_6-CO	2.18	198 (185)	0.08	1.124	56 (33)	125	-0.94	0.46	-1.39	0.94	1.79
V	Mg_4O_4-CO	2.23	197 (184)	0.01	1.121	85 (59)	88	-0.79	0.48	-1.27	1.07	2.62
	CO	-	-	-	1.130	2 169[d]	66	-0.68	0.37	-1.05	-	0.26[e]
VI	Mg_9O_9-OC	2.19	122	-0.17	1.137	-60	201	1.18	-0.17	1.34	0.51	1.44
VII	PC(001)-OC	2.19[c]	-	-	1.132	-18	99	0.83	-0.26	1.09	0.24	0.41
VIII	$NiMg_8O_9$-OC	2.06	202	-0.19	1.136	-42	206	1.20	0.09	1.10	0.58	1.92
IX	Mg_6O_6-OC	2.07	176	-0.21	1.139	-76	196	1.17	-0.18	1.35	0.49	1.82
X	Mg_4O_4-OC	2.05	197	-0.29	1.143	-98	269	1.37	-0.12	1.49	0.68	2.57
	OC	-	-	-	1.130	2 169[d]	66	0.68	-0.37	1.05	-	-0.26[e]

a) Results obtained with surrounding point charges of ±1.5 au are shown in parentheses. b) For $\partial\mu/\partial r$: 1 au = 4.80 D/Å
c) Not varied. d) Reference value for the determination of $\Delta\omega$(C-O). e) Computed dipole moment for free CO or OC.

expected from the analysis of the dynamic contribution to the C-O frequency shift with increasing coverage [10]. A detailed discussion of the reasons why this indirectly determined intensity is believed to be underestimated has been presented previously [14]. The calculated intensity enhancement can be rationalized in terms of σ and π components of the dynamical dipole moment $\partial\mu/\partial r$(C-O) which coincides here with the electronic dynamical dipole moment (the center of nuclear charges remains fixed when the C-O bond stretches). In this approach $\partial\mu_\sigma/\partial r$ and $\partial\mu_\pi/\partial r$ should be identified with the "oscillating" charge density of σ and π orbitals. An increase of $\partial\mu/\partial r$ concomitant with the C-O bond elongation corresponds to a displacement of the electronic density down to the surface, a decrease indicates an upward displacement. The effects may be caused by field, Pauli repulsion, charge transfer and polarization.

The dynamical dipole moment of CO is the sum of a small (by the absolute value) σ component and a large π component of opposite signs. Inspection of Table 1 reveals that the electrostatic field (model II) causes a decrease of I(C-O) due to a polarization of the CO σ charge directed to the (001) plane. The 1π orbital of CO is only slightly affected by the field and the σ channel is responsible for the reduction of the absolute value of $\partial\mu/\partial r$ and thus for the smaller intensity. A similar analysis may be carried out for the field-induced intensity enhancement in model VII with the O-end coordinated CO. The attraction of the σ charge by the field increases the negative $\partial\mu_\sigma/\partial r$ component and therefore also the intensity since the π contribution is negligible. In the presence of the substrate atoms (model I) which allow additional interaction mechanisms to occur the prevailing effect is a decrease of the negative $\partial\mu_\pi/\partial r$ component. This is mainly due to the interaction of the CO 1π orbital with the nearest-neighbour oxygen anions [14]. For the O-down CO (model VI), both the σ and the π interactions with the substrate contribute to the overall increase of the intensity. It is interesting to note that in this case the σ charge transfer, as measured by a Mulliken population analysis, takes place predominantly via the CO 4σ orbital. The π charge displacement, also in the direction toward the substrate, is essentially due to changes in the CO 1π orbital.

3.2 On the Ionicity of the Mg-O Bond

The results of the cluster models for isolated CO at a perfect MgO(001) surface provide some evidence for a reduced metal-oxygen charge separation [14]. On the other hand, HF calculations for bulk MgO [24] and MgO(001) slabs [25] find the Mg-O bond to be perfectly ionic. To investigate the question of possible covalent contributions, we have carried out LCGTO-LDF calculations of 2D-periodic MgO(001) slabs of up to three layers [18]. The analysis of the charge separation is based exclusively on the charge density and its ultimate goal is to define effective point charges to be used in the PC arrays surrounding the adsorption cluster models. Two procedures were employed: one following the topological

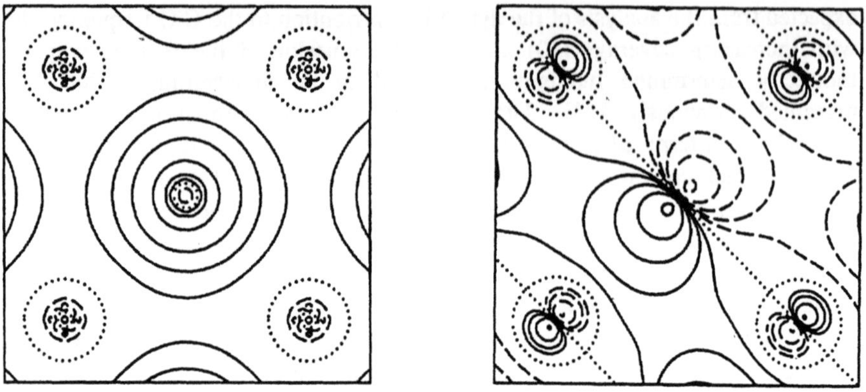

Fig. 1: Valence one-electron wave functions of the MgO(001) monolayer: O 2s dominated wave function (at $\overline{\Gamma}$ point, left); O 2p dominated wave function (at $\overline{M}$ point, right). The plots display the plane of the nuclei, oxygen at the center surrounded by four magnesium neighbours. Positive and negative values are indicated as solid and dashed lines, the nodes are shown as dotted lines. The values of the contours are 0.56, 0.32, 0.18, 0.10, and 0.056 au

atom approach suggested by Bader [26] and a direct analysis of the Madelung field near the surface.

The first method relies on a division of space into disjoint, space-filling atomic regions of reasonable shape and thus avoids the ambiguities associated with a Mulliken population analysis. Atomic boundaries are defined as zero-flux surfaces in the gradient field of the total charge density [26].

The valence electronic structure of MgO is dominated by the 2s and 2p electrons of the O^{2-} anion. Two typical examples for O 2s and 2p dominated one-electron wave functions are shown in Fig. 1. The atomic orbital contributions around the oxygen atom are clearly discernible. A significant overlap between neighbouring O 2p orbitals can also be seen. However, in the present context, the most important features are the non-vanishing s- or p-type contributions at the Mg sites. From their nodal pattern, they obviously have to be attributed to admixtures of Mg 3s and 3p orbitals. For the derivation of an effective point charge model it is irrelevant whether these contributions represent a covalent contribution to the Mg-O bond or whether they are regarded as a consequence of the orthogonalization of the rather extended oxygen orbitals with respect to the magnesium core orbitals.

To quantify the charge separation in MgO, effective atomic charges were calculated for the topologically defined atomic regions (Table 2). Their size and shape are illustrated in Fig. 2 for a triple layer (3L). The two types of ions are clearly discernible and the Mg cations are, as expected, smaller than the oxygen anions. In agreement with the view that O 2p electrons are "pushed" into the unoccupied states of the neighbouring magnesium ions, the effective charges are found to decrease with increasing coordination of the atomic regions.

Table 2: Atomic charges of the Mg cation (in au) derived from the LDF charge distribution in MgO slab models of one to three layers (L).

	1L	2L	3L	
topological atoms	1.70	1.66	1.66 (outer)	1.62 (inner)
Madelung field	1.60	1.77	1.81	

Alternatively, one may directly analyze the Madelung field around the adsorption site. If this analysis were based on an extended surface with the finite cluster removed, significant contributions to the electric field would be due to the boundary region and the results would depend crucially on the choice of the boundary. Therefore, the well-defined Madelung field of a *complete* substrate slab was analyzed. From a 2D-periodic version of the classical multipole expansion one finds [18] that the electrostatic potential above a slab which does not exhibit a total dipole moment, like MgO(001), can be expressed as

$$\varphi(r,z) = \frac{2\pi}{A_{UC}} \sum_K \frac{M_K}{|K|} e^{iKr} e^{-|K|z}, \quad \text{with } M_K = \int \rho(r,z)\, e^{-iKr} e^{|K|z}\, d^2r\, dz \,.$$

The main contribution to the Madelung field is connected to the reciprocal lattice vectors $K = (2\pi/a, 0), (0, 2\pi/a)$ and, due to the alternating phase factor $\exp(-iKr)$, provides a suitable measure for the charge separation in a MgO(001) layer. This field is essentially determined by the charge distribution in the top substrate layer.

Almost no interlayer charge redistribution occurs as can be seen from the

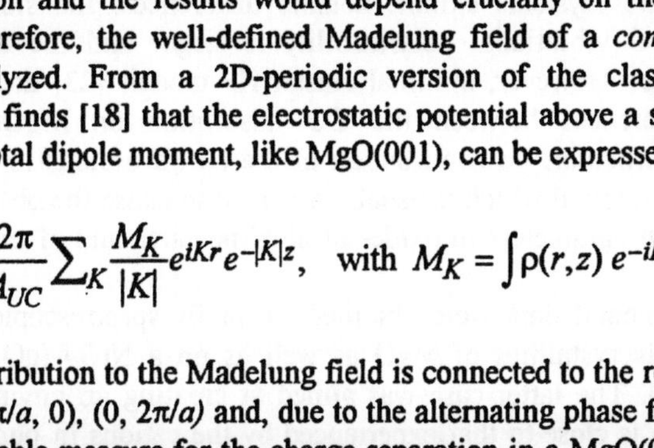

Fig. 2: Valence electron density of a MgO(001) triple layer shown in the [100] plane perpendicular to the surface. Large circles enclose the oxygen, small circles the magnesium sites. The values of the contours (solid lines) are 0.32, 0.10, 0.032, and 0.010 au, the dotted lines indicate the zero-flux boundaries of the atomic regions.

small difference between the atomic charges determined for the double and the triple layer (Table 2). The Madelung charges for a monolayer are smaller, reflecting the unique geometrical fact that the O 2p electrons can be delocalized into the vacuum at both sides of the layer.

It is important to note that both procedures yield effective charges which are significantly reduced from the nominal value ±2 au. They may be even used to construct embedding models with spatially varying point charges.

3.3 CO at a Ni Impurity Site

Characteristics of carbonyl complexes on pure MgO [11,12] have been compared to those on NiO and on Ni impurities of doped MgO [10,11]. The experimental data for the latter systems exhibit notable differences: the shift $\Delta\omega$(C-O) is smaller, the C-O vibrational polarizability is larger and the bonding to the substrate stronger. However, minimal cluster HF models [23] do not reveal any significant differences between the CO adsorption on MgO(001) and on NiO(001). In particular, no contribution of the Ni 3d orbitals to the adsorption bonding has been found which is usually assumed to cause the observed diversity of the adsorption parameters in oxides of alkaline-earth and of transition metals [10,11].

The experimental data were obtained in an IR spectroscopic study of CO adsorbed on polycrystallites of MgO as well as on a NiO/MgO solid solution (10% NiO) [11]. The latter case was aimed at creating an environment of the Ni^{2+} sites which is close to that experienced by the cations in pure MgO. These experiments were interpreted to indicate a Ni(3d)-CO(π) overlap bonding contribution besides a strong electrostatic interaction. Results of the LDF investigation [17] are given in Table 1 (models I, III, VI and VIII). A shorter Ni-CO distance and a larger adsorption energy as well as the frequency ω(Ni-CO), in particular for the C-down models I and III, clearly indicate a noticeably stronger adsorption interaction at a Ni_{5c} site compared to that at Mg_{5c}. The charge exchange between CO and a Ni_{5c} site is more intense than that in model I, as can be seen from the increase of the dynamical dipole moment $\partial\mu/\partial z$. It seems to be a classical charge transfer of σ and π type where the σ channel dominates. Mulliken populations indicate that the partial occupation of the antibonding CO $2\pi^*$ orbital is accompanied by a notable depletion of the CO 1π orbital [17]. The π interaction is mainly responsible for the reduction of the C-O frequency in the cluster $NiMg_8O_9$-CO (III) related to that for Mg_9O_9-CO (I) by 9 cm^{-1} , in line with the measured value of 13 cm^{-1} [11]. In the IR spectra of CO on the NiO/MgO solid solution [11] one finds several characteristic peaks around 2094-2060 cm^{-1}, some of which may well be a manifestation of CO adsorbed O-down at Ni_{5c} or other Ni-sites. These complexes (model VIII) should exhibit C-O vibrations red-shifted relative to that of free CO. A decisive argument for the presence of marked π contribution in the Ni_{5c}-CO bond, an intrinsic feature of d-metal complexes, can be drawn from the enhanced intensity I(C-O) in the model

III compared to I, which is in qualitative agreement with experiment [11]. A σ donation to the substrate increases $\partial\mu_\sigma/\partial r$, but the back-donation in the π channel has obviously a stronger effect on the intensity, decreasing the $\partial\mu_\pi/\partial r$ component for $NiMg_8O_9$-CO by a third compared to that in the Mg_9O_9-CO model. The present results are at variance with those of HF calculations [23] where no difference was found for CO adsorption at Ni_{5c} and Mg_{5c} centers in oxides. However, important experimental evidence [11] of specific interactions with the d-metal sites is qualitatively reproduced and rationalized at the LCGTO-LDF level of theory. Further studies are necessary to explain why the HF and LDF cluster results for CO adsorption on alkaline-earth and d-metal oxides exhibit such large differences.

3.4 Lateral Interaction between the Adsorbed CO Molecules and Propagation of the Electronic Effects through the Substrate

We will address two questions:

1. What are the reasons for coverage dependence of the static C-O frequency shift and which role does the substrate play in this phenomenon?
2. Does the lateral interaction between adsorbed molecules affect the absolute IR intensity of the C-O vibrational mode in a noticeable way?

The coverage dependence of the stretching frequency of adsorbed CO is a characteristic feature of ionic surfaces [10]. CO adsorbed on the (001) plane of sintered powder MgO samples exhibits an IR peak at 2157 cm^{-1} which shifts to 2148 cm^{-1} when the CO coverage increases to saturation [11] (where the monolayer, 1CO:2Mg_{5c}, exhibits a $c(2\times2)$ surface structure). It is possible to discriminate dynamic (due to the mutual perturbation of vibrating CO molecules) and static contributions to the adsorbate-adsorbate interaction. An isotopic dilution technique yields a positive dynamic shift $\Delta\omega$(C-O) = 3.5 cm^{-1} when the MgO(001)/CO coverage increases from zero to the maximum value. Combining the observed difference of the C-O frequency shift for isolated CO on MgO and that at saturation, 9 cm^{-1}, one derives a shift of -12.5 cm^{-1} which is attributed exclusively to the static CO-CO interaction in the MgO/CO monolayer [11].

A "solvent" and a "chemical" mechanism have been suggested to rationalize this shift. The solvent effect is a through-space interaction caused by changes in the local environment at a given site, brought about by filling the nearby sites. The chemical effect is due to a through-bond perturbation among molecules adsorbed in adjacent positions. It can occur only through the substrate because the shortest CO-CO separation, even in the monolayer (4.21 Å), is larger than the van der Waals distance. If the next molecule coming to the surface finds its adsorption site in a state of lower acceptor ability because some of the charge donated by the first CO has been transmitted through the solid, then the corresponding C-O frequency would be decreased compared to that of an isolated CO on MgO(001).

The computed frequency for the central CO moiety in a model of monolayer coverage based on the cluster $Mg_{13}O_{13}$-$(CO)_5$, 2188 cm^{-1}, is red-shifted relative to the value 2199 cm^{-1} for the cluster $Mg_{13}O_{13}$-CO (2192 cm^{-1} when corrected for the basis set superposition error) [15]. The agreement with the experimentally derived static shift of -12.5 cm^{-1} [11] may be considered reasonable since the long-range effect of a large number of the surrounding adsorbates is only partly taken into account by the cluster model.

For those models where the substrate is either absent or described as a PC array the calculated static C-O frequency shift is *positive* [15]. This implies that the through-space perturbation among the adsorbates without a charge exchange involving the substrate can not explain the observed coverage dependence of the C-O frequency. Although the CO adsorption is accompanied by some σ charge transfer to the nearest Mg_{5c} ion [14], no delocalization of this charge beyond its oxygen neighbours was found [15]. Thus, the propagation of an electronic effect through a MgO substrate may be ruled out as the dominating mechanism responsible for the coverage dependent static red shift of the C-O frequency. Rather, a change of the electrostatic field in the vicinity of a vibrating CO due to the charge transfer occurring at the adjacent sites may be invoked as a rationalization consistent with the overall picture. To test this hypothesis, each CO molecule has been replaced by point charges: 0.1 au at the CO center of mass and a change of -0.1 au in the Mg_{5c} position. The following two models were considered: (i) only the four nearest neighbours of the central CO above a PC(001) array are simulated by these dipoles; (ii) all 80 suitable cationic positions on the upper plane of this array are modified as described. The frequency shift -12 cm^{-1} (with respect to the unmodified PC(001)-CO) in model (ii) is more than twice as large as that of model (i), -5 cm^{-1}. This clearly illustrates the importance of long-range field contributions and supports the explanation of the coverage-induced static frequency shift as an electrostatic solvent effect [15].

Finally, a rather small effect of the MgO/CO coverage on the C-O vibrational intensity should be pointed out. The calculated vibrational intensities for the singleton (165 km/mol) and the monolayer models (148 km/mol) are rather similar and more than twice as large as that calculated for free CO, 66 km/mol. Therefore, the adsorption-induced intensity change may be rationalized in the same way as in section 3.1, leaving aside a coverage dependence.

4. Adsorption of CO Molecules at the Structural Defects of MgO

The calculated C-O frequencies (Table 1, models IV and V) agree qualitatively with the HF data [27] and support the assignment of the high-energy features in the carbonyl region of the IR spectra of CO on polycrystalline MgO: these features are associated with complexes of coordinatively unsaturated cations located at edge and corner positions [10,11]. As discussed in the previous sections, the models based on the surrounding point charges of ±2.0 au overestimate the Madelung field on MgO(001). The MgO/C-O frequencies for the clusters embedded in PC arrays of smaller charges ±1.5 au are in significantly better agreement with experiment ($\Delta\omega$(C-O) = 27 cm^{-1} for Mg_{4c}-CO and $\Delta\omega$(C-O) = 60 cm^{-1} for Mg_{3c}-CO complexes [11]), thus also supporting a sub-nominal value of the ionicity of MgO. The strengthened field at the more open Mg_{4c} and Mg_{3c} sites is obviously responsible for the main difference in their properties compared to the regular Mg_{5c} coordination [17]. This effect forces the adsorbate to come closer to the surface and increases the induced dipole moment as well as the adsorption energies. At the same time, a field-induced shift of the C-O frequency becomes noticeable for Mg_{3c}-CO (25 cm^{-1} versus -8 cm^{-1} in the model II). The chemical contribution to the bonding is not very different from that at Mg_{5c} centers and may be rationalized in a similar way. A partial decrease of the absolute intensity I(C-O) calculated for model V is mainly connected to the missing π interaction with the nearest-neighbour oxygen anions which results in a reduction of the $\partial\mu_\pi/\partial r$ component.

5. Conclusion

The present study demonstrates that the LCGTO-LDF cluster models are adequate to describe very sensitive parameters of weakly bonded adsorption complexes at ionic surfaces. For MgO, a typical representative of such systems, the calculations confirm the assignment of the IR spectra of various MgO/CO species and clarify the origin of the C-O frequency shift. The adsorption-induced enhancement of the absolute IR intensity of the C-O mode is predicted and rationalized in terms of the σ and π components of the CO dynamical dipole moment.

Acknowledgements

We thank J. C. Boettger and S. B. Trickey for sharing the FILMS program package and for many stimulating discussions. Financial support by the Deutsche Forschungsgemeinschaft (SFB 338) and by the Fonds der Chemischen Industrie is gratefully acknowledged. K.M.N. also thanks the Humboldt Foundation for a research fellowship.

References

[1] C. Xu, M. Hassel, H. Kuhlenbeck, H.-J. Freund: Surf. Sci. **258**, 23 (1991)

[2] H. Kuhlenbeck, C. Xu, B. Dillmann, M. Hassel, B. Adam, D. Ehrlich, S. Wohlrab, H.-J. Freund, U.A. Ditzinger, H. Neddermeyer, M. Neuber, M. Neumann: Ber. Bunsenges. Phys. Chem. **96**, 15 (1992)

[3] J. Sauer: Chem. Rev. **89**, 199 (1989)

[4] E.A. Colbourn: Surf. Sci. Rep. **15, 281 (1992)**
[5] N. Rösch, P. Knappe, P. Sandl, A. Görling, B.I. Dunlap in: The Challenge of d and f Electrons. Theory and Computation, ACS Symposium Series No. 394, eds. D. R. Salahub, M. C. Zerner, Washington 1989, p. 180
[6] B.I. Dunlap, N. Rösch: Advan. Quantum Chem. 21, 317 (1990)
[7] N. Rösch in: Cluster Models for Surface and Bulk Phenomena, NATO ASI Series B, eds. G. Pacchioni, P.S. Bagus, F. Parmigiani, Plenum Press, New York 1992, p. 251
[8] A. Svane, O. Gunnarsson: Phys. Rev. Lett. **65**, 1148 (1990)
[9] A.A. Davydov: Infrared Spectroscopy of Adsorbed Species on the Surface of Transition Metal Oxides, John Wiley, Chichester 1990
[10] E. Escalona Platero, D. Scarano, G. Spoto, A. Zecchina: Faraday Disc. Chem. Soc. **80**, 183 (1985)
[11] D. Scarano, G. Spoto, S. Bordiga, S. Coluccia, A. Zecchina: J. Chem. Soc., Faraday Trans. **88**, 291 (1992)
[12] L. Marchese, S. Coluccia, G. Martra, A. Zecchina: Surf. Sci. **269/270**, 135 (1992)
[13] D. Schmicker, J.P. Toennies, R. Vollmer, H. Weiss: J. Chem. Phys. **95**, 9412 (1991)
[14] K.M. Neyman, N. Rösch: Chem. Phys. **168**, 267 (1992)
[15] K.M. Neyman, N. Rösch: Ber. Bunsenges. Phys. Chem. **96**, 1711 (1992)
[16] K.M. Neyman, N. Rösch: J. Mol. Struct. **293**, 303 (1993)
[17] K.M. Neyman, N. Rösch: to be published
[18] U. Birkenheuer, N. Rösch, J.C. Boettger: to be published
[19] J.C. Boettger, S.B. Trickey: Phys. Rev. **B 45**, 1363 (1992)
[20] J.C. Boettger, S.B. Trickey: unpublished
[21] E.A. Colbourn, W.C. Mackrodt: Surf. Sci. **143**, 391 (1984)
[22] R.D. Amos: Advan. Chem. Phys. **67**, 99 (1987)
[23] G. Pacchioni, G. Cogliandro, P.S. Bagus: Surf. Sci. **225**, 344 (1991)
[24] M. Causà, R. Dovesi, C. Pisani, C. Roetti: Surf. Sci. **175**, 551 (1986)
[25] M. Causà, R. Dovesi, C. Pisani, C. Roetti: Phys. Rev. **B 33**, 1308 (1986)
[26] R.F.W. Bader, T.T. Nguyen-Dang: Advan. Quantum Chem. **14**, 63 (1981)
[27] G. Pacchioni, T. Minerva, P.S. Bagus: Surf. Sci. **275**, 450 (1992)

Part V

Chemical Reactivity

The Structure and Reactivity of Oxide Thin Films Grown on Transition Metal Surfaces

W. Weiß, A.B. Boffa, J.C. Dunphy, H.C. Galloway, M.B. Salmeron, and G.A. Somorjai

Department of Chemistry and Lawrence Berkeley Laboratory,
University of California, Berkeley, CA 94720, USA

Abstract: Epitaxially grown oxide and sulfide films [FeO_x and ZrO_2 on Pt(111), V_2O_3 on Au(111), S on flat and vicinal Mo(100) and Re(0001)] and oxide islands (TiO_2, ZrO_2, FeO, V_2O_3 on Rh) were produced on transition metal surfaces. Their structure has been studied by LEED-surface crystallography and by scanning tunneling microscopy. The reactivity of the oxide films is probed by chemisorption and by carrying out catalytic reactions at high pressures using these oxide-metal systems. The ordered FeO_x films are unreactive for chemisorption. The sulfur layers order into a large number of complex structures as a function of coverage, thereby inducing restructuring of the metal substrate. The periphery of the oxide islands deposited on metals proved to be very reactive for the hydrogenation of molecules with polar C-O bonds (CO, CO_2 and acetone).

1. Introduction

Studies of the structure and chemistry of transition metal oxide and sulfide surfaces is one of the frontier areas of surface science. One approach to producing well-defined surfaces of these materials is to grow them as ordered monolayers or thin films on single crystal surfaces of metals.[1] We report here our studies of producing ordered films of iron oxide, FeO_x [2], and zirconium oxide, ZrO_2 [3], on Pt(111) and (100) surfaces, and vanadium oxide, V_2O_3 [4], on Au(111). We also prepared sulfur monolayers on Mo(100), (910), (911), (28,4,1) [5,6] and Re(0001) [7] crystal faces; studying their surface structures as a function of coverage. Our main techniques of structural characterization are low energy electron diffraction (LEED)-surface crystallography and scanning tunneling microscopy (STM). This combination proved to be a uniquely powerful means to verify the arrangements of atoms in these complex oxide and sulfide thin layers.

Auger electron spectroscopy (AES) and X-ray photoelectron spectroscopy (XPS) were used to analyze the film compositions and the oxidation states. Following structural characterization, adsorption studies were conducted to establish the bonding characteristics of these films, using a variety of molecular adsorbates. The ordered oxide films only weakly adsorb most molecules and appear to be chemically inert.

Sulfur monolayers over molybdenum [8] and rhenium [9] crystal surfaces form a large variety of ordered surface structures with increasing coverage. Sulfur restructures the metal surfaces which leads to double-height steps on molybdenum.

In independent studies, we explored the catalytic activity of several transition metal oxides (TiO_2, FeO, V_2O_3, ZrO_2) that form small two-dimensional islands when they were deposited on platinum and rhodium foils [10,11] below one monolayer coverage. Most of these oxide-metal systems proved to be highly reactive with rates of selected reactions ($CO+H_2$ or CO_2+H_2) much higher than on the transition metal alone. When multilayers of oxide films covered the metal surface the reactivity declined markedly or disappeared, with the exception of FeO which became more catalytically active with increasing film thickness. The ordered oxide films make excellent chemically passive coatings, indicating they do not possess active sites, which accounts for their well-known catalytic functions. It appears that the structure and composition of the oxide films dramatically influence the chemical behavior of their surfaces.

2. Experimental

We use UHV systems equipped with AES, XPS, LEED and STM capabilities to prepare and characterize the oxide and sulfide films. After cleaning the single crystal substrates by argon ion-bombardment, they are heated by electron bombardment to order the surface atoms. Then the transition metals (Fe, Zr, V, etc.) to be oxidized are deposited from the vapor in known concentrations, heated in oxygen to appropriate temperatures to oxidize and order the deposit, layer by layer. Sulfur was deposited by either dissociating H_2S on the metal surface or by using an Ag_2S electrochemical cell that generates a beam of S_2 molecules in UHV.

The catalytic reaction studies are performed in a reaction cell that can be closed around the sample in the center of the UHV chamber and pressurized with the reactant gases. The reaction rate and the product distribution is then monitored using a gas chromatograph, as a function of time, temperature and gas pressure. The reaction cell can be evacuated and opened at any time exposing the model small area (~1 cm^2) catalyst sample to UHV and surface characterization using the techniques that are at our disposal.

3. Results

3.1 Characterization of Oxide and Sulfide Films

Table 1 lists the oxide films that were grown on platinum and gold single crystal surfaces. In all these cases the films exhibited ordered surface structures of

Table 1: The ordered oxide films that were grown on single crystal metal substrates.

Ordered Film	Substrate	Film Orientation
FeO (monolayer) Fe_3O_4 (multilayer)	Pt(111)	(111)
FeO (monolayer)	Pt(100)	(111)
V_2O_3	Au(111)	(0001)
ZrO_2	Pt(111)	(111)

hexagonal (111) or (0001) orientations. In these crystallographic directions the films may be oxygen or metal layer terminated. Our most detailed LEED-crystallography and STM studies were carried out on iron oxide films grown on Pt(111). Figure 1 shows LEED and STM pictures of the monolayer iron oxide deposits. This overlayer is stable up to temperatures of T=1300 K. LEED reveals a very stable ordered structure with a hexagonal (1x1) unit cell 15% larger than the platinum (1x1) substrate unit cell. The pattern is consistent with the presence of a (8x8) surface structure as 8 FeO(111) - (1x1) lattice spacings fit onto nine lattice spacing on the Pt(111) - (1x1) substrate surface due to the lattice mismatch. The atomic resolution STM picture confirms the periodicity of the coincidence lattice. Surface crystallography analysis of the ordered multilayer deposit that exhibits a (2x2) surface structure reveals the formation of an Fe_3O_4 structure in (111) orientation that is schematically shown in Fig. 2. Two different oxygen terminated domains are present on the surface. One has second layer Fe^{3+} ions in octahedral positions, and the other has second layer Fe^{2+} ions in tetrahedral positions. The oxide lattice lateral expansion that appears in the monolayer on Pt(111) is removed during the build-up of the multilayer iron oxide

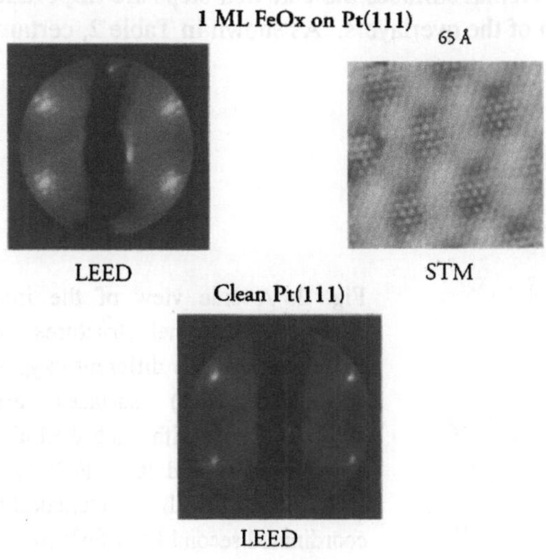

Fig. 1: LEED and STM pictures of one monolayer of iron oxide grown on Pt(111). The LEED pattern of the clean Pt(111) crystal face is also shown.

film, and an Fe_3O_4 spinel structure is formed. This is the most complex surface structure that has been solved so far by tensor-LEED, clearly showing the power of the theory. Analysis of the step structure on the Fe_3O_4 surface indicates the presence of steps of one oxygen-iron bilayer in height (2.5 Å).

On Pt(100) a monolayer deposit of iron oxide forms a c(2x10) structure in two 90°-domains. The 10 times period in the [011] direction (and in the $[0\bar{1}1]$ direction in the other domain) implies a 10% expansion or mismatch between FeO and Pt lattices in these directions and gives a periodicity of 3.05 Å for the FeO layer. Also, the 2 times period in the perpendicular direction or 5.54 Å is very close (within 5%) to the 5.28 Å that corresponds to $\sqrt{3}$ times the Fe-Fe distance in the (111) planes of FeO. These observations suggest that bilayers of (111) oriented FeO forms on the Pt(100) substrate.

Ordered films of zirconium oxide, up to seven monolayers, yield a (1x1) surface structure on Pt(111). They appear to grow in (111) orientation of the face centered cubic bulk ZrO_2 structure. The surface unit cell of the films is (2x2) times that of the bulk oxide. Annealing up to 900 K is necessary to produce long range order in the ZrO_2 films. Annealing for longer times or at higher temperatures leads to reduction and dissolution of zirconium oxide films in platinum despite the oxidative conditions maintained in the chamber.

Vanadium oxide produces a $(\sqrt{3}x\sqrt{3})R30°$ surface structure on Au(111) from submonolayer up to multilayer coverages. The structure of the oxide films is consistent with that of the hexagonal (0001) face of V_2O_3 (corundum structure). The lattice parameter of vanadium oxide (5.1 Å) and $\sqrt{3}$ x 2.88 Å, the lattice parameter of Au, differ by only 2%.

LEED and STM studies indicate that sulfur forms several ordered structures on the Mo(100) surface with increasing coverage (Table 2). A STM study of defects and the registry of the unit cells across domain boundaries indicates that all these structures consist of sulfur atoms chemisorbed at fourfold hollow sites on the Mo surface. LEED studies of vicinal surfaces indicate that steps are important sites for nucleation and growth of the overlayers. As shown in Table 2, certain

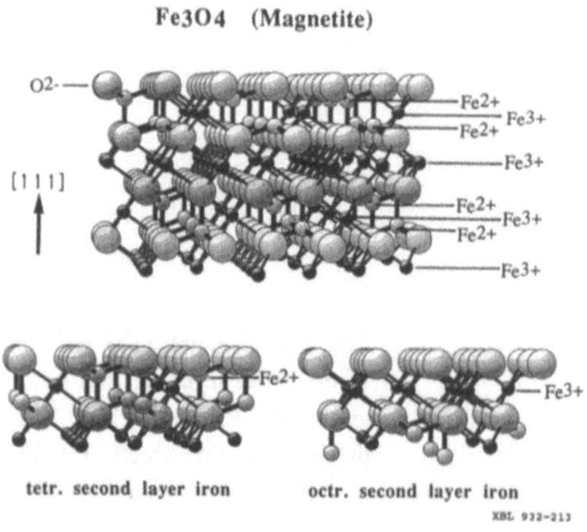

Fig. 2: A side view of the iron oxide, Fe_3O_4 spinel structures. In the lower part two different oxygen terminated (111) surfaces are shown. One with tetrahedrally coordinated second layer Fe^{2+} ions and one with octahedrally coordinated second layer Fe^{3+} ions.

Table 2: The surface structures of sulfur detected by LEED and STM on the (100) crystal face of molybdenum and on other crystal faces vicinal to Mo(100) as a function of sulfur coverage.

Sulfur Coverage (Monolayers) Surface	0.50	0.67	0.75	1.00
Mo(100)	c(2x2)	$\begin{vmatrix} 2 & \bar{1} \\ 1 & 1 \end{vmatrix}$	c(4x2)	p(2x1)
Mo(910)	c(2x2)	diffuse (1x1)	diffuse (1x1)	p(2x1) with preferred domain
Mo(911)	c(2x2)	diffuse (1x1)	diffuse (1x1)	diffuse (1x1)
Mo(28,4,1)	c(2x2)	diffuse (1x1)	diffuse (1x1)	p(2x1) with preferred domain

step directions induce or prevent ordering of particular domains of each structure. STM images indicate that at very low coverages, sulfur only orders at the lower edges of steps.

Sulfur adsorption on Mo has also been shown to have an important effect on the step structure. While the clean Mo surface contains only monoatomic steps, sulfur adsorption induces step coalescence. On the highly stepped (910), (911) and (28,4,1) surfaces sulfur adsorption, even at coverages of 0.1 monolayer, caused step doubling observed by the splitting of LEED pattern spots. On less highly stepped surfaces, STM images indicate that steps group together on the surface at low sulfur coverage with a separation of 30 Å and a much larger separation between the groups. At higher sulfur coverage the steps were observed to coalesce to form multi-layer steps. At very high coverage faceting in the (310) direction was observed. A summary of the results is shown in Fig. 3.

Several ordered surface structures of sulfur form also on the rhenium(0001) crystal face with increasing coverage (Table 3). However, on this metal sulfur coalesces to produce trimers, tetramers and hexamers as the sulfur coverage increases. These are clearly visible by STM, allowing the determination of the structure of the clusters. The formation of sulfur clusters above 0.25 monolayer is interpreted as evidence of substrate mediated interactions. The restructuring of the metal substrate alleviates the repulsive interaction between sulfur atoms that is expected at the higher coverages to permit clustering. Such behavior of sulfur monolayers is completely absent on the molybdenum surfaces that we have studied.

3.2 Adsorption Behavior and Catalytic Activity of the Oxide Films

The adsorption of water and benzene was studied on epitaxially grown iron oxide films deposited on Pt(111) and (100) crystal surfaces. Water desorbs [2] around 170 K while benzene around 400 K [12] without any sign of decomposition.

Table 3: Surface structures of sulfur on the Re(0001) crystal face detected by STM and LEED, as a function of sulfur coverage.

Sulfur Coverage (Monolayers) Surface	0.25	0.25-0.50	0.50	0.50-0.70
Re(0001)	(2x2) c($\sqrt{3}$x5)rect	$(3\sqrt{3}\times 3\sqrt{3})R30°$	$\begin{vmatrix} 3 & 1 \\ 1 & 3 \end{vmatrix}$	$(2\sqrt{3}\times 2\sqrt{3})R30°$
Sulfur Structure	monomers	trimers	tetramers	hexamers

Molecular adsorption and weak bonding to the substrate characterizes these oxide-adsorbate systems.

Thermal desorption studies of the interaction of water and a large group of polar organic molecules, ethers and fluoro-ethers have been carried out with ZrO_2 films epitaxially grown on Pt(111) surfaces [13]. Most of these molecules adsorb weakly and reversibly on the oxide film. The desorption energies correlate well with molecular polarizabilities.

These adsorption studies indicate that the ordered iron and zirconium oxide films are chemically inert.

Catalytic reaction studies carried out using several oxides deposited on rhodium foils show interesting and surprising patterns of chemical behavior [14].

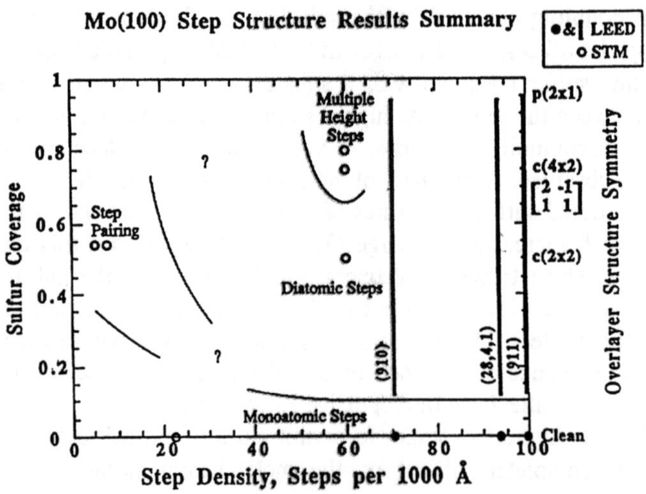

Fig. 3: A phase diagram summarizing the LEED and STM results. The horizontal axis represents the step density on the surface and the vertical axis represents the sulfur coverage. The symmetry of the sulfur overlayer structure which forms on low step density surfaces is indicated on the right axis. Dark experimental points and lines represent LEED results. Hollow points represent STM results. Roughly defined regions of different step structures are shown. The diagram shows a trend toward increasing step height as either sulfur coverage or step density is increased.

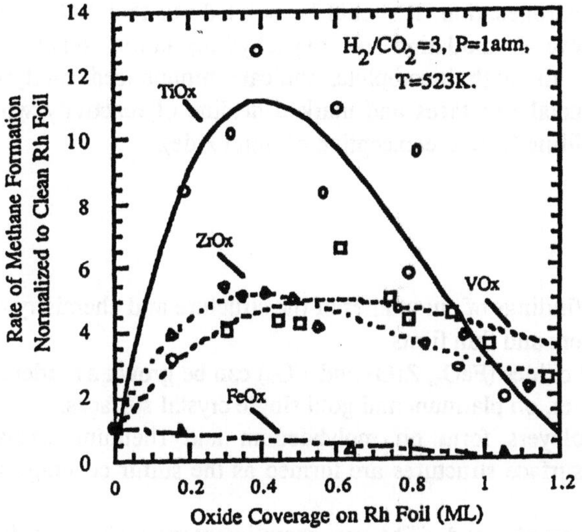

Fig. 4: Rate of methane formation as a function of oxide coverage for FeO_x, VO_x, ZrO_x and TiO_x.

In studies of the hydrogenation of CO and CO_2 to methane submonolayers of oxides, ZrO_x, TiO_x, VO_x all show increased catalytic activity as compared to the pure metal (Fig. 4). FeO_x is an exception that causes only declining activity. Surprisingly, above one monolayer oxide deposit only iron oxide exhibits increased catalytic activity for these reactions (Fig. 5). Thus, it appears that iron oxide multilayers are catalytically active, while the monolayer or submonolayer iron oxide films are totally inactive. For the other oxides, thick films proved to be

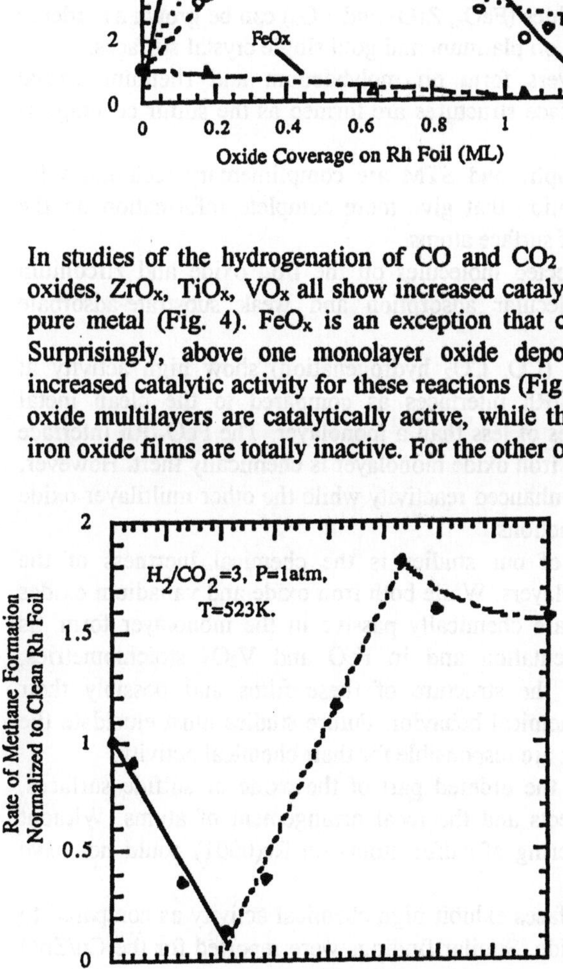

Fig. 5: Rate of methane formation as a function of iron oxide coverage. Note the sharp rise in reaction rate above one monolayer of coverage.

inactive for CO or CO_2 hydrogenation. It should be pointed out that VO_x is an excellent catalyst for several oxidation or oxydehydrogenation reactions. Nevertheless, our results, although incomplete, indicate pronounced catalytic activity at several oxide-metal interfaces and marked decline of reactivity with the thickness of the oxide films (with the exception of iron oxide).

4. Discussion

Let us summarize the key findings of our studies of the structure and chemistry of oxide and sulfide monolayers and thin films.

1) Several transition metal oxides (FeO_x, ZrO_2 and VO_x) can be grown as ordered epitaxial films, layer by layer, on platinum and gold single crystal surfaces.

2) Ordered sulfide monolayers form on molybdenum and rhenium crystal surfaces. Several ordered surface structures are formed as the sulfur coverage is altered.

3) LEED-surface crystallography and STM are complimentary techniques for surface structure characterization that give more complete information on the location and rearrangement of surface atoms.

4) Adsorption studies of selected molecules on the iron oxide and zirconium oxide surfaces indicate molecular adsorption and weak substrate-adsorbate interaction.

5) Catalytic reaction studies (CO, CO_2 hydrogenation) show high activity at TiO_x/Rh, ZrO_2/Rh and VO_x/Rh interfaces as compared to the clean metal (rhodium), for oxide coverages of less than a monolayer. The FeO_x/Rh interface is chemically inactive and the iron oxide monolayer is chemically inert. However, iron oxide multilayers show enhanced reactivity while the other multilayer oxide films are inactive for these reactions.

One surprising finding of our studies is the chemical inertness of the epitaxially grown oxide monolayers. While both iron oxide and vanadium oxides are excellent catalysts, they are chemically passive in the monolayer form, as grown with their (111) orientation and in FeO and V_2O_3 stoichiometries, respectively. It appears that the structure of these films and possibly their stoichiometry control their chemical behavior. Future studies must elucidate the surface structural features that are responsible for their chemical activity.

While LEED focuses on the ordered part of the oxide or sulfide surfaces, STM reveals the role of defects and the local arrangement of atoms. Without STM, for example, the clustering of sulfur atoms on Re(0001) could not have been identified.

Certain oxide-metal interfaces exhibit high chemical activity as compared to the pure metal or the pure oxide. Similar findings were reported for the Cu/ZnO system [14,15] that produces methanol from CO, CO_2 and H_2, and for platinum catalysts that are used under oxidizing conditions. In this circumstance there is evidence that the oxidation of ammonia or carbon monoxide occur at the interface of islands of chemisorbed oxygen on the platinum surface [16,17]. Future studies of the structure of these interfaces and their alteration during chemisorption or catalytic reactions should reveal the causes of this effect.

Acknowledgement

This work was supported by the Director, Office of Energy Research, Office of Basic Energy Science, Materials Science Division, of the U.S. Department of Energy under Contract No. DE-AC03-76SF00098. W. Weiß acknowledges support by the Deutsche Forschungsgemeinschaft.

References

[1] G.H. Vurens, V. Maurice, M. Salmeron, G.A. Somorjai: Surf. Sci. **268**, 170 (1992)

[2] W. Weiss, G.A. Somorjai: Surf. Sci., to be published (1993)

[3] V. Maurice, M. Salmeron, G.A. Somorjai: Surf. Sci. **237**, 116 (1990)

[4] K.B. Lewis, S.T. Oyama, G.A. Somorjai: Surf. Sci. **233**, 75 (1990)

[5] J.C. Dunphy, P. Sautet, D.F. Ogletree, M.B. Salmeron: J. Vac. Sci. Techn. (submitted) (1992)

[6] J.C. Dunphy, C.C. Knight, P. Sautet, D.F. Ogletree, G.A. Somorjai, M.B. Salmeron: Surf. Sci. (submitted) (1992)

[7] R.Q. Hwang, D.M. Zeglinski, A. Lopez Vazquez-de-Parga, D.F. Ogletree, G.A. Somorjai, M.B. Salmeron: Phys. Rev. **B 44**, 1914 (1991)

[8] C.C. Knight; G.A. Somorjai: Surf. Sci. **272**, 326 (1992)
C.C. Knight, G.A. Somorjai, Surf. Sci. **240**, 101 (1990)

[9] G.A. Somorjai, M.B. Salmeron, in: Surface Science Investigations in Tribology, ACS Symposium Series 485, eds. Y.-W. Chung, A.M. Homola and G.B. Street, Washington DC 1992

[10] K.J. Williams, A.B. Boffa, M.B. Salmeron, A.T. Bell, G.A. Somorjai: Catal. Lett. **9**, 415 (1991)

[11] A.B. Boffa, A.T. Bell, G.A. Somorjai: J. Catal., to be published (1993)

[12] W. Weiss, G.A. Somorjai: Surf. Sci., to be published (1993)

[13] K. Takeuchi, M.B. Salmeron, G.A. Somorjai: Surf. Sci. **279**, 328 (1992)
V. Maurice, K. Takeuchi, M.B. Salmeron, G.A. Somorjai: Surf. Sci. **250**, 99 (1991)

[14] S.S. Fu, G.A. Somorjai: J. Phys. Chem. **96**, 4542 (1992)

[15] S.S. Fu, G.A. Somorjai: Surf. Sci. **237**, 87 (1990)
S.S. Fu, G.A. Somorjai: Langmuir **8**, 518 (1992)

[16] T. Katona, L. Guczi, G.A. Somorjai: J. Catal. **136**, 434 (1992)

[17] T. Katona, G.A. Somorjai: J. Phys. Chem. **96**, 5465 (1992)

High-Resolution Electron Energy-Loss Studies of Metal Oxides

M.-Ch. Wu, Ch. Truong, and D.W. Goodman

Department of Chemistry,
Texas A&M University, College Station, TX 77843-3255, USA

Abstract. Studies of ordered MgO and NiO have been carried out by synthesizing ultra-thin films under ultra-high vacuum (UHV) conditions via evaporating Mg and Ni onto Mo(100) in the presence of oxygen. Low-energy electron diffraction (LEED) studies indicate that the films grow epitaxially with their (100) faces oriented parallel to Mo(100). The MgO and NiO films, prepared under optimum synthesis conditions, have essentially one-to-one stoichiometry, are nearly free from point-like surface defects, and have properties essentially identical to their bulk, single-crystal counterpart. Molecular adsorption onto the films has been studied using High-Resolution Electron Energy-Loss Spectroscopy (HREELS) and temperature programmed desorption (TPD). In order to circumvent the difficulty associated with intense multiple surface optical phonon (Fuchs-Kliewer modes) losses, a new approach to acquisition of HREELS data has been demonstrated. This new approach enables the direct observation of weak loss features due to excitation of the adsorbates without serious interference from multiple phonon losses.

1. Introduction

Studies of metal oxides and the characterization of adsorbates on these substrates have received considerable attention in many important areas such as catalysis, ceramics, metallurgy, and in microelectronics. In catalysis for example, magnesium oxide is known to be a typical basic oxide which catalyzes the H_2 and D_2 exchange reaction [1] and the dehydrogenation of formic acid or methanol [2]. These latter molecules, which are Brønsted acids, generally dissociate heterolytically on MgO surfaces via an acid-base interaction, forming protons adsorbed onto lattice-oxygen. It has also been demonstrated recently that Li-doped MgO catalysts can promote the oxidative coupling of CH_4 to $C_2H_6 + C_2H_4$ through a methyl radical intermediate [3].

Despite this demand for a better understanding of the surface chemistry of these materials on the atomic scale, few surface electron spectroscopic studies have been reported on well-characterized metal-oxide surfaces. Because most metal oxides are bulk insulators or wide gap semiconductors, problems with

Springer Series in Surface Sciences, Vol. 33 Eds.: E. Umbach · H.-J. Freund
Adsorption on Ordered Surfaces of Ionic Solids and Thin Films

surface charging are encountered with these materials during charged particle measurements.

However, surface charging has been circumvented in our studies by making an ultra-thin, well-characterized oxide film on the surface of a metal substrate. Any charging induced in the thin film during spectroscopic measurements is dissipated via the conducting substrate. Our low-energy electron diffraction (LEED) and surface electron spectroscopic studies [4,5] indicate that MgO and NiO films grow epitaxially with the (100) face oriented parallel to Mo(100), and that the properties of the films are essentially identical to their bulk, single-crystal counterpart.

Among the various surface spectroscopic techniques used to characterize adsorbate properties, High-Resolution Electron Energy-Loss Spectroscopy (HREELS) is a powerful tool which can be utilized to identify rather complex molecular adsorbates by their characteristic vibrational frequencies (fingerprints) and to follow step-by-step surface reactions by identifying intermediates [6]. HREELS has become a premier technique for surface vibrational studies due to its high sensitivity, wide spectral range and applications involved in the study of a wide range of molecules adsorbed on conducting substrates. However, in addition to the problem of surface charging associated with insulating oxides, this technique has encountered a second great difficulty when used to study adsorbates on ionic substrates. The difficulty arises from the fact that the accompanying vibrational spectra are dominated by losses due to excitation of surface optical phonons, known as Fuchs-Kliewer modes [7]. Since the intense multiple phonon losses generally extend over a wide vibrational frequency range of the HREELS spectra, it is not practical to directly observe losses due to the excitation of adsorbates in the 0-4000 cm^{-1} spectral range.

In order to circumvent the difficulties associated with these phonon losses, Cox et al. [8] have developed a Fourier transform method to remove the phonon-phonon combination peaks from the HREELS spectrum.This approach has recently been applied to the study of formic acid adsorbed on ZnO(0001) [9]. Clearly the Fourier deconvolution method is superior to simply acquiring difference spectra between clean and adsorbate-covered surfaces; however, it is still basically an "indirect" technique. Of particular concern are the spurious features that may be introduced in the spectrum by the Fourier deconvolution procedure itself due to difficulties inherent in deconvolution schemes [8,9].

Here a summary is presented of recent work addressing molecular adsorption on ultra-thin MgO and NiO films using a new approach to the acquisition of HREELS data. This new approach, which has been previously outlined [10,11], enables the direct measurement of losses due to the excitation of adsorbed species with little interference from surface optical phonon losses.

2. Experimental

The experiments were carried out in two UHV systems. One apparatus has capabilities for Auger electron spectroscopy (AES), LEED and TPD, and for sample heating and cooling. The MgO and NiO film characterization and thermal desorption measurements were carried out in this apparatus. HREELS was performed in a second UHV system which is also equipped with AES and LEED. A detailed description of the two systems has been given elsewhere [4,12].

The crystal cleaning procedure consisted of oxidation in 1×10^{-7} Torr oxygen at 1000-1300 K for 5-10 minutes and a flash to 2000 K. After several cycles of the above procedure, the impurities (mainly carbon and oxygen) were below the detection limit of AES. A clean surface, exhibiting a sharp (1x1) LEED pattern, was then obtained.

Spectroscopic grade samples (methanol, 99.8 %, and formic acid, 96 %) were used in this study. Further purification was achieved in the manifold via freeze-pump-thaw cycles prior to introduction into the vacuum chamber. In the TPD experiments, a directional gas doser was used to introduce methanol and formic acid to the crystal surface; however, the exposure of the crystal to the adsorbates was carried out via backfilling the UHV chamber for HREELS measurements. Surface coverages of the adsorbates are reported with respect to the saturation or monolayer coverages. The monolayer coverages (ML) of methanol (or formic acid) on MgO(100) (or NiO(100)) deduced from TPD spectra is assumed to be $\theta = 1$. This value is used to calibrate a specific coverage deduced by measuring the area below the adsorbate-TPD peak.

The HREELS spectra were acquired both in the specular direction and in an off-specular direction. The incident angle of the electron beam was set at 60^{0} with respect to the surface normal. The primary energy of the electron beam of the spectrometer (LK-2000, Larry Kesmodel Technologies) can be varied between 0-250 eV. While an elastic-peak count rate of 200,000 cps could be easily obtained with a low primary energy electron beam, acquisition of the spectra at high primary energies resulted in a decrease in the count rate. The spectral resolution (full-width at half maximum of the elastic peak) of the data presented here was typically 70-90 cm^{-1}.

3. Results and Discussion

3.1 Preparation and Characterization of Ultra-Thin MgO and NiO Films on Mo(100)

Ultra-thin MgO and NiO films were synthesized in a UHV environment by evaporating Mg and Ni onto a clean Mo(100) surface in the presence of oxygen. Deposition of Mg and Ni was performed by evaporation of the corresponding high-purity metal wire tightly wrapped around a tungsten filament and mounted inside a doser. The stabilization and calibration of the flux of metal evaporation have been described in previous publications [4,5]. Briefly, Auger spectra were recorded following successive doses of the metal onto Mo(100). The metal evaporation rate can be determined using the correlation between the first break point in the AES plot (Auger signal versus time) and the completion of the first Mg monolayer. This evaporation rate was independently confirmed by TPD measurements. Our thermal desorption spectra have shown that the initial appearance of the multi-layer desorption peak of Mg or Ni from Mo(100) corresponds very closely to the monolayer break point in AES. The thickness of the MgO or NiO film was then determined from the calibrated metal evaporation rate and the deposition time.

Auger and XPS measurements suggest that the MgO and NiO films have essentially a one-to-one stoichiometry. No evidence for the formation of intermediate sub-oxide precursors is indicated. Details of these studies can be found in references [4,5] and [13].

The growth of MgO and NiO films on Mo(100) has been examined by LEED in the 200-600 K substrate temperature range. It was found that the films grow epitaxially with their (100) faces oriented parallel to Mo(100). At MgO or NiO coverages, θ_{MgO} or θ_{NiO}, below one monolayer, the (1x1) LEED pattern remained essentially unchanged with the exception that the background intensity increased slightly. Distinct (1x1) LEED spots were observed from epitaxial MgO or NiO films whose coverage was varied from 2 to 30 ML. The detailed evolution of the LEED pattern as a function of film thickness has been described previously [4,5].

3.2 Temperature Programmed Desorption Studies

Methanol on MgO(100). Figure 1 presents TPD spectra as a function of methanol exposure. Methanol multilayers desorb between 137-166 K, with a maximum at 146 K. A second desorption feature appears in the 200-400 K temperature range. This feature, which can be interpreted to arise from methanol formed via recombination of a surface methoxy with hydrogen from surface hydroxyls (see section 3.4 for details), increases with increasing exposures and approaches saturation when the methanol multi-layer desorption peak appears. No desorption signal was observed for mass channels other than that of CH_3OH, e.g. m/e = 2 or 28, implying that the parent molecule is the only species desorbed from the surface.

Formic Acid on NiO(100). The adsorption of HCOOH on NiO(100) films was carried out at 90 K and has been examined using TPD. The evolution of various desorption features as a function of formic acid coverage, θ, has been monitored in several mass channels (m/e = 2, 18, 28, 44, 45, 46) and the desorption signal of the parent molecules is shown in Fig. 2. No desorption signal was observed for mass channels other than that of HCOOH and its fragments, e.g. m/e = 45 or 46, implying that the parent molecules are the only species

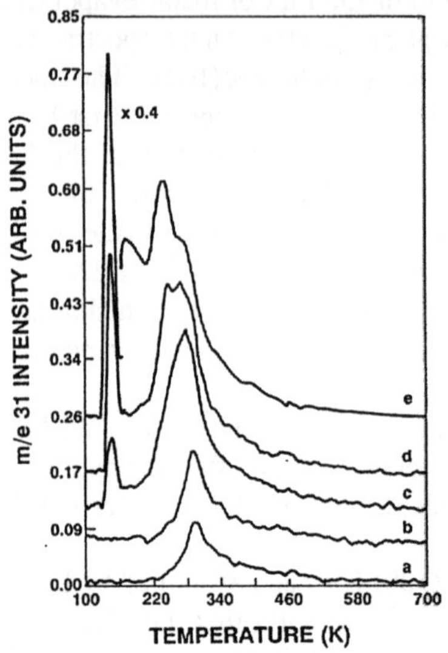

Fig. 1: TPD spectra of methanol adsorbed on MgO(100) surfaces at 90 K as a function of coverage (ML): (a) 0.25; (b) 0.34; (c) 0.90; (d) 1.25; (e) 2.04. The linear heating rate for the TPD spectra is 7 K/s.

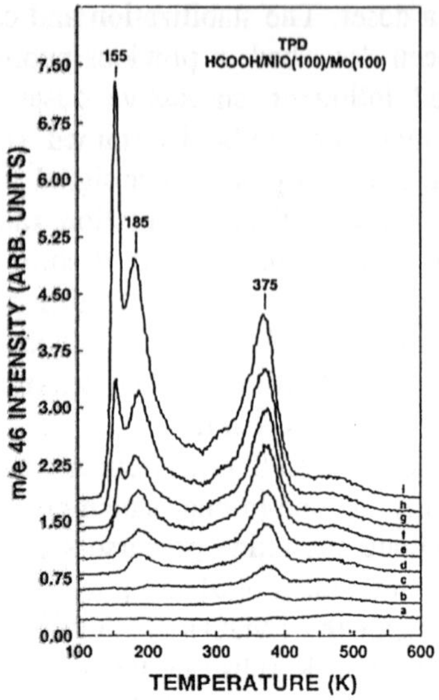

Fig. 2: TPD spectra of formic acid adsorbed on NiO(100) surfaces at 90 K as a function of coverage (ML): (a) 0.07; (b) 0.12; (c) 0.22; (d) 0.43; (e) 0.58; (f) 0.90; (g) 1.09; (h) 1.58 and (i) 2.37. The linear heating rate for the TPD spectra is 5 K/s.

desorbed from the surface. TPD spectra (m/e = 46) at low coverages exhibited a desorption feature with a maximum at 375 K and a broad weak feature at ~ 450 K. Further exposure to formic acid resulted in the population of the 375 K desorption state and the appearance of two new features at lower temperatures. These desorption states appear to fill in non-sequentially with increasing exposure. While other desorption features remained saturated at $\theta \geq 2$ monolayer, the desorption peak at 155 K increased in intensity as θ is increased, indicating that it is from condensed multilayers of formic acid. The peak at 185 K, which occurred near the maximum of multilayer desorption, can be interpreted as arising from desorption of bilayers of the molecules. Assuming the first order kinetics and a prefactor $\nu = 8 \times 10^{14}$ Hz [14], the activation energy of desorption of this bilayer is deduced to be 12.7 kcal/mol using Redhead's peak maximum method [15], which is in excellent agreement with the previously reported value (12.6 kcal/mol) for the decomposition of dimeric formic acid [16]. The desorption of monolayer formic acid formed via recombination of surface formate and surface hydrogen occurred between 300-400 K with a maximum at 375 K. The heterolytic dissociation of HCOOH on NiO(100) at coverages below ML upon heating has been confirmed spectroscopically by HREELS which follows.

3.3 HREELS Studies: Theoretical Considerations

Modern quantum theory allows the development of a general theory for the analysis of the scattering of electrons from a surface; however, it is extremely difficult in practice to formulate a single expression for the loss processes in HREELS. Accordingly, two distinct regimes have been outlined to describe the loss processes. One involves small-angle scattering (with reference to the specularly reflected beam direction) induced by long-range dipolar fields generated either by adsorbate vibrations or by surface atomic vibrations. The second is the impact scattering mechanism which involves large-angle electron deflection (with reference to the specularly reflected beam direction). HREELS spectra acquired using small-angle scattering geometry can be interpreted semi-classically within the framework of the so-called dielectric theory [6,17]. The energy-loss probability, Γ, which describes a single event corresponding to either the excitation or emission of one surface optical phonon, is given by [17]

$$\Gamma(\omega)=\frac{4e^2}{\pi^2\hbar V_\perp}\int d^2k\frac{1}{k^2}\frac{(kV_\perp)^3}{\left[(\omega-k\cdot V_\parallel)^2+(kV_\perp)^2\right]^2}\cdot \mathrm{Im}\frac{-1}{\varepsilon(\omega)+1} \qquad (1)$$

The above integration is computed over a region defined by the angular acceptance of the energy analyzer, where $\mathrm{Im}(-1/(\varepsilon(\omega)+1))$ is the loss function for surface excitation in an isotropic medium and $\varepsilon(\omega)$ the dielectric function.

It should be pointed out that the above integral function has a spatial distribution which peaks sharply very close to the specularly reflected beam direction. The angular width of this distribution has been deduced to be approximately $\hbar\omega/2E_0$, where $\hbar\omega$ and E_0 are the phonon energy and electron primary energy, respectively. As E_0 is increased, the distribution narrows and, accordingly, the energy analyzer samples a larger fraction of the distribution. Consequently, for emission of a surface optical phonon, the scattering probability integrated over the total solid angle can be computed without loss of accuracy and is given as follows: $\Gamma = KE_0^{-1/2}$, where K is a constant.

It has been demonstrated both theoretically and experimentally [6,18] that the total probability (Γ_N) for an electron to excite N quanta as it passes by a surface, losing the energy $N\hbar\omega$ in the process, can be expressed using a Poisson distribution $\Gamma_N = (\Gamma^N/N!)\exp(-\Gamma)$. The intensity of the Nth multiple loss in the Poisson distribution decreases by a factor of Γ/N ($\Gamma < 1$) relative to that of the (N-1)th loss. This relationship, in conjunction with the above characteristic energy dependence of $E_0^{-1/2}$, suggests that the intensities of multiple phonon losses can be modulated greatly by using a high-energy electron beam. Figure 3 demonstrates how the multiple phonon losses diminish as E_0 is increased. At $E_0 = 46.2$ eV, all losses due to excitation of the multiple quanta of the MgO surface phonon virtually disappear at frequencies above the third multiple loss. Indeed, the intensity of this third loss decreases more than two orders of magnitude relative to the comparable losses measured with $E_0 = 3.2$ eV.

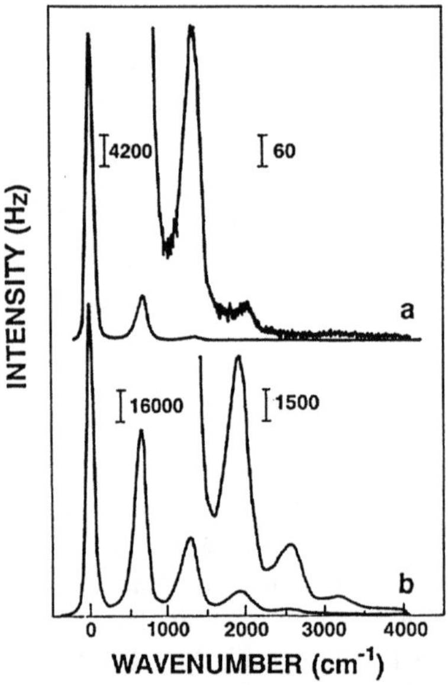

Fig. 3: Demonstration of diminution of the multiple surface optical phonon losses with increasing primary energy, Eo, of the electron beam. The HREELS spectra were acquired in the specular direction and at 90 K with primary energies of (a) 46.2 eV; (b)3.2 eV. The incidence of the electron beam was at 60^0 with respect to the surface normal. The vertical scales indicated represent electron counts.

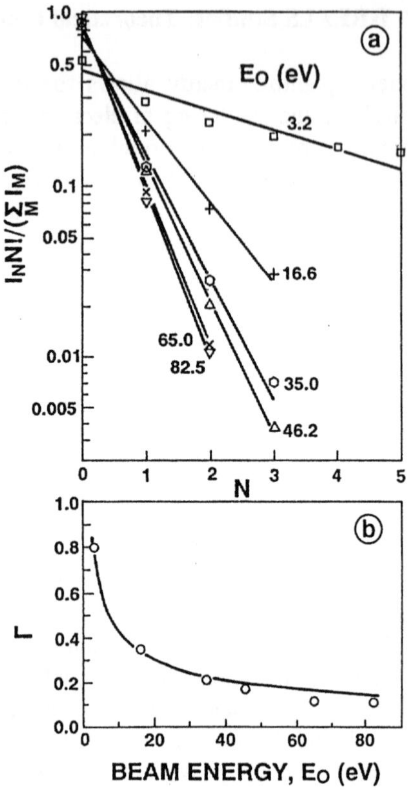

Fig. 4a: Intensity distribution of multiple losses due to excitation of N quanta of the MgO surface phonon at various primary energies indicated. The experimental plots match the theoretically predicated Poisson distribution excellently.

Fig. 4b: The scattering probability, Γ, corresponding to excitation of the fundamental mode of the MgO surface phonon as a function of Eo.

Figure 4a gives the intensity distribution of multiple losses due to excitation of N quanta of the MgO surface phonon at various impact energies. With the exception of $E_0 = 3.2$ eV, where the experimental points exhibit a deviation from the theoretically predicated Poisson distribution (straight line), the overall agreement between our experiment and theory is excellent. In Fig. 4b, the scattering probability, Γ, corresponding to the excitation of the fundamental mode of the MgO surface phonon, normalized to that of the specularly reflected beam, is plotted as a function of E_0. The experimental points obtained from the slope of each line in Fig. 4a are in excellent agreement with the energy dependence of $E_0^{-1/2}$.

It is known that the Fuchs-Kliewer surface optical phonon has a small wave-vector component parallel to the surface and has a long characteristic penetration depth [6]. It is therefore feasible to further decrease the phonon-loss peaks by acquiring the spectra in an off-angle scattering geometry, i.e. in the impact scattering regime.

3.4 HREELS Studies: Experimental Results

Methanol on MgO(100). Adsorption of methanol on thin MgO(100) films has also been examined using HREELS [10,11], as shown in Fig. 5. Exposing the specimen to 180 L CH_3OH produces four distinct peaks at 1140, 1440, 2950 and

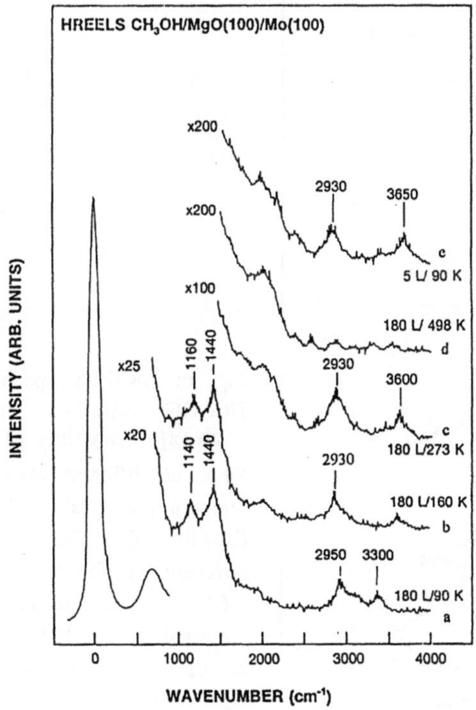

Fig. 5: HREELS spectra of methanol adsorbed on a ~30 ML MgO(100) film at 90 K. (a) 180 L CH_3OH at 90 K; (b) flash to 160 K; (c) flash to 273 K; (d) flash to 498 K; (e) 5 L CH_3OH at 90 K. 1 Langmuir (L) = 1×10^{-6} Torr·s. The spectra were collected at $E_0 \cong 46$eV and an angle 8.5^0 off the specularly reflected beam direction.

3300 cm⁻¹, respectively. These losses can be attributed to C-O stretching (ν (CO)), CH_3 bending ($\delta(CH_3)$), C-H stretching (ν (CH)) and O-H stretching (ν (OH) modes, respectively, of methanol multilayers [12]. These loss features were essentially preserved in the 90-145 K annealing temperature range; however, the spectrum (not shown here) obtained using a low-energy beam ($E_0 = 3.2$ eV), following an anneal to 145 K, was indistinguishable from the spectrum of the clean surface. Spectrum b of Fig. 5 acquired following an anneal to 160 K, exhibits the loss features of the methoxy monolayer (see the following discussion); the desorption of methanol multilayers occurs in the 137-166 K range, as indicated by our TPD measurements. Therefore, the new approach outlined here does facilitate a direct observation of relatively weak adsorbate losses in a wide spectral range (in this case, 1000-4000 cm^{-1}) without serious interference from the intense multiple phonon losses.

Annealing to T ≥ 160 K resulted in the disappearance of the 3300 cm^{-1} loss, and the appearance of a new peak at 3600 cm^{-1}. The 2950 cm^{-1} loss remained essentially unperturbed. The appearance of the new peak, which is identical to the one assigned to OH species, indicates the desorption of methanol multilayers, exposing the dissociatively adsorbed methanol monolayer. Further annealing to higher temperatures leads to concurrent diminution of the two losses. At 498 K, features due to the excitation of the adsorbates completely disappear (Fig. 5e).

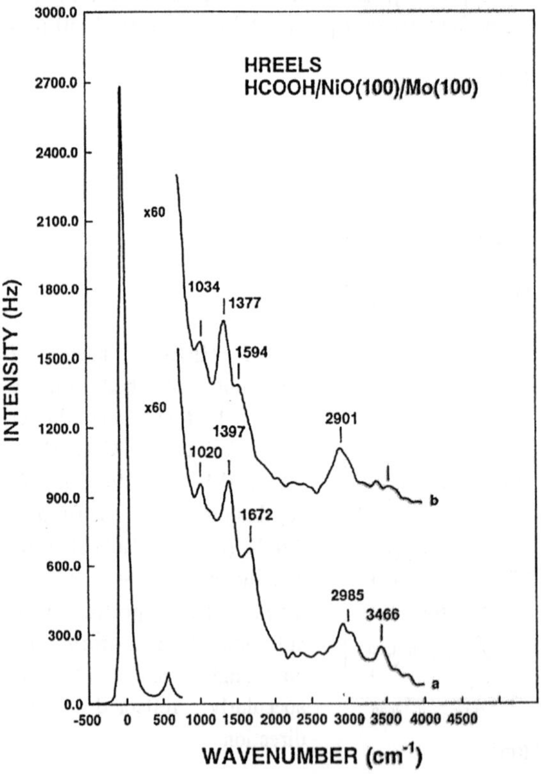

Fig. 6: HREELS spectra of HCOOH adsorbed on a ~20 ML NiO(100) film at 90 K at different annealing tempera-ture: (a) 90 K; (b) 220 K. The surface coverage of formic acid is 0.64 ML. The spectra were collected at $E_0 \cong 49.8$ eV and at the specularly reflected beam direction

The above behavior indicates that a direct recombination of adsorbed methoxy with atomic hydrogen occurs at temperatures above 273 K. The desorption of methanol is completed at 498 K. The dissociative adsorption of methanol was also observed at low exposures and at 90 K, as shown in Fig. 5e.

Formic Acid on NiO(100). Adsorption of formic acid on thin NiO(100) films has been further examined using HREELS [19]. Displayed in Fig. 6 is a set of HREELS spectra of formic acid adsorbed on a ~ 20 monolayer NiO film at 90 K as a function of annealing temperature. Exposing the sample to HCOOH at 90 K gave rise to a number of distinct loss features in the 750-4000 cm^{-1} region. At coverages below monolayer, the losses observed at 1020, 1397, 1672, 2985 and 3466 cm^{-1} are due to excitations of the C-O stretching (ν (C-O)), the C-H deformation (δ(CH)), the C=O stretching (ν (C=O)), the C-H stretching (ν(CH)) and the O-H stretching (ν (OH)) modes, respectively [9,16,20]. All loss features remained essentially unchanged at T < 200 K. However, dramatic changes occurred upon annealing over 200 K: the C=O stretching mode completely disappeared and a new loss at 1594 cm^{-1} appeared. The shape and relative intensity of the losses at 1397 and 1020 cm^{-1} also changed as T exceeded 200 K. The changes of these vibrational modes clearly indicate the dissociation of formic acid and the formation of a surface formate species [9,16,21,22]. The losses observed at 1034, 1377 and 1594 cm^{-1} are thus attributed to the out-of-plane C-H deformation (π(CH)), the symmetric and asymmetric O-C-O stretching modes, respectively, of the surface formate [9,16,21,22]. The in-plane C-H deformation (δ(CH)) mode has a loss energy very near that of the ν_s(OCO) mode; however, these two modes cannot be differentiated with the present energy resolution of the instrument (~ 80 cm^{-1}).

4. Conclusions

The present study indicates that highly ordered, stoichiometric ultra-thin MgO and NiO films can be formed on Mo(100), and that the properties of the films are essentially identical to their bulk, single-crystal counterparts. Because of the intrinsic merits of thin films, one can perform surface electron spectroscopy on this insulator without difficulties associated with surface charging. Model studies of oxide thin films have another advantage in the avoidance of a sample temperature gradient while heating, caused by poor bulk thermal conductivity of oxides. It has been further demonstrated that surface chemistry on this model surface can be followed in great details using modern surface analytical techniques.

A new approach to the acquisition of HREELS data has also been demonstrated. By utilizing a high-energy incident electron beam in combination with an off-specular scattering geometry, this new approach enables a direct observation of relatively weak loss features due to excitation of the adsorbates without serious interference from intense multiple surface optical phonon losses. Combined HREELS/TPD spectroscopies together with the thin film preparation technique facilitate the acquisition of a variety of data on oxide materials which have been heretofore precluded.

Acknowledgements

We acknowledge with pleasure the support of this work by the Department of Energy, Office of Basic Energy Sciences, Division of Chemical Sciences and the Gas Research Institute.

References

[1] M. Boudart, A. Delbouile, E.G. Derouane, V. Indovina, A.B. Walters: J. Am. Chem. Soc. **94**, 6622 (1973)

[2] P. Mars, J.J.F. Scholten, P. Zwietering: Adv. Catal. **14**, 35 (1963)

[3] T. Ito, J.H. Lunsford: Nature **314**, 721 (1985)

[4] M.-C. Wu, J.S. Corneille, C.A. Estrada, J.-W. He, D.W. Goodman: Chem. Phys. Lett. **182**, 472 (1991)

[5] M.-C. Wu, D.W. Goodman: in preparation

[6] H. Ibach, D.L. Mills: Electron Energy Loss Spectroscopy and Surface Vibrations, Academic Press, New York 1982

[7] R. Fuchs, K.L. Kliewer: Phys. Rev. **140**, A2076 (1965)

[8] P.A. Cox, W.R. Flavell, A.A. Williams, R.G. Egdel: Surf. Sci. **152/153**, 784 (1985)

[9] W.T. Petrie, J.M. Vohs: Surf. Sci. **245**, 315 (1991)

[10] M.-C. Wu, C.A. Estrada, D.W. Goodman: Phys. Rev. Lett., **67**, 2910 (1991)

[11] M.-C. Wu, C.A. Estrada, J.S. Corneille, D.W. Goodman: J. Chem. Phys. **96**, 3892 (1992)

[12] J.E. Parmeter, X. Jiang, D.W. Goodman: Surf. Sci. **240**, 85 (1990)

[13] J.-W. He, J.S. Corneille, C.A. Estrada, M.-C. Wu, D.W. Goodman: J. Vac. Sci. Technol. **A 10**, 2248 (1992)

[14] L.A. Larson, J.T. Dickinson: Surf. Sci. **84**, 17 (1979)

[15] P.A. Redhead: Vacuum **12**, 203 (1962)

[16] G. Herzberg: Molecular Spectra and Molecular Structure, Vol II, Van Nostrand Reinhold, New York 1960, p 535

[17] Ph. Lambin, J.P. Vigneron, A.A. Lucas: Phys. Rev. **B 32**, 8203 (1985)

[18] H. Ibach: Phys. Rev. Lett. **12**, 1416 (1970)

[19] C. Truong, M.-C. Wu, D.W. Goodman: J. Chem. Phys. in press

[20] K. Nakamoto, S. Kishida: J. Chem. Phys. **41**, 1554 (1964)

[21] W.T. Petrie, J.M. Vohs: Surf. Sci. **259** L750 (1991)

[22] B.A. Sexton: Surf. Sci. **88** 319 (1979)

Surface Oxygen and Chemical Reactivity

C.T. Au, A.F. Carley, A. Pashuski, S. Read, M.W. Roberts, and A. Zeini-Isfahan

School of Chemistry & Applied Chemistry, University of Wales, Cardiff, UK

Abstract: The probe-molecule or coadsorption approach has emphasised some unique aspects of the chemistry of dioxygen at metal surfaces at low temperatures. Models have been developed making use of XP and HREEL spectroscopies for the chemisorption of dioxygen where the chemical reactivity is dominated by the transients $O^{\delta-}(s)$ and $O_2^{\delta-}(s)$ - which of these controls the chemistry depending on the metal. By comparison the oxide overlayer is unreactive under the same conditions.

Dynamic or real time studies provided the evidence for the transient or precursor states and estimates have been made of their surface life-times and concentrations with support for the concepts coming from recent scanning tunnelling microscope results. The charge on the surface oxygen and whether it exists as isolated atoms or as a component of an 'oxide' island - even at low coverage - are suggested to be the crucial factors in determining its chemistry. Theoretical calculations support this view and analogies drawn with 'defective' oxide surfaces where both O^- and O^{2-}-like species can exist.

1. Introduction

Although there was initial scepticism expressed regarding the early claims that chemisorbed oxygen present at metal surfaces could activate such adsorbates as H_2O and H_2S at low temperatures [1] - the oxygen acting as a base - there is now substantial evidence [2] that it can indeed exhibit very diverse chemistry depending both on the state of the surface oxygen and the nature of the adsorbent. Furthermore the experimental data has been supported by theoretical studies [3]. We emphasise in the present discussion how the experimental strategy adopted allows us to attribute specific chemistry to oxygen as its state transforms from a transient species to oxygen adatoms to that associated with surface "oxide" - however the latter is defined at the atomic level. Although the more conventional approach to investigate oxygen surface dynamics is through scattering studies, the use of molecular probes - a chemical trapping approach - offers the possibility to comment on the chemical specificity of surface oxygen transients present during the transition of an atomically clean single crystal metal surface, and involving reconstruction, to an 'oxide' overlayer. The coadsorption approach can therefore be seen as a bridge between the more

Springer Series in Surface Sciences, Vol. 33 Eds.: E. Umbach · H.-J. Freund
Adsorption on Ordered Surfaces of Ionic Solids and Thin Films

traditional static approach to chemisorption and dynamic molecular-beam studies. It is the coadsorption approach that led us to highlight the unique reactivity of oxygen adatom transients in the chemistry of coadsorbed molecules - dioxygen and carbon monoxide [4], dioxygen and ammonia [5] and dioxygen and propene [6] - and also in the dissociative chemisorption of oxygen containing molecules such as nitric oxide and nitrous oxide [5].

When dioxygen dissociation is kinetically slow - as it is at a Zn(0001) surface [7] where the sticking probability is ~ 0.001 at 80 K - the chemistry can be dominated by a dioxygen transient $O_2^{\delta-}(s)$ i.e. oxygen present at a predissociation stage. For example low energy pathways become available when dioxygen is coadsorbed with ammonia leading to facile cleavage of the dioxygen bond and amide formation with the kinetics showing all the characteristics of an ammonia-dioxygen precursor mediated reaction. Both the clean metal and the oxide overlayer are under the same conditions unreactive.

In this paper we revisit some metal-oxygen systems that first led us to consider the possibility that in the formation of the chemisorbed oxygen overlayer at metal surfaces oxygen transients could exhibit chemistry quite distinct from the stable chemisorbed layer. These XPS studies drew attention to the high chemical reactivity of oxygen species present at low temperature (80 K) but which became inactive when they became incorporated into the oxide overlayer at higher temperature. The oxygen species were considered to be transient and metastable with respect to the thermodynamically stable oxide overlayer. Two metal oxygen systems - Ni(210) and Pb(pc) - exhibited [8] specific oxygen reactivity to $H_2O(a)$ at or close to 80 K and in this paper we return to the lead-oxygen system but studying the Pb(110) single crystal with in addition to XPS the advantage of HREEL spectroscopy. The earlier work using water as the adsorbate is also extended to include ammonia and deuterated ammonia. The role of hot oxygen adatoms in catalytic chemistry - established [4] for the carbon monoxide-dioxygen-aluminium system is explored further with Mg(1010) surfaces and analogies drawn with recent studies by Boronin et al. [9] over silver surfaces and also by Ertl and his colleagues [10]. These more recent studies are discussed in the context of earlier work at Cardiff and the role of defective oxygen at the surfaces of oxide overlayers as recently reported by Wu and Goodman [11].

2. Experimental

The core-level and vibrational spectra were obtained using a single spectrometer with X-ray and LEED facilities manufactured by VG Scientific (now Fisons) and HREEL facilities provided by VSW. For the XP spectra the pass energy used was usually 60 eV and the binding energies calibrated against the Cu(2p) line at 932.8 eV. Electron energy loss spectra were collected in the specular mode with a beam energy of 5 eV. The spectrometer is capable of ultimate pressures of

5×10^{-11} mbar and has facilities for single crystal cleaning through sputtering by Ar^+ ions.

3. Results and Discussion

Activation of ammonia and water at Pb(110)-oxygen surfaces. In XPS studies of the Ni(210) and Pb(polycrystalline)-oxygen-water systems it was concluded on the basis of O(1s) spectra that the oxygen species active in water activation at close to 80 K were coordinately unsaturated, only present at low temperatures and distinct from the oxygen present as the oxide overlayer [8]. We designated the former as $O^{\delta-}(s)$ and the oxygen of the chemisorbed oxide overlayer as $O^{2-}(a)$; in other words the $O^{\delta-}(s)$ species have not developed the full ionic charge associated with an oxide where the Madelung term plays a significant role in determining the charge. The following pathways were proposed for the interaction of water vapour with oxygen chemisorbed at a Ni(210) surface at 77 K and subsequent warming to 295 K.

(1)	$O_2(g)$	$\Rightarrow$	$2O^{\delta-}(s)$	: oxygen transient
(2)	$O^{\delta-}(s)$	$\Rightarrow$	$O^{2-}(a)$	: surface oxide
(3)	$O^{\delta-}(s) + H_2O(a)$	$\Rightarrow$	$O^{\delta-}(s)$----H-OH(a)	: H-bonded water
(4)	$O^{\delta-}(s)$----H-OH(a)	$\Rightarrow$	OH(a) + OH(a)	: potentially reactive radicals

The relative rates of steps (1)-(4) and their respective temperature dependencies will determine the chemistry - if step (2) is kinetically fast and step (4) requires some activation then water desorption is preferred with no H-abstraction occurring. In the case of the coadsorption of $O_2(g)$ and $H_2O(g)$ at a Zn(0001) surface at 180 K only OH(a) species are formed.

With the advantage of combining XPS with vibrational spectroscopy we returned to the lead-oxygen system but with a single crystal Pb(110) substrate. We also explored the activation of both ammonia and deuterated ammonia at low temperatures. The adsorption of water at a Pb(110) surface at 80 K gives rise to a single O(1s) peak with a binding energy of 534 eV and four loss-peaks in the vibrational spectrum (Fig. 1). The latter are at 230 cm^{-1} (ν_{Pb-H_2O}); 1660 cm^{-1} (δ_{HOH}) and a broad feature centred at 3420 (ν_{H-OH}). Desorption of the molecularly adsorbed water is complete at 150 K leaving an atomically clean Pb(110) surface.

The chemisorption of oxygen at a Pb(110) surface at 80 K is characterised by an O(1s) peak at 530.1 eV and a single loss peak at 510 cm^{-1} (ν_{Pb-O}). When the chemisorbed oxygen adlayer is exposed to water vapour at 80 K and the adlayer warmed to 160 K, the $\nu_{(H-OH)}$ loss feature now centred at 3600 cm^{-1} narrows at 140 K to give at 160 K a particularly sharp loss centred at 3640 cm^{-1} (Fig. 2). The O(1s) peak at 160 K is asymmetric towards higher binding energy but narrows to that characteristic of chemisorbed oxygen on warming to 300 K.

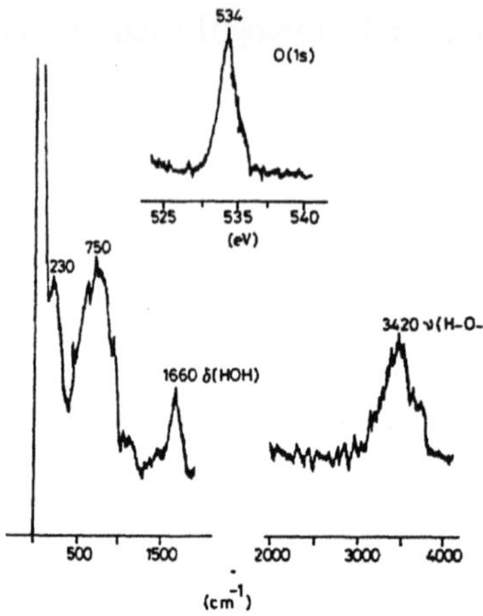

Fig. 1: O(1s) and HREEL spectra for molecularly adsorbed water at Pb(110) at 80 K.

These observations are consistent with H-abstraction occurring at or close to 80 K to give hydroxyl species OH(a), followed by desorption of molecularly adsorbed water on raising the temperature and reflected in both the O(1s) and HREEL spectra.

Following our observations of H-abstraction from ammonia by chemisorbed oxygen present at Cu(111) surfaces [12], and also in the coadsorption of dioxygen-ammonia mixtures under dynamic conditions [13], we studied the activity of $O^{\delta-}$(s) species at Pb(110) surfaces (Fig. 3). Both ammonia and

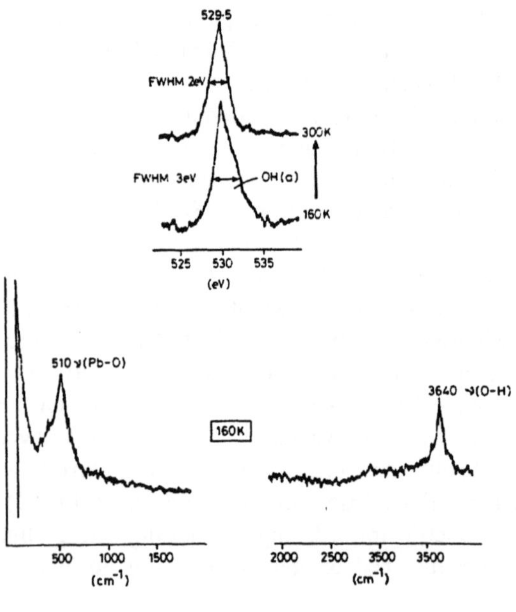

Fig. 2: O(1s) and HREEL spectra after exposing chemisorbed oxygen at 80 K to H_2O(g) and warming the adlayer to 160 K; also the O(1s) spectrum on further warming to 300 K. Note the formation of OH(a).

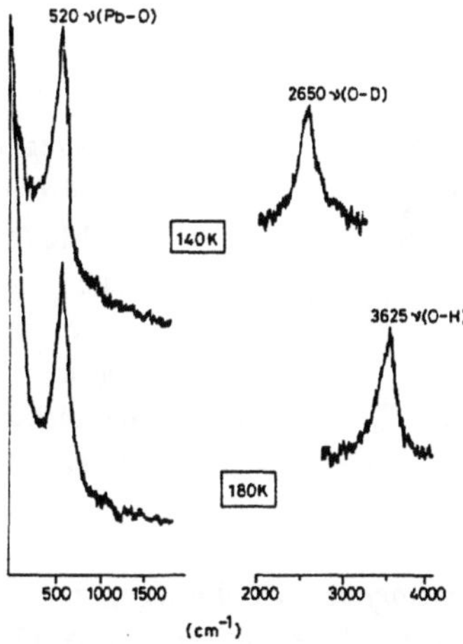

Fig. 3: HREEL spectra for the coadsorption of dioxygen-ammonia mixtures (NH_3 and ND_3) at 80 K and warming to 140 K (ND_3) and 180 K (NH_3). Note the presence of OD(a) and OH(a).

deuterated ammonia were used thus enabling unambiguous interpretation of the spectra and also ruling out any possibility of the participation of traces of water that conceivably might be present as an artefact at low temperatures.

The loss features for molecularly adsorbed NH_3 and ND_3 at 80 K are shown in Fig. 4; the main points to note are the two loss features at 2540 cm^{-1} and 3390 cm^{-1} assigned to $\nu_{N\text{-}D}$ and $\nu_{N\text{-}H}$, respectively, and the loss features characteristic of molecularly adsorbed ammonia (discussed elsewhere) in the 300 cm^{-1} to 1600 cm^{-1} region. Figures 3 and 5 show electron energy loss features

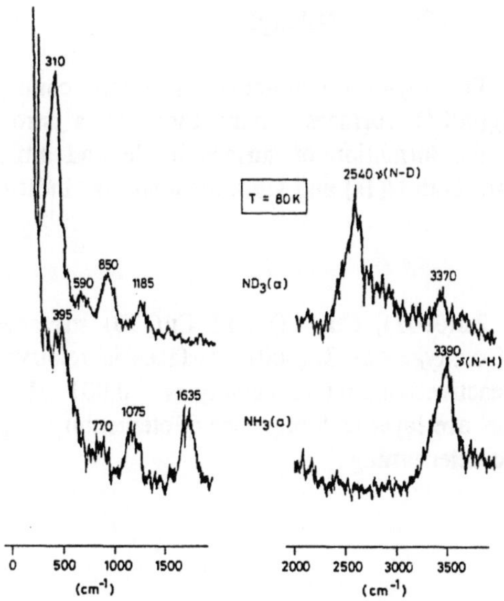

Fig. 4: HREEL spectra for NH_3(a) and ND_3(a) at Pb(110) surface at 80 K.

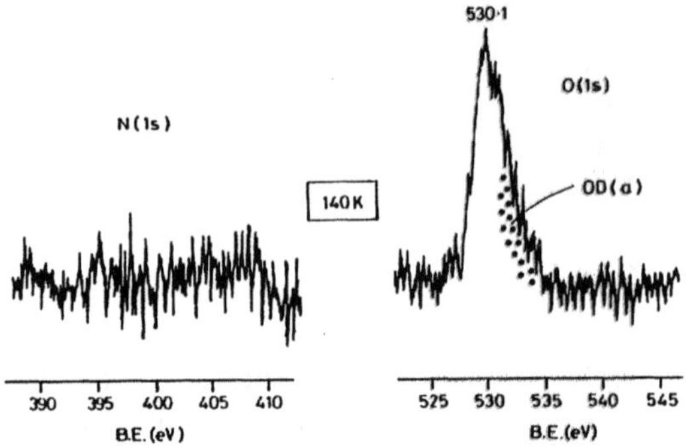

Fig. 5: N(1s) and O(1s) spectra regions for the ND_3-O_2 coadsorption experiment at 140 K (compare HREEL spectra Fig. 3). Note the absence of any surface nitrogen species although H-abstraction from NH_3 had occurred.

and an O(1s) spectrum for the *coadsorption* of dioxygen and ammonia at a Pb(110) surface at 80 K and warming to 140 K (in the case of ND_3) and 180 K (in the case of NH_3). The vibrational spectra provide unambiguous evidence for the formation of surface hydroxyl species - stable up to at least 180 K - with $\nu_{(O\text{-}D)}$ and $\nu_{(O\text{-}H)}$ values of 2650 cm^{-1} and 3625 cm^{-1}, respectively. In the O(1s) spectrum (Fig. 5) the hydroxyl species are reflected by the asymmetry to higher binding energy. Other than the loss at 520 cm^{-1} assigned to $\nu_{(Pb\text{-}O)}$ there are no other features present in the HREEL spectra. It is also particularly significant, and in accordance with the vibrational spectra, that there is no evidence for any intensity in the N(1s) region of the core-level spectrum. The spectra therefore support a mechanism where transient $O^{\delta-}$(s)-type species effect H(D)-abstraction from ammonia to generate relatively stable surface hydroxyls with the radical NH_2 species desorbing possibly after surface recombination:

$$O^{\delta-}(s) + NH_3(s) \Rightarrow OH(a) + NH_2(s)$$

$$2NH_2(s) \Rightarrow N_2H_4(g)$$

The molecular events in the Pb(110)-dioxygen-ammonia system contrast with Cu(111), Cu(110) and Mg(0001) surfaces where there is a strong thermodynamic driving force for the formation of surface imide and amide species - and indeed in these systems both N(1s) and vibrational spectra indicate their formation.

The ammonia-dioxygen system: Zn(0001), Cu(111) and Cu(110) surfaces. The dissociative chemisorption of dioxygen at Zn(0001) surfaces is relatively inefficient [7] and at 80 K the reactive sticking probability is ~0.001. If we envisage the formation of the 'oxide' overlayer as a sequence of elementary steps and that one of the following is rate-determining

$$O_2(g) \Rightarrow O_2(s) \overset{e}{\Rightarrow} O_2^{\delta-}(s) \Rightarrow O^{\delta-}(s) \overset{e}{\Rightarrow} O^{2-}(a)$$

then by using a carefully chosen probe-molecule - unreactive at both the clean Zn(0001) surface and the oxide overlayer - there is the possibility to explore the existence of surface transients such as $O_2^{\delta-}(s)$ and $O^{\delta-}(s)$ through a chemical trapping approach [5,7]. Taking note of the high reactivity of oxygen ions and ammonia in the gas phase and the inactivity of both the oxide overlayer and the clean Zn(0001) metal in ammonia chemisorption at low temperatures we chose to investigate the chemistry of the coadsorption of mixtures of $NH_3(g)$ and $O_2(g)$ at Zn(0001) and subsequently Cu(111) and Cu(110) surfaces.

In the case of Zn(0001) there was clear evidence with ammonia rich ammonia-dioxygen mixtures for the participation of a surface complex with the kinetic behaviour consistent with the ammonia-dioxygen complex participating as a precursor in the formation of surface hydroxyl, amide and chemisorbed oxygen species. Furthermore the reaction rate *increased* dramatically as the temperature *decreased* in the range 300 K to 80 K. The mechanism of the surface reaction was suggested to be

$$O_2^{\delta-}(s) + NH_3(s) \Rightarrow [O_2^{\delta-}(s){-}{-}{-}{-}NH_3](s) \Rightarrow OH(a) + NH_2(a) + O^{2-}(a)$$

with both XPS and HREELS establishing the nature of the chemisorbed species. This is essentially a reaction between transient species and analogous to a two dimensional gas reaction with a strong thermodynamic driving force from the formation of the chemisorbed species.

At a Cu(110) surface [13,14] exposed to an ammonia rich NH_3-O_2 mixture at 295 K the formation of surface imide species NH(a) occurred efficiently but without any evidence during the reaction for the presence of chemisorbed oxygen. Oxydehydrogenation occurred at an effectively atomically clean Cu(110) surface with loss-peaks at 3240 cm^{-1} ($\nu_{N\text{-}H}$), 780 cm^{-1} (δ_{NH}) and 430 cm^{-1} ($\nu_{Cu\text{-}NH}$) and an N(1s) peak at 397.4 eV. When the concentration of NH(a) approached 4×10^{14} cm^{-2} then evidence for chemisorbed oxygen emerged with an O(1s) peak at a binding energy of 530 eV. The Cu(111) surface exhibited similar chemistry and in this case Boronin et al. [13] established that the rates of surface imide formation and dissociative chemisorption of dioxygen were very similar suggesting that oxydehydrogenation was effected by a hot transient oxygen atom - the chemisorbed oxygen overlayer (for θ approaching unity) being less active and also restricted in the extent in which it could participate in the oxydehydrogenation reaction. The evidence is clear in that a distinction has to be made between oxygen atoms generated by dioxygen dissociation - and present at low coverage - and those of a chemisorbed layer and present at higher coverages. The occurrence of surface reconstruction with increasing oxygen coverage enables, through contributions from the Madelung potential, the oxygen atom to develop its full O^{2-}-like charge whereas at low oxygen coverages ($\theta \rightarrow 0$) we designate the oxygen adatom as $O^{\delta-}$ or O^{-}-like - and inherently highly reactive.

Activation of carbon monoxide at a Mg(1010) surface. Carbon monoxide is only weakly adsorbed (dispersion-type forces) at sp-metal surfaces and at a Mg surface there is no evidence for adsorbed carbon monoxide at 80 K after extensive exposure at 80 K (Fig. 6). However, when a mixture of carbon monoxide and dioxygen is exposed to the clean surface at 80 K - the oxygen content being about 1% - multiple peaks are observed in both the O(1s) and C(1s) regions of the spectrum for the same exposure to CO(g) (Fig. 6). On heating the adlayer, both the O(1s) and C(1s) spectra become less complex with essentially a single peak in each case at 473 K. The C(1s) peak at a binding energy of 291 eV we assign to a surface carbonate species CO_3(a) the intensity of which decreases with increasing temperature in the range 80 K to 493 K. That at 284.6 eV we attribute to surface carbon while that at ~287 eV to a $CO_2^{\delta-}$-type species. The corresponding O(1s) spectra support these assignments with the 533.7 eV peak corresponding to CO_3(a) which decomposes to surface oxide (531 eV) on heating to 473 K. These O(1s) and C(1s) peaks are virtually identical to those observed when carbon dioxide interacts with Mg(0001) at 295 K [15].

These observations are similar to those reported in the *coadsorption* of carbon monoxide and dioxygen at an aluminium [4] surface, the main difference being that with aluminium an additional peak was present due to a carbidic-type species $C^{\delta-}$(a) with a C(1s) binding energy of 282 eV. We suggest - as for aluminium - a step-wise mechanism involving a weakly adsorbed CO molecule present at very low surface coverage ($\theta < 0.1$) undergoing surface diffusion and

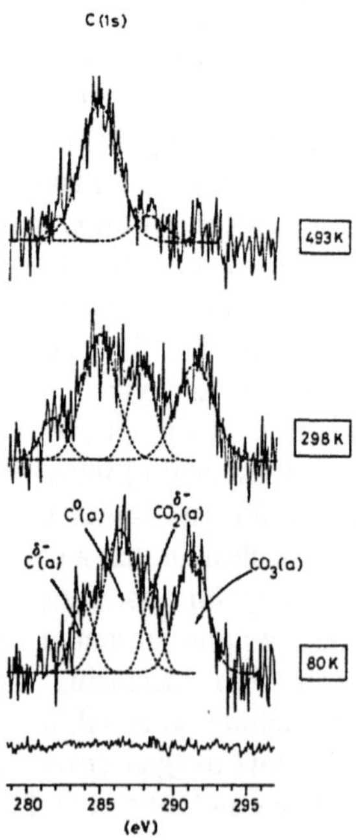

Fig. 6: C(1s) spectra for coadsorption of carbon monoxide and dioxygen (1%) at a Mg(1010) surface at 80 K followed by warming to 295 K and 493 K. Note the formation of CO_3(a), CO_2^-(a) and C(a) with binding energies of 291.7 eV, 286.6 eV and 284.6 eV, respectively. The Mg(1010) surface does not adsorb CO(g) at 80 K.

interacting with the transient oxygen adatom $O^{\delta-}(s)$.

$CO(g)$	$\Rightarrow$	$CO(s)$	: CO undergoing surface diffusion
$O_2(g)$	$\Rightarrow$	$O^{\delta-}(s)$	: transient O-adatom
$O^{\delta-}(s) + CO(s)$	$\Rightarrow$	$CO_2(s)$	: "two-dimensional gas" surface reaction
$CO_2(s) + O^{\delta-}(s)$	$\Rightarrow$	$CO_3(a)$	: carbonate formation
$O^{\delta-}(s)$	$\Rightarrow$	$O^{2-}(a)$	: oxide overlayer formation
$CO_3(a)$	$\Rightarrow$	$C^{o}(a) + 3O^{2-}(a)$	: carbonate decomposition

We have therefore a mechanism by which the strong carbon-oxygen bond in the CO molecule can be cleaved through the participation of an oxygen surface transient $O^{\delta-}(s)$ and under conditions where the thermodynamically stable MgO overlayer is unreactive to carbon monoxide. We make the important distinction - which we suggest is central to this and many other analogous reactions - between the $\delta-$ charge on the oxygen transient and the full 2e charge on the oxide-oxygen. This has been the central feature of our experimental studies of the reactivity of chemisorbed oxygen at atomically clean single crystal metal surfaces. A corollary of this is that studies of the metal-chemisorbed oxygen overlayer *per se* can be misleading in the context of attempts to understand the catalytic chemistry of oxygen at metal surfaces and illustrates the limitations of the static surface science approach.

4. The Implications of Coadsorption Studies

The evidence for the existence of oxygen transients: the methodology of chemical trapping. When a molecule collides with a surface it may undergo an elastic collision, alternatively it may transfer energy to the surface for some finite time. The time a molecule is trapped at the surface τ is an important concept since it can lead to adsorption - physical and chemisorption - surface diffusion, desorption or reaction with other species. In discussing the dynamics of adsorption we require such information as the enthalpy of adsorption, the kinetics of desorption, information on surface diffusion and structure. The more recently developed surface sensitive spectroscopies have in general, but not exclusively, been used in the static mode. In general we then extrapolate from a post-reaction type analysis to the situation under more realistic dynamic conditions.

An intuitively appealing picture of the adsorption process can be developed [5,7,16] in terms of the surface residence time τ. Two different residence times have to be distinguished. First we define τ the surface residence time as $\tau = \tau_o \exp(E_{des}/RT)$ where $\tau_o \sim 10^{-13}$ sec and E_{des} is the activation energy for desorption and secondly we have a "surface site" residence time $\tau_s = \tau_o \exp(E_{diff}/RT)$ where E_{diff} is the activation energy for surface diffusion or molecular hopping. The number of molecular hops N_τ made during time τ is $R_s\tau$ where R_s is the hopping rate given by τ_s^{-1} and which may be very large even though the finite surface coverage θ of the impinging molecule is "negligibly small". The value of θ is $N\tau$ where N is the molecular impingement rate and for $T = 295$ K, $E_{diff} = 5$ kJ mol^{-1} and $E_{des} = 20$ kJ mol^{-1} the value of τ is 3×10^{-6} sec and N_τ, the number of molecular hops in time τ, about 400. So several hundred surface sites may be visited by the hopping molecule during its residence time τ at the surface. The site residence time is therefore (in this case) about 10^{-8} sec but during this time an efficient reaction may occur (see below). It is the relationship between the activation energy of surface diffusion (hopping) and that of desorption that may determine the chemistry. In the coadsorption of dioxygen and ammonia it is the reaction

$$NH_3(s) + O^{\delta-}(s) \Rightarrow NH(a) + H_2O(g)$$

that leads to the formation of imide species at the Cu(110) surface, while at the Mg(0001) surface the reaction is $NH_3(s) + O^{\delta-}(s) \Rightarrow NH_2(a) + OH(a)$ with subsequent dehydrogenation of hydroxyl species to give chemisorbed oxygen. Detailed kinetic analysis making use of steady state assumptions for the transients designated by (s) are given elsewhere [5].

Recently scanning tunnelling microscopy has established [10], in an elegant and unequivocal way, the existence of transient or 'hot' oxygen adatoms. Dioxygen exposed to Al(111) surfaces has been shown to dissociate leading to isolated oxygen adatoms no closer than 80 Å apart at very low coverages ($\theta \sim 0.0014$). The life-time of these translationally hot adatoms - analogous to those we designated as $O^{\delta-}(s)$ - is estimated to be $> 10^{-12}$ sec. Recent studies [14] at Cardiff of the reactivity of oxygen adatoms at Cu(111) surfaces suggest that for concentrations as small as 5×10^{13} cm^{-2} ($\theta \sim 0.1$) not all the oxygen atoms are active in hydrogen abstraction from $NH_3(g)$ at 295 K. Yet up to 4×10^{14} cm^{-2} (i.e. $\theta \rightarrow 1.0$) of NH(a) species will form under dynamic or coadsorption conditions. Clearly for coverages as small as 0.1 a fraction of the oxygen adatoms are inactive in H-abstraction presumably since they have developed the full 2e charge associated with 'oxide' islands. Although we would not expect Cu(111) and Al(111) to behave identically it is interesting to note that for $\theta = 0.18$ (O_2 exposure of 72 L at 300 K) very few of the oxygen atoms are present at Al(111) as isolated adatoms whereas for an exposure of 20 L the largest fraction are present as isolated adatoms but with the rest mainly present as nuclei consisting of 2 and 3 oxygen adatoms. The evidence therefore points to a decrease in chemical reactivity (H-abstraction from NH_3) as the oxygen adatoms aggregate into relatively small islands.

What are the concentrations of the surface transients? Computer simulation studies [17] for the coadsorption of $NH_3(g)$ and $O_2(g)$ at a Mg(0001) surface have shown that the reaction can be modelled with a $NH_3(s)$ concentration of 5×10^8 molecules cm^{-2} - invariant during the reaction - and an $O^{\delta-}(s)$ concentration of about 2×10^9 cm^{-2}. For an E_{diff} value close to zero the value of $\tau_{O^{\delta-}(s)}$ is about 10^{-8}

sec under the experimental (pressure and temperature) conditions; the $O^{\delta-}(s)$ species become inactive when they aggregate into islands and develop the full 2e charge associated with the 'oxide' overlayer [5]. Analysis of the Zn(0001)-dioxygen-ammonia experimental data [18], where a precursor (NH_3---$O_2^{\delta-}$) complex provided a low energy pathway to dioxygen bond cleavage, suggests that the concentration θ_c of the precursor necessary to sustain the overall kinetics can be exceptionally small - as low as 10^5 molecules cm^{-2}! The analysis, however, is only illustrative as there are no good estimates available for kinetic parameters for surface diffusion or for ν, the vibrational frequency for decomposition of the precursor complex. According to the model [18] as ν decreases or E_{diff} increases, the value of θ_c would be expected to increase.

Heterogeneous Catalysis and Oxygen Transients. Selective oxidation is an area of immense industrial significance, and inherently complex with a variety of mechanisms proposed. Weng and Delmon [19] have discussed recently the role that individual components of multiphase oxide catalysts may have in oxidative dehydrogenation reactions. They discuss the concept of synergy arising from remote control - where one phase, the donor, dissociates dioxygen to generate a rapidly diffusing oxygen adatom which spills over to the second phase, the acceptor. It is these mobile oxygen adatoms that Delmon has established are active in dehydrogenation catalysis and where their surface life-time and concentration will be determined by such factors as those considered in our kinetic analysis of the coadsorption of gas mixtures where oxygen is one of the components.

A further illustration is the recent study of Boronin et al. at Novosibirsk of the silver-carbon monoxide-oxygen system [9]. These authors have investigated the chemistry of CO-O_2 (1%) mixtures over a wide range of temperature and pressure using a combination of thermal desorption spectroscopy and XPS. There are strong similarities between these studies and those of the aluminium-CO-O_2 [4] and magnesium-CO-O_2 systems (Fig. 6). The main conclusion is the high reactivity of oxygen transients to form surface carbonate prior to forming the relatively unreactive oxide overlayer. By changing the temperature, pressure and the composition of CO-O_2 gas mixture evidence for the participation of oxygen transients is established.

5. Review of the Present Position

There are compelling arguments now emerging from both experimental and theoretical studies for considering the chemistry of dioxygen at atomically clean metal surfaces as involving oxygen transients. Chemical reactivity can be related closely to the charge on the oxygen which depends on the surface coverage and whether or not the oxygen has become incorporated through place-exchange and surface reconstruction to develop an O^{2-}-like oxide species with the full negative charge through Madelung stabilisation. Oxygen adatoms - isolated or transients - we have designated as $O^{\delta-}(s)$ or $O^{-}(s)$. As Colbourn [20] has emphasised recently in a theoretical review of bulk oxide surfaces it is the charge on the surface oxygen that distinguishes edge or coordinatively unsaturated defective oxygen from lattice oxygen, O^{-}-like and O^{2-}-like, respectively, with the former exhibiting the higher chemical reactivity. We can therefore draw an analogy between the very reactive transient oxygen species present at metal surfaces and the coordinatively unsaturated oxygen present at defective oxide surfaces as studied by both Zechina [21] and his colleagues and more recently by Wu and Goodman [11]. There is, however, one essential difference in that the metal is a good thermodynamic sink for radicals generated by H-abstraction and indeed they can be characterised spectroscopically as stable chemisorbed species, whereas in the case of an oxide, radical recombination followed by desorption is frequently - but not always - the preferred pathway.

From the experimental evidence now available, and discussed in this paper, it is clear that oxygen present at "very low" surface coverages at metal surfaces can exhibit unique chemical reactivity. The Cu(110)-dioxygen-ammonia coadsorption system illustrates this principle, an oxydehydrogenation reaction to give a monolayer of NH(a) species being highly efficient and occurring at effectively an oxygen-free surface with no evidence for chemisorbed oxygen being present during the reaction. By comparison preadsorbed oxygen (for example $\theta > 0.2$) is very much less reactive. Brune et al. report [10] similarly unique reactivity to what they describe as "hot" oxygen atoms in the Al(111)-dioxygen-carbon dioxide system, results analogous to the earlier observations at Cardiff with polycrystalline aluminium surfaces [4]. Whether or not the chemistry observed with for example CO-O_2 (Fig. 6) or CO_2-O_2 mixtures [10] is uniquely attributable to the hot $O^{\delta-}(s)$ atom while undergoing surface diffusion or due to the *isolated oxygen adatom* after coming to rest in the chemisorbed state, is a question still to be resolved. In our kinetic analysis of the Mg(0001)-dioxygen-ammonia system [5] we defined $\tau_{O^{-}(s)}$ the surface life-time of the $O^{-}(s)$ transient as the time for which the transient existed at the surface under reaction conditions before becoming unreactive as an "oxide" island i.e. O^{2-}-like.

The examples discussed in this paper illustrate the need to take cognisance of the role that species present at unmeasurably small coverages can have in providing efficient routes in heterogeneous catalysis, reactions occurring in what is effectively a two-dimensional gas. Predictions based on the known reactivities of the individual components of a gaseous reactant mixture involving dioxygen will be at the best restrictive and at the worse misleading which emphasises a need to bridge the gap between the static surface science and the dynamic molecular beam approaches to the chemistry of metal surfaces.

Acknowledgements

We are grateful to SERC, the Royal Society and the British Council for their support.

References

[1] K. Kishi, M.W. Roberts: J. Chem. Soc. Faraday Trans. **71**, 1721 (1975); R.W. Joyner, K. Kishi, M.W. Roberts: Proc. Roy. Soc. (London) **A358**, 223 (1977); C.T. Au, J. Breza, M.W. Roberts: Chem. Phys. Lett. **66**, 340 (1979)

[2] For example: P.A. Thiel, T.E. Madey: Surf. Sci. Reports **7**, 211 (1987); E.M. Stuve, R.J. Madix, B.A. Sexton: Surf. Sci. **111**, 11 (1981); K. Bange, D.E. Grider, T.E. Madey, J.K. Sass: Surf. Sci. **137**, 18 (1984)

[3] E. Shustorovich, A.T. Bell: Surf. Sci. **268**, 397 (1992)

[4] A.F. Carley, M.W. Roberts: J. Chem. Soc. Chem. Comm. **355** (1987)

[5] C.T. Au, M.W. Roberts: Nature **319**, 206 (1986); J. Chem. Soc. Faraday Trans. I **83**, 2047 (1987)

[6] C.T. Au, Li Xing-Chang, Tang Ji-an, M.W. Roberts: J. Catal. **106**, 538 (1987)

[7] A.F. Carley, Song Yan, M.W. Roberts: J. Chem. Soc. Chem. Comm. **267** (1988); J. Chem. Soc. Faraday Trans. I **86**, 2701 (1990); M.W. Roberts: Chem. Soc. Rev. 18, 451 (1989); J. Mol. Catal. **74**, 11 (1992)

[8] A.F. Carley, S. Rassias, M.W. Roberts: Surf. Sci. **135**, 35 (1983)

[9] V.I. Bukhtiyarov, A.I. Boronin: (personal communication)

[10] H. Brune, J. Wintterlin, R.J. Behm, G. Ertl: Phys. Rev. Lett. **68**, 624 (1992)

[11] M.-C. Wu, D.W. Goodman: Catal. Lett. **15**, 1 (1992)

[12] C.T. Au, M.W. Roberts: Chem. Phys. Lett. **29**, 55 (1980)

[13] B. Afsin, P.R. Davies, A. Pashuski, M.W. Roberts: Surf. Sci. **259**, L724 (1991); A. Boronin, A. Pashuski, M.W. Roberts: Catal. Lett. **16**, 345 (1992)

[14] B. Afsin, P.R. Davies, A. Pashuski, M.W. Roberts, D. Vincent: Surf. Sci. **284**, 109 (1993)

[15] R.G. Copperthwaite, P.R. Davies, M.A. Morris, M.W. Roberts, R.A. Ryder: Catal. Lett. **1**, 11 (1988)

[16] M.W. Roberts: Catal. Today **12**, 501 (1992)

[17] P.G. Blake, M.W. Roberts: Catal. Lett. **3**, 379 (1989)

[18] A.F. Carley, M.W. Roberts, M. Tomellini: J. Chem. Soc. Faraday Trans. **87(21)**, 3563 (1991)

[19] T. Weng, B. Delmon: Appl. Catal. A: **81**,141 (1992)

[20] E.A. Colbourn: Surf. Sci. Reports **15**, 284 (1992)

[21] E. Borello, S. Coluccia, A. Zechina: J. Catal. **93**, 331 (1985)

Part VI

Application-Oriented Studies

Adsorption of Carbon Monoxide on Catalyst Surfaces: Oxides, Sulfides and Zeolites

H. Knözinger

Institut für Physikalische Chemie, Universität München, Sophienstrasse 11, D-80333 München 2, Germany

Abstract. Examples are discussed for the application of low-temperature FT infrared transmission absorption spectroscopy of adsorbed CO for the characterization of adsorption sites on catalysts such as binary oxides, supported oxides and sulfides, and zeolites. The nature of cation sites, their oxidation state and their coordination environment can be analyzed. The H-bonding characteristics, particularly the O-H stretching frequency shifts induced by CO adsorption on surface hydroxyl groups permits a relative acidity scale of these groups to be established.

1. Introduction

Real heterogeneous catalysts as applied in technical processes, are typically complex polydisperse materials that develop high specific surface areas and micro- and macroporosity [1-3]. As a consequence, many surface science techniques can either not at all or only with limited success be applied for the characterization of heterogeneous catalysts.

A challenge of great importance in the science of catalysis is the elucidation of the nature and properties of adsorption and potential active sites on the surfaces of heterogeneous catalysts. Among such sites on non-metallic catalysts are the acid sites that are responsible for the extremely important class of acid-catalyzed reactions. Two types of acid sites are distinguished, namely aprotonic (Lewis) acid sites and protonic (Brönsted) acid sites. The former consist of coordinatively unsaturated (cus) cations which are exposed e.g. on oxide or sulfide surfaces, while the latter are typically described as surface hydroxyl groups with sufficiently high proton donor strength. The acid properties of heterogeneous catalysts can conveniently be studied by transmission infrared absorption spectroscopy using probe molecules [4-6]. The most widely used probe molecules for this approach are certainly ammonia and pyridine, which are very strong bases and which are therefore not very specific [6]. An alternative probe molecule for the characterization of acid sites and coordination sites in general on oxides and sulfides is carbon monoxide CO [6]. This molecule - when adsorbed at low temperature, e.g. liquid N_2 temperature - readily coordinates to cus cation sites and forms hydrogen bonds with surface OH groups. It thus provides information via vibrational spectroscopy about the nature and properties (including acid properties) of these centers. A few examples of catalytic relevance including simple binary oxides, supported oxides, supported sulfides, and zeolites, will be described in the following.

2. Adsorption of CO on Cationic Sites

General Considerations. The frequency shift of the C-O stretching mode induced by coordination of the CO molecule to metal centers can be interpreted in terms of the bonding mechanism: The 5σ orbital of CO forms a donor bond to the metal with prevalent σ character. Considerable contributions to the stabilization of the metal-carbon bond come from the participation of metal d-orbitals and the antibonding π* orbitals of the ligand (π-back donation). It is the increased electron density in the π* orbital due to this back donation which leads to a decrease in the C-O force constant and stretching frequency, this effect overcompensating for an increase in the force constant by formation of the σ-bond. When, however, metal centers with zero or negligible d-electron densities (cations) are involved, the σ-bond becomes the dominating bond. Comparably weak bond energies and carbonyl stretching frequencies higher than the gas phase frequency are the consequence. Theoretical considerations by Pacchioni et al. [7,8] have shown that the bonding of CO on MgO and NiO is entirely electrostatic in nature and the positive frequency shift of the C-O stretching mode was attributed to a wall effect (stretching motion in the presence of a rigid surface) with only minor electric field contributions.

It has also been shown that adsorbate-adsorbate interactions between CO oscillators can affect the C-O stretching frequency of adsorbed CO molecules on oxide surfaces provided the coordination sites are located in sufficiently close proximity [9,10]. These interactions are principally due to both dynamic dipole-dipole interactions and static effects. In the absence of image charges (non-metals), the dynamic effect is restricted to direct (through space) dipole-dipole coupling causing high-frequency shifts. These are significant only when d-electrons are involved in the bonding (d-π* contributions). In many oxide systems the effect is therefore negligible. The static adsorbate-adsorbate interactions contain two contributions, namely solvent-like effects [9,11] and "chemical" effects. The solvent effect leads to comparably small shifts of the carbonyl stretching frequency to lower wavenumbers [9] whereas the "chemical" effect can be substantial also producing negative shifts. It was suggested that this latter effect is brought about by inductive effects, i.e. reduction of the electron-acceptor strength of a cus surface metal ion after coordination of CO onto adjacent sites [9]. Again, this effect must be a short-range effect and indicates, if observed, the close proximity of adjacent sites. Both, the solvent-like and the "chemical" interactions are insensitive toward isotope dilution and can thus be separated from dynamic dipole-dipole coupling.

For studies of the intrinsic properties of the coordination sites, spectral data are desirable which are free of contributions from these adsorbate-adsorbate interactions. Measurements at low coverage are therefore recommendable [12].

Adsorption of CO on Oxides. Adsorption of CO on oxide surfaces (preferably at low temperatures) typically shows this behaviour with the carbonyl frequencies varying between the gas phase frequency and values as high as 2240 cm^{-1}, depending on the type of oxide and its pretreatment. It can thus be anticipated that CO is in fact a useful probe for the nature of the coordination site (acid site) and the strength of the coordination bond (acid strength). Among the systems studied are the binary oxides Al_2O_3 [9,13-18], MgO [17,18], TiO_2 [9,17-21], Cr_2O_3 [10,12,22] and more complex multicomponent systems such as SiO_2-Al_2O_3 [13,17], vanadium oxide supported on TiO_2 [23], tungsten oxide supported on TiO_2 [24] and silica-titania mixed oxides [25].

The potential of CO for discrimination of adsorption sites in different coordinations was nicely demonstrated for transition aluminas , which expose cus Al^{3+} in octahedral and tetrahedral sites. A carbonyl band observed near 2190 cm^{-1} was attributed to CO coordinated to an octahedral Al^{3+} center [12,13] while carbonyl bands beyond 2200 cm^{-1} were related to CO coordinated to tetrahedral Al^{3+} sites [12,13,15,26]. Only the high frequency species was observed on SiO_2 - Al_2O_3 [12,13,17] which exclusively exposes tetrahedral Al^{3+} cations. As the Lewis acid strength of a cation likely increases with decreasing coordination, the relative acid strength of these sites can also be deduced from the C-O frequency shift relative to the gas phase frequency at 2143 cm^{-1}.

Similarly, Cr^{3+}_{3c} and Cr^{3+}_{5c} sites with three (subscript 3c) and five (subscript 5c) O^{2-} ligands, respectively, on reduced α-Cr_2O_3 surfaces were characterized by C-O stretching frequencies of 2184 and 2162 cm^{-1}, respectively [10,27].

Although high surface area porous oxides are usually not well-defined as regards the surface structure and hence, the local environment of a particular site, some oxide materials can be prepared in microcrystalline form. The microcrystals may then develop crystallographically well-defined surface faces as can be shown by transmission electron microscopy. Knowledge of the most abundant faces and hence, of the coordination environment of an exposed cation site, can be used advantageously for the interpretation of infrared spectra of chemisorbed species. This approach has been applied very successfully by Zecchina and his coworkers [10,27].

An example of microcrystalline α-Cr_2O_3 is shown in Fig. 1. The particle morphology is similar to that reported earlier by Scarano et al. [9] and by Zaki and Knözinger [12]. The exposed faces appear to be flat and well developed and should preferentially consist of the basal hexagonal (0001) and the prismatic $(10\bar{1}0)$ faces.

The potential of CO for discrimination of coordination sites consisting of metal ions in various oxidation states, can also be demonstrated for chromia when it is treated in reducing and oxidizing atmospheres [12]. Carbonyl stretching frequencies that have been observed on low-temperature adsorption of

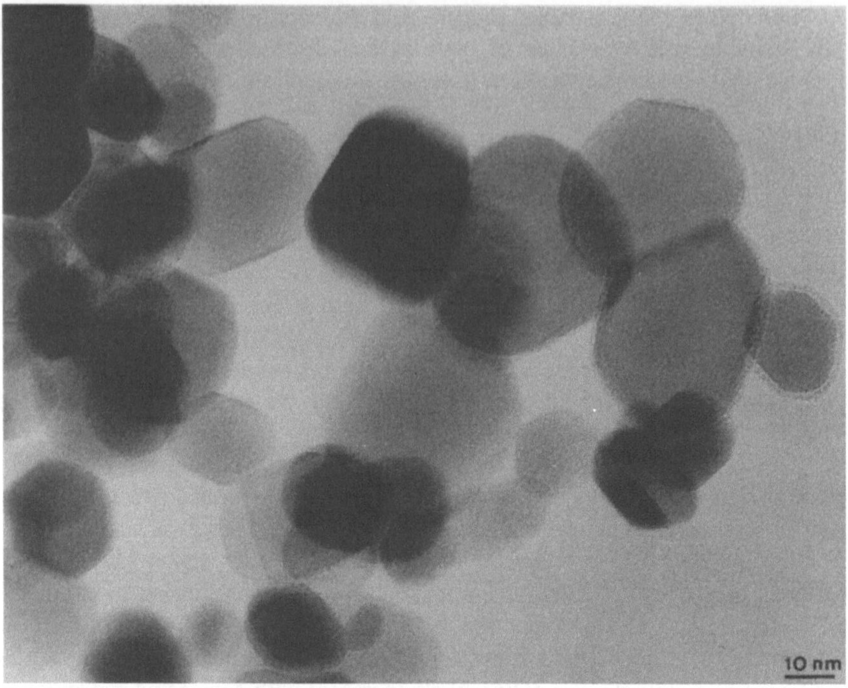

Fig. 1: Transmission electron micrograph of α-Cr_2O_3.

CO on these α-Cr_2O_3 samples are summarized in Table 1 together with the suggested assignments for the nature of the coordination site. Hence, the dominant chromium oxidation state turns out to be Cr^{3+} in different coordination environments on the reduced surface. In contrast, thermally rather stable adsorption complexes with Cr^{5+}_{5c} sites are formed on the oxygen-rich surface. In this study it was assumed that the CO ligand is chemisorbed end-on on the Cr^{x+} sites. It should be noted that Kuhlenbeck et al. [28] observed a C-O stretching mode at 2167±12 cm^{-1} in EELS spectra when CO was adsorbed on an epitaxially grown Cr_2O_3 (111) surface. Based on ARUPS results, these authors were able to demonstrate that the corresponding adsorbed species should be tilted toward the surface with the molecular axis oriented approximately parallel to the surface.

Table 1: Infrared carbonyl frequencies of CO on reduced and oxidized α-Cr_2O_3.

CO Streching frequency / cm^{-1}		Coordination
Reduced (H_2 / 770 K)	Oxidized (O_2 / 570 K)	Site
2164	2162	Cr^{3+}_{5c}
2176 - 2187	2182	Cr^{3+}_{3c}
	2200	Cr^{5+}_{5c}

It has been shown by theoretical considerations [29] that the carbonyl stretching frequency should monotonously increase with the electric field strength at the cation (d^0 electron configuration) center. Several examples for such correlations have in fact been reported in the literature [12,13,17,30-32]. Zaki and Knözinger [12,13] have pointed out that oxidation states and coordination numbers of cation centers can be estimated from experimental carbonyl stretching frequencies by means of a universal frequency/field strength correlation, provided the carbonyl stretching frequencies are free of shifts due to coverage-dependent dynamic and static lateral interactions. This can most easily be achieved by using frequencies which are determined by extrapolation to zero-coverage [12,13].

Studies of the multicomponent systems, e.g. tungsten oxide supported on TiO_2, are particularly interesting, since here the infared spectra of chemisorbed CO permit the differentiation between different elements being exposed, e.g. W^{6+} and Ti^{4+} [24]. Figure 2 shows carbonyl infrared spectra of CO adsorbed on TiO_2 and on WO_x/TiO_2. The band at 2180 cm^{-1} is characteristic of $Ti^{4+}\leftarrow CO$ complexes, that at 2196 cm^{-1} for $W^{6+}\leftarrow CO$ complexes [24]. The relative acid strength of the two sites can be qualitatively characterized by the temperature and pressure dependence of the carbonyl band intensities, and the relative abundance of the two types of sites is correlated with their intensity ratios at saturation coverage.

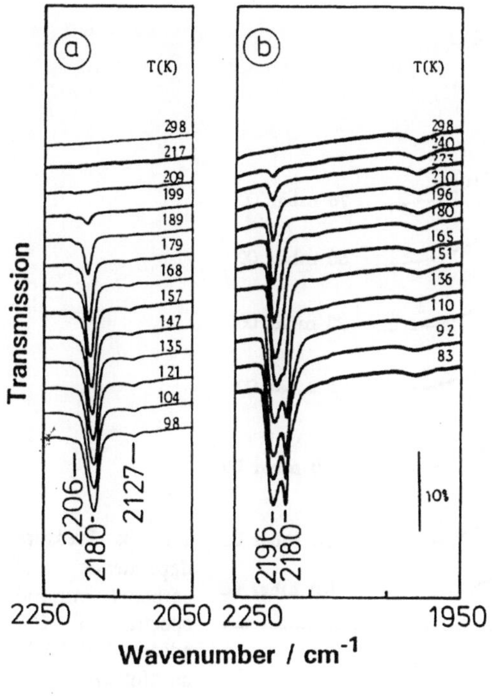

Fig. 2: Temperature dependence of carbonyl infrared spectra of CO (70 mbar) adsorbed on (a) TiO (anatase), and (b) 9 wt% WO_3/TiO_2.

Adsorption of CO on Sulfides. Sulfided molybdena-on-alumina catalysts are technologically important materials which are used at large scale in hydrotreating processes for upgrading crude oil [1,33]. The supported sulfide phases are described as small MoS_2 platelets. Adsorption sites on the supported sulfide phases can be probed by CO adsorption. Figure 3 shows transmission-absorption infrared spectra in the carbonyl stretching region of CO adsorbed at 80 K on a sulfided Mo/Al_2O_3 catalyst [34-36]. The dominant band in the spectra appears at 2100 cm^{-1}. This band and a second less intense band near 2191 cm^{-1} saturate at low CO pressures while the central band at 2156 cm^{-1} is pressure dependent and can easily be removed on evacuation at 80 K. The latter band is due to H-bonded CO molecules (vide infra). The high frequency component at 2191 cm^{-1} is due to Al^{3+}←CO complexes on the support (vide supra) and may contain contributions from Mo^{4+}←CO complexes [34-36]. The same band position is observed on H_2-reduced Mo/Al_2O_3 catalysts for which XPS shows the presence of Mo^{4+} sites. It is interesting to note that the frequency of the stretching mode of this CO ligand being coordinated to a $Mo^{4+}(d^2)$ center is still greater than the gas phase frequency at 2143 cm^{-1}, although some π-back donation may occur. The strongest carbonyl band at 2100 cm^{-1} in the spectra of Fig. 3 is in contrast clearly shifted to a position below the gas phase frequency ($\Delta\nu$ = -43 cm^{-1}). This suggests that the

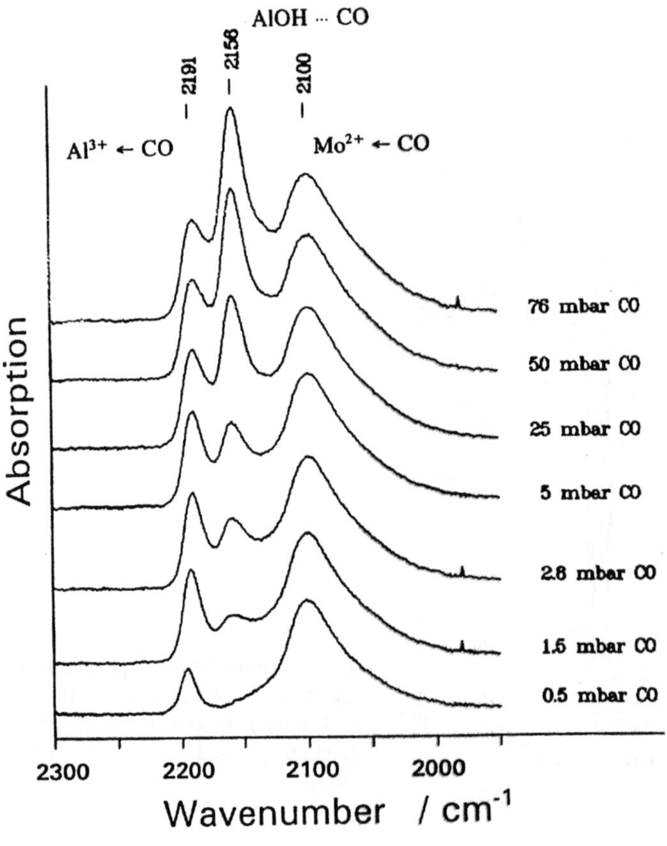

Fig. 3: Pressure dependence of carbonyl infrared spectra of CO adsorbed at 80 K on MoS_2/Al_2O_3.

corresponding coordination site must be a low-valent Mo site providing significant d-electron density which leads to an appreciable π-back donation in the coordination bond. It has been suggested [34-36] that the coordination site is to be identified as a $Mo^{2+}(d^4)$ center which is presumably located along the edges of MoS_2-slabs.

3. Adsorption of CO on Protonic Sites

Binary and multicomponent oxides typically contain surface hydroxyl groups [4] and zeolites may contain protons as charge-compensating ions [37,38]. The O-H stretching frequency of unperturbed hydroxyl groups depends on the electronegativity of the coordination center of the group, the number of cationic coordination centers it is bonded to, the oxygen coordination number of these centers and the overall bond character of the particular material. The proton donor strength of a hydroxyl group should increase with increasing polarizing power and number of coordination centers and, hence, with decreasing O-H stretching frequency. However, the absolute frequency cannot be taken as a direct measure of the proton donor strength or acid strength of OH groups when comparing different types of materials, although the frequencies of different OH configurations on one oxide material may with some caution be taken as a relative measure of an acid strength sequence, as e.g. on aluminas [4,39].

A possible approach toward the analysis of acid properties of surface OH groups uses their H-bonding ability. H-bonding will induce characteristic shifts of internal vibrations of the probe, but more importantly, typically rather significant shifts occur in the O-H stretching region $\Delta\nu_{OH}$ accompanied by the characteristic broadening of the O-H band and increase in integrated band intensity [40,41]. Carbon monoxide does form O-H...CO H-bonding complexes at low temperatures although the H-bond energies are low [11-13].

CO adsorption on silica has been studied extensively by Ghiotti et al. [42] and by Beebe et al. [11]. An H-bonded complex of the type

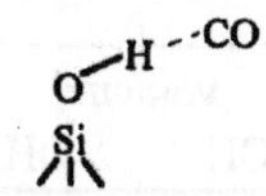

is undoubtedly formed at 77 K as indicated by the decrease in intensity of the band of unperturbed OH groups and the appearance of an intense band shifted to lower frequency by 93 cm^{-1} [11]. Simultaneously, the CO stretching mode was observed at 2158 cm^{-1}, i.e. shifted by 15 cm^{-1} to higher frequency as compared to the gas phase frequency of CO.

Aluminas are amphoteric and due to their structural characteristics with Al^{3+} cations in octahedral and tetrahedral coordination, their surfaces contain five different OH configurations which give rise to five individual O-H stretching bands [4,39]. Table 2 summarizes the frequencies of these five bands, which are attributed to isolated OH groups of types I, II and III being coordinated to one,

Table 2: Hydroxyl band positions (cm^{-1}) of free and associated OH groups on γ-Al_2O_3 before and after CO adsorption at 77 K.

Isolated OH's Type I	Type II	Type III	Associated OH's (presence of CO)	$\Delta\nu_{OH}$
3785 (b)			3785	-
3775 (a)			3775	-
	3725 (b)		3655	70
	3715 (a)		3635	80
		3695	3600	95

two and three Al^{3+} ions, respectively [4,39]. Due to the presence of Al^{3+} ions in octahedral and tetrahedral sites, there are two forms of type I (a and b) and of type II (a and b) OH groups. Simple estimations based on Pauling's electrostatic valence rule [39] suggested increasingly positive charge on the OH groups with decreasing O-H stretching frequency.

The data of Table 2 indicate that after admission of CO the two bands at 3785 and 3775 cm^{-1} remain entirely unaffected, whereas the three bands at lower frequency, namely at 3725, 3715 and 3695 cm^{-1}, are shifted to lower frequencies with band positions for the associated OH groups at 3655, 3635 and 3600 cm^{-1}. The relative intensities of individual bands in the two triplets permit a correlation to be made between the unperturbed and associated OH groups. The observed frequency shifts $\Delta\nu_{OH}$ are also summarized in Table 2. The results indicate - in agreement with the sequence of acid strengths predicted on the basis of the estimated net charges of the free hydroxyls - that the low frequency bands are representing the stronger H-bond donor groups. Moreover, the weak interaction with CO permits the distinction between AlOH...CO surface complexes of only slightly different H-bond energies. The increasing frequency shift for the individual associated complexes as the O-H stretching frequency of the

Table 3: O-H stretching frequencies of zeolites prior to and after CO adsorption at 77 K and the corresponding frequency shifts $\Delta\nu_{OH}$

Zeolite	SiO_2/Al_2O_3	ν_{OH}/cm^{-1} free OH	OH...CO	$\Delta\nu_{OH}/cm^{-1}$	Ref.
[Al]-HZSM-5	24	3620	3305	315	43
	13.6	3617	3305	312	44
	26.8	3619	3290	339	45
	41.5	3617	3295	322	45
[Ga]-HZSM-5	-	3622	3330	282	45
$H_{70}Na_{30}$-Y		3652	3377	275	43
$H_{95}Na_5$-Y		3643	3347	296	44
$H_{70}Na_{30}$-Y		3647	3369	278	44
$H_{20}Na_{80}$-Y		3650	3385	265	44
$H_{90}Na_{10}$-Y		3640	3355	285	46

unperturbed groups decreases, provides strong support for the increasing H-bond donor strength with increasing coordination number of the OH group. The $\Delta\nu_{OH}$-value 95 cm^{-1} for the low-frequency type III hydroxyl groups is almost identical with the frequency shift for silanol groups and, hence, suggests that the corresponding OH groups should possess an H-bond donor strength (acid strength) comparable to that of SiOH groups.

An interesting case is CO adsorption on zeolites. The acidic OH groups in zeolites are believed to bridge an Al^{3+} and a Si^{4+} center: Si(OH)Al [37,38]. H-bonding occurs when CO is adsorbed at 77 K and leads to $\Delta\nu_{OH}$ shifts of 250 to 340 cm^{-1}, depending on the type of zeolite and its composition. Simultaneously, the CO stretching mode shifts to higher frequency relative to the gas phase frequency by as much as 30 to 35 cm^{-1}. Table 3 summarizes reported data on CO adsorption on HNa-Y, [Al]-HZSM-5 and [Ga]-HZSM-5. The spectra in the O-H stretching regime of [Al]-HZSM-5 and [Ga]-HZSM-5 are compared in Fig. 4, which - together with the data in Table 3 - clearly demonstrates that the $\Delta\nu_{OH}$-value measured upon CO adsorption for the gallium - form of ZSM-5 is significantly lower than that obtained for the aluminium-form consistent with the known relative acid strength of these materials.

The examples presented clearly demonstrate the potential of CO as a highly specific probe molecule for the characterization of protonic acidity on surfaces, which permits a ranking of acid strength to be established on the basis of $\Delta\nu_{OH}$ shifts.

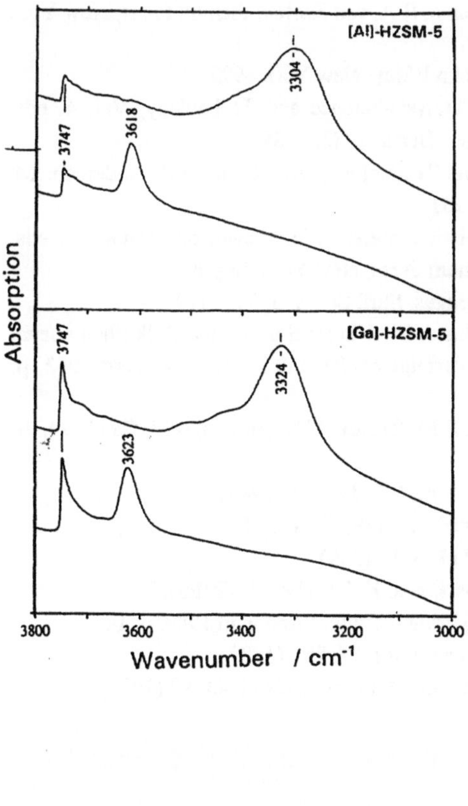

Fig. 4: Hydroxyl infrared spectra of [Al]-HZSM-5 and [Ga]-HZSM-5 at 80 K prior to (lower spectra) and after (upper-spectra) adsorption of CO.

4. Conclusions

Carbon monoxide is a probe molecule of high potential for the characterization of real catalyst surfaces. It provides information on the nature of cus cations on complex oxide and sulfide catalysts, on the oxidation state and coordination environment of these coordination sites. Furthermore, CO undergoes H-bonding interactions with surface hydroxyl groups, and frequency shifts of O-H stretching bands resulting from this interaction can be used as a measure of the relative Brönsted acid strength of OH groups on various catalyst surfaces.

Acknowledgements

The studies reported in this review were financially supported by the Deutsche Forschungsmeinschaft (Sonderforschungsbereich 338) and by the Fonds der Chemischen Industrie. The electron microscopy (Fig.1) was carried out by Dr. B. Tesche, Fritz-Haber-Institut der Max-Planck-Gesellschaft, Berlin.

References

[1] B.C. Gates, J.R. Katzer, and G.C.A. Schuit: Chemistry of Catalytic Processes, McGraw-Hill, New York 1979

[2] H. Knözinger in: Fundamental Aspects of Heterogeneous Catalysis Studied by Particle Beams, eds. H.H. Brongersma, R.A. van Santen, Plenum Press, New York 1991, p. 7

[3] B.C. Gates: Catalytic Chemistry, John Wiley, New York 1992

[4] H.-P. Boehm, H. Knözinger in: Catalysis-Science and Technology, Vol. 4, eds. J.R. Anderson, M. Boudart, Springer, Berlin 1983, p. 39

[5] K. Tanabe in: Catalysis-Science and Technology, Vol. 2, eds. J.R. Anderson, M. Boudart, Springer, Berlin 1981, p. 231

[6] H. Knözinger in: Elementary Reaction Steps in Heterogeneous Catalysis, eds. R.W. Joyner, R.A. van Santen, Plenum Press, New York, in print

[7] G. Pacchioni, G. Cogliandro, P.S. Bagus: Surf. Sci. **225**, 344 (1991)

[8] G. Pacchioni and P.S. Bagus in: Cluster Models for Surface and Bulk Phenomena, eds. G. Pacchioni, P.S. Bagus, F. Parmigiani, Plenum Press, New York 1992, p. 305

[9] A.A. Tsyganenko, L.A. Denisenko, S.M. Zverev, V.N. Filimonov: J. Catal. **94**, 10 (1985)

[10] D. Scarano, A. Zecchina, A. Reller: Surf. Sci. **198**, 11 (1988)

[11] T.P. Beebe, P. Gelin, J.T. Yates: Surf. Sci. **148**, 526 (1984)

[12] M.I. Zaki, H. Knözinger: J. Catal. **119**, 311 (1989)

[13] H. Knözinger in: Acid-Base Catalysis, eds. K. Tanabe, H. Hattori, T. Yamaguchi, T. Tanaka, Kodansha, Tokyo, and Verlag Chemie, Weinheim 1989, p. 147

[14] M.I. Zaki, H. Knözinger: Mater. Chem. Phys. **17**, 201 (1989)

[15] G. Della Gatta, B. Fubini, G. Ghiotti, C. Morterra: J. Catal. **43**, 90 (1976)

[16] A. Zecchina, E. Escalona Platero, D. Otero Arean: J. Catal. **107**, 244 (1987)
[17] M.I. Zaki, H. Knözinger: Spectrochim. Acta **43 A**, 1455 (1987)
[18] R.I. Soltanov, E.A. Paukshtis, E.N. Yurchenko: Kinet. Catal. (Engl.) **23**, 135 (1982)
[19] K. Tanaka, J.R. White: J. Phys. Chem. **86**, 4708 (1982)
[20] G. Busca, H. Saussey, O. Saur, J.-C. Lavalley, V. Lorenzelli: Appl. Catal. **14**, 245 (1985)
[21] C. Morterra: J. Chem. Soc. Faraday Trans. I **84**, 1617 (1988)
[22] A. Zecchina, S. Coluccia, E. Guglielminotti, G. Ghiotti: J. Phys. Chem. **75**, 2774 (1971)
[23] G. Busca: Langmuir **2**, 577 (1986)
[24] F. Hilbrig, H. Schmelz, H. Knözinger: Proc. 10th Intern. Congr. Catal., Budapest 1992, in print
[25] C.U.I. Odenbrand, S.L.T. Andersson, L.A.H. Andersson, J.G.M. Brandin, G.Busca: J. Catal. **125**, 541 (1990)
[26] C. Morterra, S. Coluccia, E. Garrone, G. Ghiotti: J. Chem. Soc. Faraday Trans. I **75**, 289 (1979)
[27] D. Scarano, A. Zecchina: Spectrochim. Acta **43 A**, 1441 (1987)
[28] K. Kuhlenbeck, C. Xu, B. Dillmann, M. Haßel, B. Adam, D. Ehrlich, S. Wohlrab, H.-J. Freund, U.A. Ditzinger, H. Neddermeyer, M. Neuber, M. Neumann: Ber. Bunsenges. Phys. Chem. **96**, 15 (1992)
[29] N.S. Hush, M.L. Williams: J. Mol. Spectrosc. **50**, 349 (1974)
[30] C.L. Angell, P.C. Schaffer: J. Phys. Chem. **70**, 1413 (1966)
[31] P.G. Harrison, E.W. Thornton: J. Chem. Soc. Faraday Trans. I **74**, 2703 (1978); and **75**, 1487 (1979)
[32] R. Larsson, R. Lykvist, B. Rebenstorf: Z. Phys. Chem. (Leipzig) **263**, 1089 (1982)
[33] H. Knözinger: Proc. 9th Intern. Congr. Catal., Calgary 1988, Vol. 5, eds. M.J. Phillips, M. Ternan, The Chemical Institute of Canada, Ottawa 1989, p. 20
[34] E. Delgado, G.A. Fuentes, C. Hermann, G. Kunzmann, H. Knözinger: Bull Soc. Chim. Belg. **93**, 735 (1984)
[35] M.I. Zaki, B. Vielhaber, H. Knözinger: J. Phys. Chem. **90**, 3176 (1986)
[36] B. Müller, A.D. van Langeveld, J.A. Moulijn, H. Knözinger: submitted to J. Phys. Chem.
[37] P.A. Jacobs: Catal. Rev.-Sci. Eng. **24**, 415 (1982)
[38] J.A. Rabo, G.J. Gajda: Catal. Rev.-Sci. Eng. **31**, 385 (1989/90)
[39] H. Knözinger, P. Ratnasamy: Catal. Rev.-Sci. Eng. **17**, 31 (1978)
[40] G.C. Pimentel, A. McClellan: The Hydrogen Bond, Freeman, San Francisco 1960
[41] H. Knözinger in: The Hydrogen Bond. Vol. 3, eds. P. Schuster, G. Zundel, C. Sandorfy, North Holland, Amsterdam 1976, p. 1263
[42] G. Ghiotti, E. Garrone, C. Morterra, F. Boccuzzi: J. Phys. Chem. **83**, 2863 (1979)
[43] L.M. Kustov, V.B. Kazansky, S. Beran, L. Kubelkova, P. Jiru: J. Phys. Chem. **91**, 5247 (1987)
[44] L. Kubelkova, S. Beran, J.A. Lercher: Zeolites **9**, 539 (1989)
[45] I. Mirsojew, S. Ernst, J. Weitkamp, H. Knözinger: submitted to J. Phys. Chem.
[46] N. Echoufi, P. Gélin: J. Chem. Soc. Faraday Trans. I **88**, 1067 (1992)

Schottky Barriers and Ohmic Contacts with Pt/TiO_2(110): Implications to Control Gas Sensor Properties

K.D. Schierbaum, X. Wei-Xing, S. Fischer, and W. Göpel*

Institut für Physikalische and Theoretische Chemie, Universität Tübingen, Auf der Morgenstelle 8, D-72076 Tübingen, Germany
*Department of Chemistry, Nanjing University, Nanjing, China

Abstract. We have studied Pt overlayers on a thermodynamically stable oxide surface between submono- and multilayer coverages for different oxygen stoichiometries. The latter were adjusted by different reducing and oxidizing conditions in the gas phase. Surface spectroscopic techniques and current-voltage measurements have been used to determine the structures, stabilities, and electrical interface properties of Pt adatoms, clusters, contacts, and Pt ions at and below the TiO_2(110) surface. At temperatures below 700 K, metallic Pt exhibits electron acceptor properties and causes Schottky barrier formation at the Pt/TiO_2 interface. At temperatures above 1000 K, Pt ions migrate into subsurface layers. The ions act as bulk donors and cause a switch from Schottky barrier towards Ohmic contact behaviour of electronic transport across the Pt/TiO_2 interface.

1. Introduction

Semiconducting gas sensors based upon oxides such as SnO_2 and TiO_2 monitor changes in the conductance during their interaction with molecules. Figure 1 shows schematically different possible elementary steps of such "electronic conductance sensors". The specific detection of certain molecules is possible by a careful choice of the oxide with a certain geometric structure and doping, of the contact geometry, of the operation temperature, and of the frequency in AC measurements [1].

Low-temperature reactions consist of adsorption and catalytic reactions at active sites. The latter may involve *intrinsic* point defects, e.g., oxygen vacancies and/or *extrinsic* point defects, e.g., metal adatoms at the surface. Similar reactions occur at grain boundaries or at three-phase boundaries, e.g., at metal contacts or clusters. Most of these reactions involve adsorbed negatively charged molecular (O_2^-) or atomic (O^-) oxygen species. All of these reactions occur at the surface [2].

In contrast to this, *high-temperature reactions* involve the bulk of the oxide

Springer Series in Surface Sciences, Vol. 33 Eds.: E. Umbach · H.-J. Freund
Adsorption on Ordered Surfaces of Ionic Solids and Thin Films

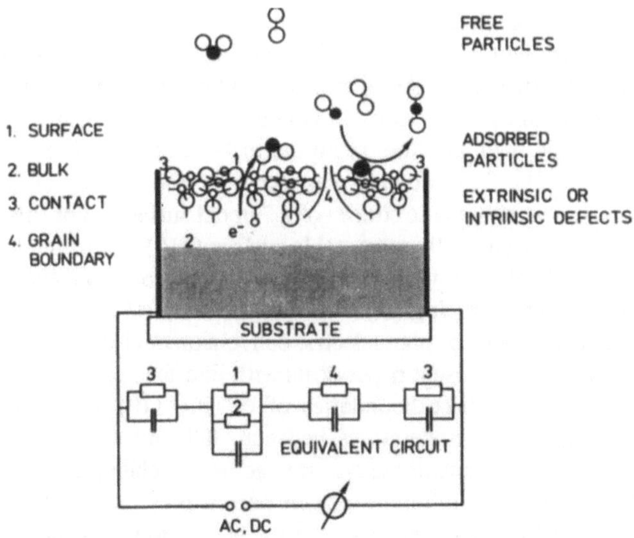

Fig. 1: Schematic representation of elementary steps during the detection of molecules ("free particles") in the gas phase with electronic conductance sensors. Surface and bulk reactions lead to changes of the overall DC (or AC) conductance, which contains (frequency-dependent) contributions from the surface (1), bulk (2), three-phase boundary or contacts (3), and grain boundaries (4). Usually two Ohmic contacts are used. In Schottky diode sensors, however, one contact/oxide interface is a Schottky barrier with non-linear current-voltage behaviour [2,3]. The formation of Pt/TiO_2 Schottky barriers and their changes towards Ohmic contact behaviour is treated in detail in the present study.

and in particular the chemical equilibria between point defects in the crystal (e.g. positively (2+) charged oxygen vacancies $V_O^{\bullet\bullet}$) and neutral oxygen (O_2) in the gas phase.

Complete reversibility of the sensor response is mandatory for practical applications. This requires that all the reactions involved are thermodynamically or kinetically controlled. For low-temperature sensor operation, this requires that the concentration of all adsorbed species is unequivocally given by the temperature and the corresponding partial pressures in the gas phase. It requires in addition, that the surface atom composition of the substrate including the doping or metal contacts does not change irreversibly. For high-temperature operation, this requires that bulk equilibria are adjusted. This concerns in particular the concentration of vacancies $V_O^{\bullet\bullet}$ in the oxygen sublattice of the oxide as function of the oxygen partial pressure and temperature and it concerns the segregation distribution of the dopants such as Pt and Pd and of the metal atoms forming the contacts. Many of these reactions lead to changes in the concentration of free electrons and hence to changes in the overall electronic conductance. The latter may contain changes in the different contributions from the surface, bulk, contacts, or grain boundaries.

With the aim of clarifying the origin of electronic conductances in such systems, we performed systematic studies on undoped oxides and obtained very detailed atomistic models to describe physisorption, chemisorption, surface and bulk defect formation as well as grain boundary effects [1,2]. The present paper concentrates at the next step, i.e. at an atomistic understanding of doping with sensor active metals. This doping may occur at the surface or in subsurface regions. We focus at systematic studies of a "model surface". For this purpose, we choose the thermodynamically most stable surface (110) of a technologically most interesting oxide (TiO_2). In view of a variety of practical applications, we choose a most important metal doping (Pt). It is used in chemical sensors, heterogeneous catalysts, electrocatalysts, or photoelectric conversion devices.

In this context, the present paper deals with interface properties and reactions at the two- and three-phase boundaries of Pt/TiO_2(110) and Pt/TiO_2(110)/gas systems investigated at temperatures between 300 and 1070 K. To study the influence of oxygen stoichiometry, we adjusted chemically reducing and oxidizing conditions, i.e. different oxygen partial pressures $p(O_2)$. Temperature and oxygen pressure determine the stability of Pt atoms, clusters, and ions at different TiO_2 surface and bulk lattice sites and also determine the stability of Pt contacts at TiO_2. In addition, they determine the electrical and catalytic properties of corresponding Pt/TiO_2(110) structures.

Three different types of Pt/TiO_2 structures will be discussed: 1) TiO_2(110) with invidual platinum adatoms, 2) TiO_2(110) with metallic Pt clusters and electrical Pt contacts, and 3) TiO_2(110) with segregated platinum ions which are formed by oxidation of platinum overlayers at elevated temperatures.

2. Experimental

We used TiO_2 single crystals (Industries de Pierres Scientifiques, Hrand Djévahirjian S.A., CH-1870 Monthey, Valais, Suisse) with only small concentrations of the bulk impurities Si (< 0.05 %), Fe (< 0.04 %) and Al (< 0.02 %). The samples were oriented in (110)-direction by Laue-diffractometry.

Experimental results on chemical compositions, stabilities, electronic structures, and charge transfer reactions are obtained from X-ray and ultraviolet photoelectron spectroscopy (XPS and UPS), Auger electron depth profiling, and current-voltage- (I-V) measurements of Pt overlayers with submono- to multilayer coverages on TiO_2(110) single crystal surfaces [4]. The I-V-measurements in particular are extremely sensitive to monitor indirectly three-phase boundary reactions for electrode arrangements perpendicular to the Pt/TiO_2(110) interface provided that ohmic back contacts (metallic Zr or Pt) are applied. Experimental details of the spectroscopic and electrical setups have been published elsewhere [4].

3. Results and Discussion

3.1 Electronic and Geometric Structure of the Ideal $TiO_2(110)$ Surface

TiO_2 is a wide-band gap ionic semiconductor (E_{gap} = 3.1 eV [5]). The electronic structure of the ideal $TiO_2(110)$ surface is characterized by a valence band mainly derived from O 2p-states, a conduction band mainly derived from Ti 3d-states, and negligible band gap states. This follows from both, experiments [6] and calculations. The latter include discrete variational $X\alpha$-methods [7] and scattering methods with tight-binding linear combinations of atomic orbitals which are based on a Hamiltonian including second-neighbour interactions [8]. For thermodynamic reasons, the bulk contains oxygen vacancies at $T \neq 0$ K and $p(O_2)$ values usually adjusted between UHV and atmospheric pressure conditions. These defects act as donors which lead to n-type conductance of TiO_2 [9].

The geometric structure of clean $TiO_2(110)$ is well known from low energy electron diffraction (LEED) and recently from scanning tunnelling microscopy (STM) [10,11]. Ordered surfaces with (1x1) LEED patterns are found if ion-sputtered $TiO_2(110)$ single crystals are annealed at high temperatures in the presence of oxygen (typically T = 823 K and $p(O_2) = 10^{-5}$ Pa). This procedure leads to stoichiometric and electrostatically neutral ("ideal") surfaces which are theoretically expected to be formed by a cleavage of TiO_2 bulk bonds with a minimum number of Ti-O broken bonds (Fig. 2). In reality, this cleaving of single crystals is not possible and therefore the above-mentioned preparation is required.

The following calculations were performed to find possible structures and to estimate related energies attributed to the Pt adsorption on $TiO_2(110)$. We represent the free surface to a first approximation by placing the ions at surface positions which are taken as identical with their bulk positions. The total excess charge $\Sigma\, q^{\pm}$ of the ideal surface is zero [12]. As the basis for the following

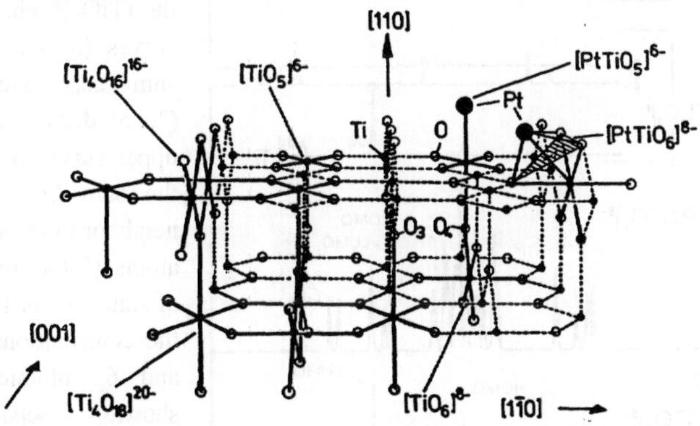

Fig. 2: Geometric structure of the ideal $TiO_2(110)$ surface with different Ti and O sites. $[TiO_5]^{6-}$, $[Ti_4O_{16}]^{18-}$, $[TiO_6]^{8-}$, and $[Ti_4O_{18}]^{20-}$ clusters are shown which represent surface and bulk lattice units. The distance between two Ti-O-layers is $3.26 \cdot 10^{-10}$ m. The Pt adpositions at a Ti- and an O-site are represented by $[PtTiO_5]^{6-}$ and $[PtTiO_6]^{8-}$ clusters.

discussion of Pt sites at $TiO_2(110)$, Fig. 2 shows the geometric structure with possible Ti and O adpositions as well as corresponding structures of free clusters which are used in our ab-initio molecular orbital calculations.

3.2 Invidual Pt Adatoms at the Ideal $TiO_2(110)$ Surface

Invidual platinum atoms at the $TiO_2(110)$ surface are formed by submonolayer evaporation of Pt onto the stoichiometric $TiO_2(110)$ surface at low temperatures.

Their stability as compared with the stability of metallic Pt clusters (which may form by subsequent surface diffusion) is to a first approximation determined by the bond energy between Pt and the Ti and/or O ions as compared with the energy of Pt-Pt bonds in the metal cluster. The latter is assumed to have a value between the dissociation energy of the gaseous Pt_2 dimer ($E_{diss} = 3.7$ eV [13]) and the sublimation enthalpy of bulk Pt at T = 298 K ($\Delta H° = 5.85$ eV [14]).

We calculate covalent bond energies of 2.27 eV ($[PtTiO_5]^0$, $[PtTiO_5]^{6-}$), 6.87 eV ($[PtTiO_6]^0$), 8.44 eV ($[PtTiO_6]^{8-}$), 3.65 eV ($[PtTi_4O_{15}]^{16-}$) in ab-initio molecular orbital calculations of free Ti-O- and platinum containing Ti-O-clusters using the GAUSSIAN 90 program [15]. In these calculations we consider contributions from Ti 3s, Ti 3p, Ti 3d, Ti 4s, O 2s, O 2p, Pt 5d, Pt 6s, and Pt 6p atomic orbitals to the molecular orbitals (MO) of the clusters. The MO

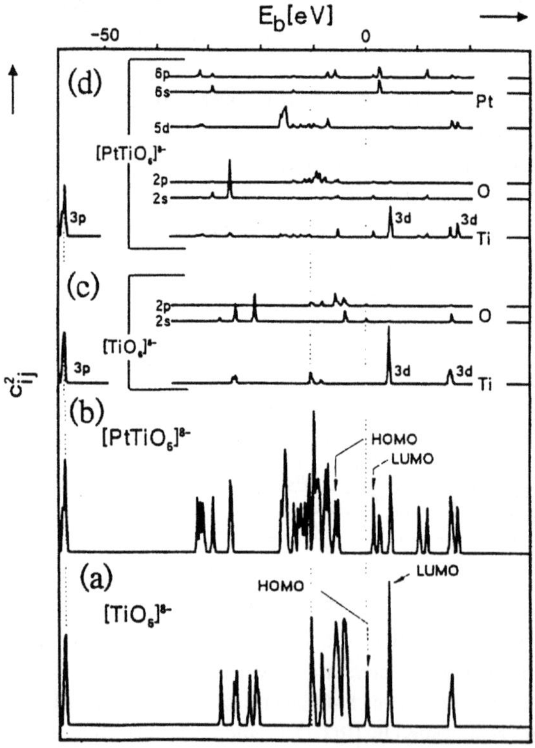

Fig. 3. Square of molecular orbital coefficients c_{ij}^2 vs. binding energy E_b of $[TiO_6]^{8-}$ and $[PtTiO_6]^{8-}$ clusters. E_b is referred to the highest occupied molecular orbital (HOMO) of the $[TiO_6]^{8-}$ cluster. The lower curves (a) and (b) show the sum Σc_{ij}^2 over all centers ("total density of states"), the upper curves (c) and (d) show the sum Σc_{ij}^2 over O (next neighbor to Pt only), Ti, and Pt atoms ("atom-resolved density of states"). For Pt in particular, the contributions from 5d, 6s, and 6p atomic orbitals are shown separately. All eigenstates are broadened with Gauss curves by 1 eV. Further explanations are given in the text.

calculation is also used to determine the electronic structure including the total and atom-resolved density of states as a function of the electron binding energy E_b. A characteristic result is shown in Fig. 3. These data correspond to the minimum total energy of the $[PtTiO_6]^{8-}$ cluster obtained by an optimized Pt-Ti distance of 0.238 nm.

Curve (a) shows the total density of states of the $[TiO_6]^{8-}$ cluster. States below the highest occupied molecular orbital (HOMO) are mainly derived from O 2p states. They correspond to the valence band of TiO_2. States above the lowest unoccupied molecular orbital (LUMO) are mainly derived from Ti 3d states. They correspond to the conduction band of TiO_2. Curve (b) shows the total density of states of the $[PtTiO_6]^{8-}$ cluster. Unoccupied Pt 6s and 6p states are found between the HOMO and LUMO of the $[TiO_6]^{8-}$ cluster. We find occupied states at and below the O 2p states of $[TiO_6]^{8-}$ which are mainly derived from Pt 5d states. They overlap partly with O 2p atomic orbitals. This leads to a covalent

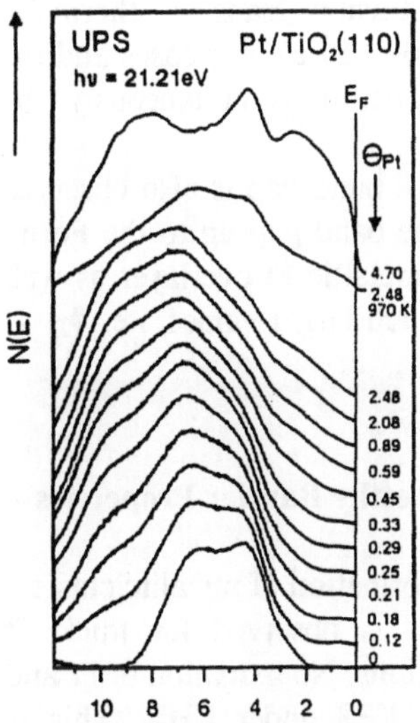

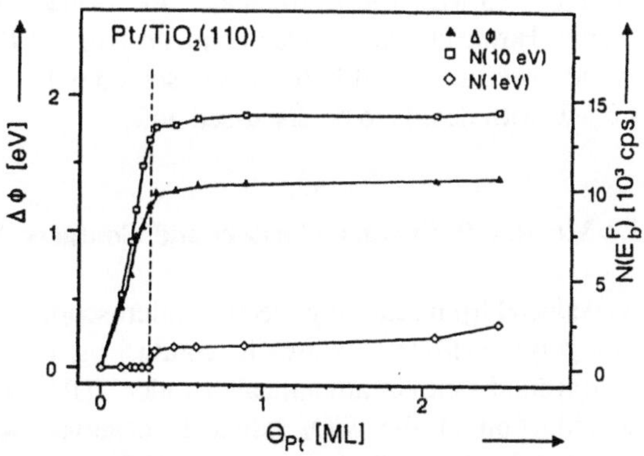

Fig. 4 (left): UP valence band spectra of ideal $TiO_2(110)$ surfaces with a work function $\Phi_{TiO_2} = 5.4$ eV after stepwise evaporation of Pt at T = 330 K with increasing Pt coverages θ_{Pt} (in monolayers). The latter is determined from corresponding XPS intensities. The upper curves show the valence band spectra after heating the sample at T = 970 K in UHV (t = 60 min) with a workfunction Φ(300 K) = 5 eV and after evaporation of a thick metallic Pt layer with a workfunction Φ_{Pt}(300 K) = 5.6 eV.

Fig. 5 (right): Change of the work function $\Delta\Phi$ (as determined from the onset of the secondary electron emission) and photoemission intensities $N(E_b{}^F)$ at binding energies $E_b{}^F$ = 10 eV (i.e., below the TiO_2 valence band) and 1 eV (i.e, in the band gap of TiO_2). Binding energies are related to the Fermi energy E_F.

bond between platinum and oxygen atoms. These findings are qualitatively in accordance with UP valence band spectra of $Pt/TiO_2(110)$ (Fig. 4).

The theoretical results are in line with experimental findings: A significant increase is observed at low coverages in the photoemission intensity below the O 2p valence band of TiO_2. Simultaneously, a decrease of the workfunction Φ occurs (Fig. 5).

Under these conditions, a relatively high concentration of invidual Pt atoms is adjusted during the Knudsen evaporation. This follows from simple statistical arguments of a Monte Carlo statistics with negligible particle/particle interaction because of negligible surface diffusion [16].

For $\Phi_{Pt} \leq 0.3$ ML, the position of the valence band maximum E_V relative to the Fermi level E_F is constant. This indicates, that no delocalized charge transfer and hence band bending occurs at the surface. The latter would change the concentration of free electrons in the TiO_2 conduction band. Under these conditions, the decrease in the workfunction Φ results totally from the decreased electron affinity. However, subsequent surface diffusion leads to thermally induced clustering of Pt adatoms. The intensity at $E_b{}^F = 10$ eV decreases and the UP spectrum shows again the TiO_2 valence band structure. Simultaneously, the workfunction increases.

An increase of the photoemission intensity in the band gap is also observed for $\Phi_{Pt} > 0.3$ ML. Electronic states are formed in the band gap up to the Fermi energy. However, the valence band structure of the metallic Pt overlayer as well as the workfunction of bulk Pt is observed only at much larger values, i.e. $\Phi_{Pt} \geq 4.7$ ML with details to be discussed now.

3.3 Metallic Pt Surface Clusters and Contacts: Schottky Barrier Properties

As deduced from scanning electron microscopy, the formation of metallic clusters and polycrystalline electrically conducting contacts is observed for thick Pt overlayers if samples are annealed under UHV conditions. No oxidation of Pt and no reduction of the TiO_2 lattice is observed with XPS under UHV. This is expected for thermodynamic reasons [17]. It is also expected for simple energetic reasons: The oxidation of Pt would require the ionisation energy and the reduction Ti^{4+} to Ti^{3+} which is correlated with changes $\Delta z_{Ti} e \Phi_{Ti}$. Here, $\Delta z_{Ti} e$ is the change of the charge and $\Phi_{Ti} = 29.53$ eV is the Madelung potential of the Ti ions [18].

Due to the larger workfunction $\Phi_{Pt} = 5.6$ eV compared to $\Phi_{TiO_2} = 5.0...5.4$ eV (as determined for different samples with different concentrations of intrinsic point defects), the conduction band electrons are transferred from TiO_2 to Pt (Fig. 6). This leads to the formation of a Schottky barrier at the Pt/TiO_2 interface and to a band bending $-e\Delta V_s$ in the TiO_2 which can be deduced from I-V measurements of Pt-coated $TiO_2(110)$ single crystals with ohmic back-side contacts (Fig. 7a,b) [20]. In the forward-voltage region (i.e. for a positive voltage

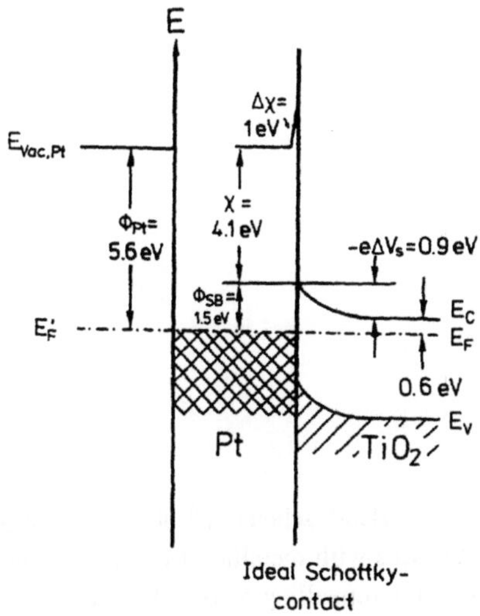

Fig. 6: Band scheme of Schottky-type Pt/TiO_2 interfaces with metallic Pt overlayers. Data are obtained from UPS and I-V measurements. The bulk position of the Fermi energy relative to the conduction band edge $(E_C\text{-}E_F)_b$ = 0.6eV is determined from temperature-dependent conductivities [19].

applied to the platinum) we determine a Schottky barrier height Φ_{SB} = 1.5 eV and an ideality factor n = 1 under UHV conditions (for the definition of n, see [21].

The value Φ_{SB} is higher than theoretically expected from

$$\Phi_{SB} = \Phi_{Pt} - \chi \tag{1}$$

[22] for an electron affinity χ_{TiO_2} = 4.8 eV of pure TiO_2 which is calculated from

$$\chi_{TiO_2} = \Phi_{TiO_2} - (E_C - E_F)_b \tag{2}$$

The difference is formally attributed to an additional dipole contribution $\Delta\chi$ of the first Pt layer which corresponds to $\Delta\Phi$-effects observed in UPS (compare Fig. 6).

A strong effect on the I-V-curves is observed if we expose Pt/TiO_2

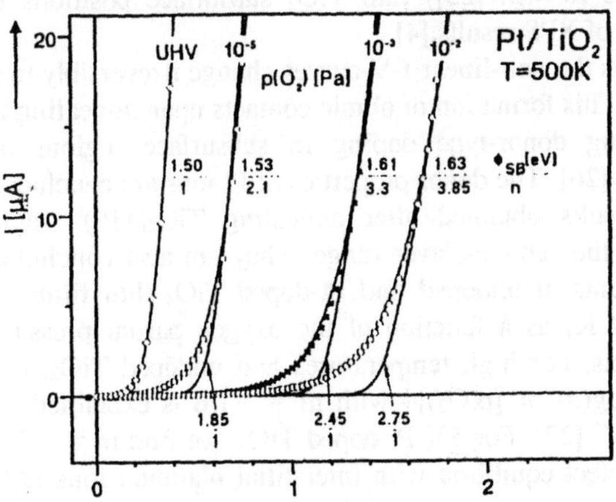

Fig. 7: Typical results of I-V-curves of polycrystalline Pt/TiO_2 contacts in UHV and for different oxygen partial pressures at T = 500 K. Schottky-barrier heights Φ_{SB} and ideality factors n are determined by fitting calculated to experimental values [18]. The I-V-curves determined after O_2 exposure lead to ideality values n > 1.

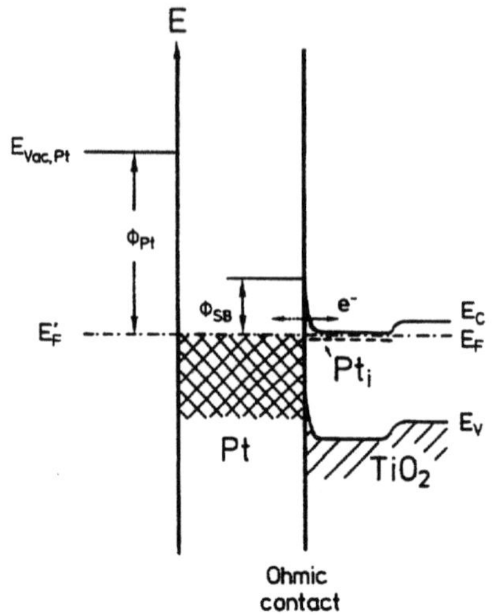

Fig. 8: Band scheme of ohmic Pt/TiO_2 interfaces with metallic Pt overlayers and very thin tunnelling barriers for electrons.

polycrystalline contacts to acceptor- and donor-type molecules such as O_2, CO, and H_2 [23,24] around T = 500 K. As an example, we determine changed values $n = f(p(O_2)) > 1$ and only small increases of the Schottky-barrier height Φ_{SB} during interaction with O_2. This is attributed to the formation of interface states at the three-phase boundary which leads to a pinning of the Fermi energy in Pt/TiO_2 Schottky diode sensors at high oxygen partial pressures.

3.4 Ionic Pt Sites at Subsurface and Bulk Positions of TiO_2: Ohmic Contact Properties

Pt overlayers become unstable upon exposure to O_2 at higher temperatures T > 970 K. The formation of Pt-oxides and the segregation of the small Pt ions (with large radii of the Pt atoms $r_{Pt} = 1.75 \cdot 10^{-10}$ m and small ones of the free ions $r_{Pt^{2+}} = 0.80 \cdot 10^{-10}$ m and $r_{Pt^{4+}} = 0.65 \cdot 10^{-10}$ m [25]) into TiO_2 subsurface positions is concluded from an evaluation of XPS results [4].

Under the same conditions the non-linear I-V-curves change irreversibly to a linear behaviour with $I \propto V$. This formation of ohmic contacts upon annealing is usually attributed to a strong donor-type doping in subsurface regions as schematically shown in Fig. 8 [26]. The donor properties of Pt ions are concluded independently from UPS results obtained after annealing $TiO_2(110)$ single crystals with Pt coverages in the submonolayer range. They are also concluded from conductance measurements of undoped and Pt-doped TiO_2 thin films at high temperatures (T = 1020 K) as a function of the oxygen partial pressure $p(O_2)$ by using Ohmic contacts. For high temperatures and undoped TiO_2, the characteristic slope of plots $\log(\sigma) \propto [p(O_2)]^m$ with $m = -1/6$ is explained by bulk defect equilibria with $V_O^{\bullet\bullet}$ [27]. For 5% Pt-doped TiO_2, we find $m = -1/5$ which is explained by bulk defect equilibria with interstitial platinum ions $Pt_i^{4\bullet}$

[28]. This conductivity effect is also made use of in O_2 gas sensors based upon TiO_2 [27].

4. Conclusions and Summary

Conditions are studied under which invidual Pt adatoms, Pt clusters, and Pt contacts are stable at $TiO_2(110)$, and under which Pt ions are stable at interstitial positions in the bulk of TiO_2. The formation of metallic Pt contacts at the surface leads to a delocalized charge transfer involving conduction band electrons of TiO_2 which are transferred to Pt. Hence, Pt/TiO_2 interfaces show Schottky-diode behavior with the barrier height and ideality value sensitively determined by the surrounding atmosphere. This is made use of in *Schottky diode gas sensors.*

At higher temperatures, Pt-oxides are formed in the presence of O_2 at the interface. Subsequently as-prepared Pt ions diffuse into subsurface regions and form donor-type dopants. This process leads to Ohmic contacts at TiO_2 which are used in TiO_2-based *Ohmic conductance sensors* to determine O_2 partial pressures at high temperatures.

References

[1] W. Göpel, K.D. Schierbaum, Electronic Conductance and Capacitance Sensors, in: Sensors - A Comprehensive Survey, Vol. 2, Part I, ed. by W. Göpel, J. Hesse, J.N. Zemel, VCH Verlagsgesellschaft, Weinheim 1991, p. 430

[2] W. Göpel, K.D. Schierbaum, Specific Molecular Interactions and Detection Principles, in: Sensors - A Comprehensive Survey Vol.2, Part I, ed. by W. Göpel, J. Hesse, J.N. Zemel, VCH Verlagsgesellschaft, Weinheim 1991, p. 120

[3] K.D. Schierbaum, J. Geiger, U. Weimar, W. Göpel, Proc.4th Int. Meeting on Chemical Sensors, Tokyo 1992;. Sensors and Actuators B, in press.

[4] U.K. Kirner, K.D. Schierbaum, W. Göpel: Fresenius' Z. Anal. Chem. **341,** 416 (1991)

[5] C.J. Kevan: Phys. Rev. **133**, A1431 (1964)

[6] D.W. Fischer: Phys.Rev. **B 5**, 4219 (1972)

[7] S. Munnix, M. Schmeits: Phys. Rev. **B 30**, 2202 (1984)

[8] M. Tsukada, H. Adachi, C. Satoko: Progr. Surf. Sci. 113 (1983)

[9] U. Balchandram, W.G. Eror: J. Mater. Sci. **23**, 2676 (1988)

[10] P.J. Møller, M.-C. Wu: Surf. Sci. **224**, 265 (1989)

[11] G. Rohrer, V.E. Henrich, D.A. Bonnel: Surf. Sci. **278**, 146 (1992)

[12] R.A. van Santen, Theoretical Heterogeneous Catalysis, ed. by S.H. Lin, World Scientific Publishing Co., Singapore 1991, p. 290

[13] Gmelins Handbook of Inorganic Chemistry, Platinum, Suppl. Vol. A2, 8th Ed., Springer, Berlin 1989

[14] D'Ans-Lax, Taschenbuch für Chemiker und Physiker, Bd. I, Makroskopische

physikalisch-chemische Eigenschaften, Springer, Heidelberg 1967
[15] Gaussian 90, Revision H.M.J. Frisch, M. Head-Gordon, G.W. Trucks, J.B. Foresman, H.B. Schlegel, K. Raghavchari, M. Robb, J.S. Binkley, O. Gonzales, D.J. Defrees, D.J. Fox, R.A. Whiteside, R. Seeger, J.J.P. Stewart, S. Topiol, J.A. Pople, Gaussian Inc., Pittsburg PA, 1990
[16] H. Mayer, Physik dünner Schichten, Teil II, Wiss. Verlagsges., Stuttgart 1955, p. 83
[17] J. Barin, O. Knacke, Thermochemical Properties of Inorganic Substances, Springer, Berlin 1973
[18] E. Iguchi, H. Sawatari, R.J.D. Tilley: Phys. Stat. Sol. B **101**, 333 (1980)
[19] Here, bulk concentrations of electrons are calculated using mobility data taken from the literature, compare G. Rocker, W. Göpel: Surf. Sci. **181**, 530 (1987)
[20] K.D. Schierbaum, X. Wei-Xing, W. Göpel: Ber. Bunsenges. Phys. Chem. **97**, 363 (1993)
[21] H.C. Card, E.H. Rhoderick: J.Phys. D: Appl. Phys. **4**, 1589 (1971)
[22] L.J. Brillson: Surf. Sci. Reps. **2**, 123 (1982)
[23] N. Yamamoto, S. Tonomura, T. Matsuoka, H. Tsubomura: Surf. Sci. **92**, 400 (1980)
[24] K.D. Schierbaum, U.K. Kirner, J.F. Geiger, W. Göpel: Sensors and Actuators B **4**, 87 (1991)
[25] D'Ans-Lax, Taschenbuch fü Chemiker und Physiker, Bd. III, Eigenschaften von Atomen und Molekeln, Springer, Heidelberg 1970, p. 3
[26] S.M. Sze, Physics of Semiconductor Devices, 2nd Ed., John Wiley, New York, p. 306
[27] U.K. Kirner, K.D. Schierbaum, W. Göpel, B. Leibold, N. Nicoloso, W. Weppner, D. Fischer, W F. Chu: Sensors and Actuators B **1**, 103 (1990)
[28] compare R.T. Dirstine, C.J. Rosa: Z. Metallkde. **70**, 322 (1979)

Applications of Lithium Niobate Acoustic Plate Mode Devices as Sensors for Liquids

R. Dahint, K. Bierbaum, and M. Grunze

Lehrstuhl für Angewandte Physikalische Chemie, Universität Heidelberg, Im Neuenheimer Feld 253, D-69120 Heidelberg, Germany

Abstract. The operation principles of acoustic wave sensors for liquids are summarized and results are reported on the use of ZX-$LiNbO_3$ acoustic plate mode (APM) devices as detectors for dilute electrolyte and metal ion solutions. The APM is a slow shear wave having a relatively strong acousto-ionic interaction, and thus allows to measure the liquid electrical properties by the perturbation of the wave propagation characteristics.

To utilize the APM device as a *selective* chemical sensor for liquids, the surface is derivatized in order to covalently couple a chemical selective film onto it. For the development of a selective sensor, the surface cleaning and derivatization of the $LiNbO_3$ surface proved to be a major obstacle. We will summarize our recent results on the derivatization of $LiNbO_3$ with alkylaminosilanes. To understand the relationship between the state of the oxide substrate surface and the homogeneity and structural aspects of aminosilane monolayers further, model experiments with different aminosilanes on a Si(100) surface covered with a native oxide layer are described.

1. Introduction

Fast detection of specific molecules is important in the areas of medical diagnostics, environmental sciences, food analysis and biological research. In particular, there is an interest in diagnostic methods which require only small volumes of analyte and allow an on-line analysis of the process under study.

A very important technique in the field of immuno chemistry is the detection of antigens by monoclonal antibodies. Their high equilibrium constants (10^6 - 10^{12} liter/mole) and their high selectivity make these antibodies an appropriate tool to separate specific antigens from a mixture of molecules. The antigen-antibody reaction is usually detected by time-consuming multi-stage processes using radioactive, enzyme or fluorescence labeled antibodies. However, these techniques - though very sensitive - are not suitable for on-line measurements [1].

Contrary to the above methods, ultrasonic waves offer the opportunity to detect antigen-antibody reactions in a one-step process without using labeled molecules. Only a small amount of analyte is required. If acoustic waves are excited in a solid, immuno reactions at the surface will change the propagation characteristics of the wave. Antigen-antibody reactions have already been observed on a quartz microbalance by a decrease in the frequency of oscillation, due to the binding of antigen [2].

However, acoustic techniques have not been able to compete with traditional methods in terms of sensitivity so far. Problems arise from corrosion caused by the contact between the liquid phase analyte and the sensor electrodes, baseline drifts, the use of unsuitable wave types, operating frequencies being too low and a poor quality of the immobilized antibody films. It is to be expected that the above difficulties can be overcome by optimizing both sensor design and film quality. The great advantages of acoustic wave sensors, such as small size and fast response times, definetely justify this effort.

One emphasis of our work is the improvement of the derivatization technique for the sensor surface. The activation of hydroxylated substrates (lithium niobate ($LiNbO_3$), silicon (Si)) with antibody films is usually carried out by using the standard immobilization technique, i.e. silanization with 3-aminopropyltriethoxysilane (APTS) prior to linkage of antibody molecules via a glutaraldehyde crosslinker [3]. In biochemical and immunological analysis, e.g. for applications in biosensors, a very high reproducibility of film characteristics such as homogeneity, thickness, long term stability and activity is required. In particular, the quality of the aminosilane films is very important, since they are directly linked to the substrate and therefore have a great influence on further layer deposition.

Recent studies [3] have shown that silanization of $LiNbO_3$ with APTS in liquid and from the gas-phase, e.g. by chemical vapour deposition, did not result in homogenous aminosilane monolayers. For this reason a variety of different organosilane films on hydroxylated surfaces like Si and $LiNbO_3$ have been characterized using X-Ray Photoelectron Spectroscopy (XPS), Near Edge X-Ray Absorption Fine Structure (NEXAFS) and Atomic Force Microscopy (AFM) to determine whether a uniform ordered monolayer is in fact achieved.

In the following we first discuss different types and operation principles of acoustic wave liquid sensors. In particular, we will focus on the sensor type used by our group and discuss ways of optimizing the acoustic properties for biosensor applications. We compare short and long alkyl chain silane films with differing type and number of reactive groups and present XPS, AFM and NEXAFS data and results. This will be followed by a discussion relating these results to enhanced film quality for sensing applications.

2. Acoustic Wave Sensors

Acoustic wave devices have been used in signal processing within the last few decades [4]. Compared to electromagnetic devices they offer dramatic reductions in size and weight. They are compatible with integrated circuit technology and can be fabricated by photolithographic techniques. In the field of sensor applications acoustic wave devices have demonstrated the capability of performing as detectors in both gas and liquid environments, requiring only small volumes of analyte for analysis [5]. The different types of waves used for acoustic sensors can be divided into two classes: bulk acoustic waves (BAWs) and surface acoustic waves (SAWs). Their characteristic properties will be discussed below.

2.1. Common Sensor Types

The use of BAWs as mass detectors is based on Sauerbrey's work in 1959 [6]. He formed a quartz bulk wave resonator by contacting the upper and lower surface of a thin crystal plate with metal electrodes. The frequency of operation f_n of such a device is determined by the necessity to form standing waves between the crystal surfaces. This yields $f_n = (n + 1) \cdot v / (2\, d)$, where n is an integral number and v the crystal sound velocity. Thus, f_n is related to the thickness of the crystal. Using resonances with $n > 0$ increases resonator bandwidth and degrades device performance.

Thin films deposited on the crystal surface change the lowest order resonant frequency f_0 of the crystal by $\Delta f = - C\, f_0^2\, \Delta m_S$. Here, Δm_S denotes the surface mass density of the deposited film and C a constant, which is determined by the material properties of the crystal. For quartz C equals $2.3 \cdot 10^{-6}$ s cm^2 g^{-1}. As typical frequencies of operation are about 10 MHz, a mass sensitivity of $\Delta f / \Delta m_S = -230$ Hz µg^{-1} cm^2 is achieved for a quartz device.

Another class of waves being widely used for sensor applications are SAWs. In contrast to BAWs, the particle displacements of SAWs are confined to a small region near the crystal surface. This fact makes them promising candidates for the study of surface reactions. As the frequency of operation is increased, the penetration depth of the wave into the crystal is further reduced.

SAWs became of particular interest in 1965 when White developed a simple and effective method of exciting them on piezoelectric substrates [7]. The technique used is to excite and detect SAWs by interdigital transducers (IDTs). An IDT consists of two interlocked combs of metal electrodes deposited on the surface of a piezoelectric crystal (Fig. 1). SAWs are excited by applying an alternating voltage to the electrodes. The waves propagate along the surface until they reach the receiver IDT. Here, the acoustic energy is converted back to an electrical signal by the reverse mechanism.

The frequency of excitation is determined by the periodicity of the metal electrodes. Resonance occurs if the partial waves excited by each finger pair interfere constructively. Therefore, $f_0 = k \cdot v / (2\, l)$, where l is the distance between two metal fingers of different polarity and k an odd integer. As photolithographic techniques provide finger widths of several microns and less, frequencies up to the GHz region can be achieved.

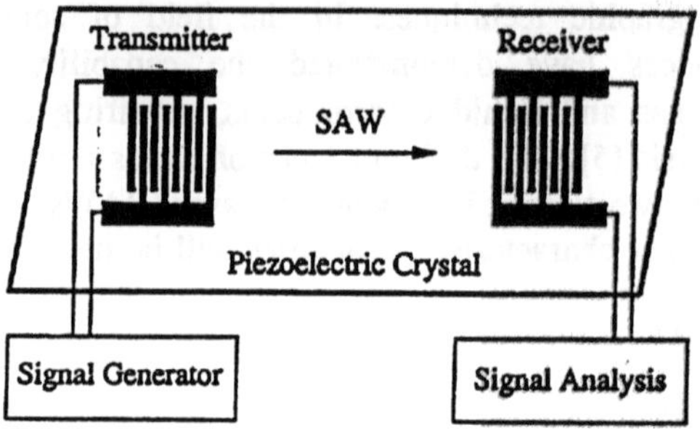

Fig. 1: Surface acoustic wave (SAW) device. Thin metal electrodes are deposited on top of a piezoelectric crystal. SAWs are excited by applying an alternating voltage to the electrical contacts.

Perturbation theory [8] allows to calculate the frequency shift of a SAW due to mass loading. Making some simplifying assumptions it is found that $\Delta f = -K f_0^2 \Delta m_S$. The constant K depends on crystal type, crystal cut and direction of wave propagation. For Y-cut, X-propagating quartz K is $1.3 \cdot 10^{-6}$ s cm^2 g^{-1}. Thus, mass sensitivity is $\Delta f / \Delta m_S = -130$ Hz µg^{-1} cm^2 at 10 MHz and $\Delta f / \Delta m_S = -13$ kHz µg^{-1} cm^2 at 100 MHz. A comparison of these data with the corresponding values for bulk resonators clearly shows that the advantages of SAW sensors are the high frequencies of operation and hence the intrinsically higher mass sensitivity that can be achieved.

2.2. Acoustic Plate Mode Sensors

Liquid sensors have to make use of special wave types which can propagate at a solid/liquid interface without drastic losses. Acoustic waves having considerable particle displacements normal to the crystal surface will suffer high attenuation as compressional waves are radiated into the fluid. For this reason the most common wave type for acoustic gas sensors - the Rayleigh wave - cannot be used in liquid sensing. To minimize energy transfer into the liquid horizontally polarized waves are to be preferred.

However, polarization is not the only problem to be solved. For both types of sensors discussed in section 2.1. the analyte is in close contact with an electroded crystal surface (Fig. 2a). Therefore, severe corrosion problems may occur if the sensor is operated in the liquid phase, resulting in instabilities, drifts and short lifetimes of the devices. In addition, surface cleaning and coating have to be done very carefully, as the electrodes are easily destroyed.

To overcome these obstacles, we decided to use a different kind of device for our studies, a so-called acoustic plate mode (APM) sensor with predominantly shear horizontal (SH) particle displacements (Fig. 2b). An APM sensor consists of a thin piezoelectric crystal plate with two IDTs deposited on the bottom surface. These serve as a transmitter and a receiver, respectively, just as for SAW devices. If an alternating voltage is applied to the transmitter, acoustic waves are radiated into the bulk of the crystal. After multiple reflection at the crystal

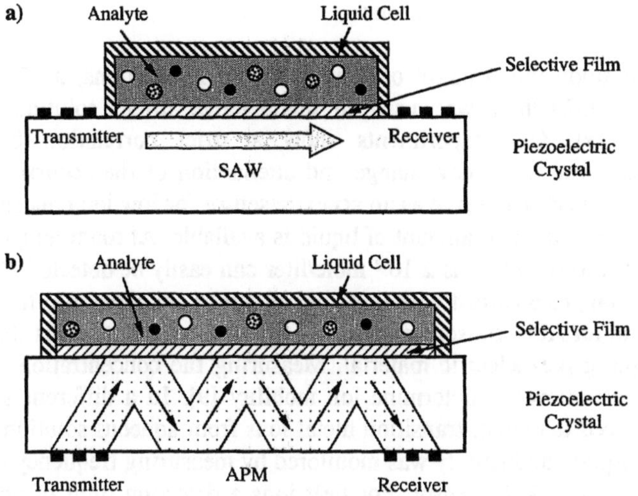

Fig. 2: Acoustic wave liquid sensors. Binding of specific molecules to a selective film changes the propagation characteristics of the acoustic device. a) For SAW sensors the electroded crystal surface is exposed to the liquid. b) For APM sensors liquid and electrical contacts are strictly separated.

surfaces the waves are detected by the receiver and converted back to electrical signals. For thin crystal plates particle displacements form standing waves along the thickness dimension. The number of nodes in the displacement profile is called the order of the mode. Due to their special geometry APM sensors allow to expose the non-electroded crystal surface to the liquid under study. Because of its high piezoelectric coupling constant Z-cut, X-propagating $LiNbO_3$ was selected for the crystal plate material.

For sensors fabricated on strong piezoelectric crystals like $LiNbO_3$, a further mechanism of interaction between sensor and liquid may become very important. In addition to mass loading effects, acoustoelectric interactions are noticed. This is due to the fact that an acoustic wave propagating in a piezoelectric material is always accompanied by an electric field. The field extends outside the surface and interacts with the adjacent medium. Changes in the electric or dielectric properties of the adjacent medium will, therefore, lead to damping processes and a frequency change of the acoustic wave. The degree of interaction is a function of the strength of the electric field, which directly determines the sensitivity of the device. For high coupling piezoelectric materials like $LiNbO_3$ a strong interaction is noticed as discussed in the following paragraph.

2.2.1. Acoustoelectric Interactions

In order to study the strength of acoustoelectric interactions, a 13 MHz APM sensor on ZX-$LiNbO_3$ was used to characterize dilute solutions of strong electrolytes [9]. Our experiments show a strict correlation between ion concentration and frequency change and attenuation of the acoustic wave (Fig. 3). Thus, the device can serve as an accurate sensor for low ion concentrations, in particular if only a small amount of liquid is available. At room temperature KCl ion concentrations as low as $2 \cdot 10^{-4}$ mole/liter can easily be detected. The relative frequency changes obtained with this kind of device are about 500 times stronger than those of an APM sensor on quartz [10]. This is due to the fact that quartz is a low coupling piezoelectric material. Measuring the concentration dependence of attenuation allows to determine ion mobility [9]. In a different study silane films were used to extract transition metal ions from aqueous solutions [11]. The change in liquid conductivity was monitored by measuring frequency change and attenuation of the APM sensor. For Fe^{3+} ions a detection limit of about 2 ppm was observed.

The above experiments also yield valuable results for biosensor applications, as antigens are usually detected from electrolytic solutions. Thus, the data tell how properly liquid parameters have to be controlled in order to avoid experimental errors. As has been shown by Shana et al. [12], at high ionic concentrations the sensor saturates with respect to its acoustoelectric response, i.e. only mass loading effects or changes in the viscosity of the adjacent medium lead to a signal response.

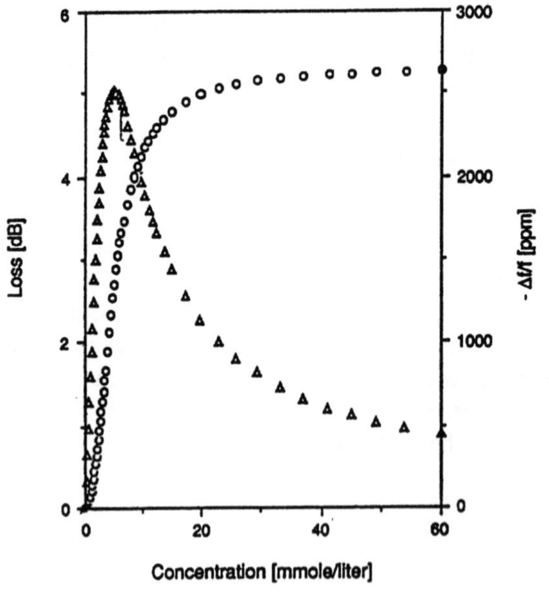

Fig. 3: APM sensor response due to KCl at 25 °C. Loss (Δ) and relative frequency change (o) of a 13 MHz APM device are plotted as a function of ion concentration. The position of the peak in the loss curve is related to ion mobility [9].

2.2.2. Mass Loading

Acoustic wave sensors for molecule detection are commonly based on mass loading effects. To evaluate the potential of APMs as mass-sensitive biosensors five APM devices with operating frequencies ranging from 13 to 42 MHz have been designed and fabricated [13]. The top surface of the crystals was metallized with a thin aluminum film to shorten the electric field and eliminate acoustoelectric interactions. Afterwards photoresist of well-defined thickness and density was deposited on top of the aluminum layer by spin coating. Now, mass sensitivity was determined by dissolving the photoresist in liquid environment and monitoring sensor response.

Theoretical calculations treating the piezoelectric crystal plate simply as an isotropic substrate predict that mass sensitivity $s = \lim_{dm \to 0} df / dm$ is proportional to f_0/d, where d is the thickness of the plate [10]. However, if the anisotopic structure of $ZX\text{-}LiNbO_3$ is taken into account, a $f^{1.7}$ dependence is found [13]. According to our experimental data mass sensitivity increases as f^2. This enhanced mass sensitivity is probably due to the fact that APMs couple to surface waves.

The most sensitive device was the 42 MHz sensor fabricated on a 0.02 inch thick $ZX\text{-}LiNbO_3$ plate. Mass sensitivity was found to be -660 Hz μg^{-1} cm^2. For the 13 MHz device fabricated on a crystal plate of 0.04 inch thickness -64 Hz μg^{-1} cm^2 have been measured. The data show that the opportunity to work at higher frequencies makes APM devices more sensitive than conventional bulk wave resonators. However, the sensitivity of SAW devices cannot be achieved.

On a piezoelectric crystal plate a wide spectrum of acoustic plate modes can propagate, each of them having a different frequency of excitation. The fact that mode separation decreases with increasing frequency limits the operation frequency of APM sensors. For $ZX\text{-}LiNbO_3$ values of $f_0 \cdot d$ up to about 26 MHz mm seem to be practical. It should be noted that the use of very thin crystal plates improves mass sensitivity and allows to design high frequency devices. However, severe mechanical problems are to be expected in sensing applications.

3. Film Preparation and Characterization

The above discussion was related to the interaction mechanism of the sensor with the adjacent liquid medium. For the sensing of a particular element or molecule, however, the sensor surface has to be derivatized for selective adsorption of the element or molecule searched for.

In the following we discuss the possibilities to use the APM device as an immunosensor. A standard technique for covalent immobilization of proteins, e.g. antibodies, is silanization of the substrate after cleaning prior to the reaction with a crosslinker (e.g. glutardialdehyde) and the protein. The detailed experimental procedures have been described by us elsewhere [3]. However, the amount of covalently linked antibodies (IgG POD-conjugate) was only ~ 6 % of the theoretical monolayer concentration and hence too small to be detected by the APM sensor when 3-aminopropyltriethoxysilane (APTS) was used as the silanization agent.

In the following we describe our results of the analysis of the short alkane chain silane films and present recent data on the adsorption of long alkane chain organosilanes, which give much better defined films and hence may lead to improved coupling layers for antibodies [14].

The data shown below were taken on oxidized and subsequently hydroxylized Si(100) wafers, which were used as model substrates for the reaction of organosilanes with surface hydroxyl groups. Because APTS with three reactive groups resulted in multilayers due to polymerization of the molecules [3], 4-aminobutyldimethylmethoxysilane ($H_2N(CH_2)_4Si(CH_3)_2(OCH_3)$, ABDMS) having only one reactive methoxy group seemed to be more promising to form monolayers, since the probability for polycondensation on the surface is strongly reduced. However, the decrease in the number of reactive methoxy groups also results in a reduced reaction probability with the surface, so that the substrates were heated in ambient atmosphere at 110 °C for 1h after removal from the ABDMS solution to accelerate covalent coupling.

The C(1s) NEXAFS spectra of ABDMS show the expected C-H* resonance at 287 eV and a C-C σ* resonance at 293 eV (Fig. 4a) [15]. From the variation of

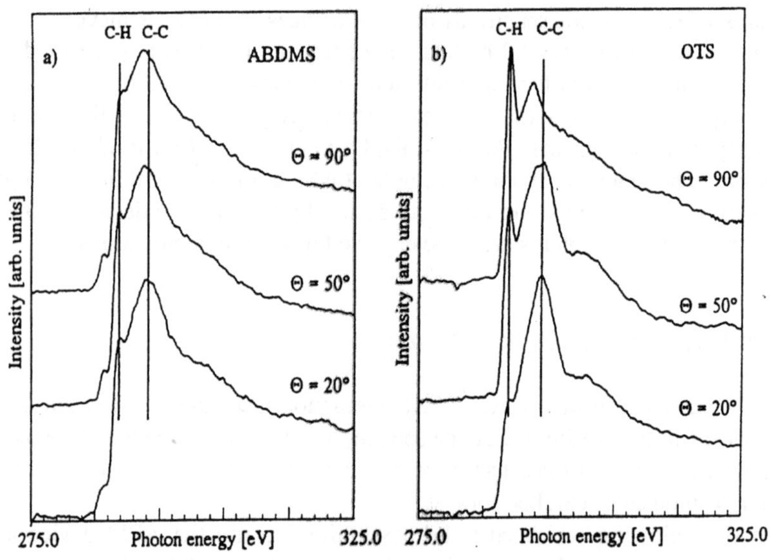

Fig. 4: NEXAFS spectra of ABDMS and OTS. a) Series of NEXAFS spectra for ABDMS for different angles of incidence θ ranging from 20° to 90°. b) Series of NEXAFS spectra for OTS for different angles of incidence θ ranging from 20° to 90°.

the two resonances with the angle of incidence of the photons the orientation of the molecules can be determined, if the direction of the transition dipole moments governing these transitions are known [17]. The NEXAFS spectra for the ABDMS samples displayed in Fig. 4a do not vary with the angle of incidence of the radiation. Only a slight change in the C-C σ^* resonance could be observed (strongest intensity for grazing incidence), while the C-H* resonance showed almost no change at all. This could be due to a tilt of 55° (magic angle) or to a random orientation of the molecular axes in the films and/or a high concentration of gauche conformations in the chains. Supported by the XPS results (not discussed here) we conclude that the film is not well ordered.

In addition, the ABDMS films are inhomogenous, which is also expected to reduce the availability of covalent bonding sites for antibodies. In the AFM micrographs of the silicon sample with ABDMS three topographically different areas could be clearly distinguished (Fig. 5a). In the upper right part isolated bumps of approximately 0.25 μm height are detected, in the lower left part irregularly shaped islands of film material which are up to 20 μm in diameter and 0.3 μm in height are present.

Compared to short alkane chain aminosilanes, an increase in the number of methylene groups leads to homogenous films. This observation is analogue to the ordering of thiols of different alkane chain lengths reported by us previously [17]. Films of OTS (CH_3-$(CH_2)_{17}$-$SiCl_3$) were prepared on Si and $LiNbO_3$ substrates by adsorption from solution [14]. XPS spectra gave a C(1s):Si(2p) intensity ratio of 18:1, which is in good agreement with the theoretical carbon/silicon ratio in OTS molecules. No chlorine signal was detected, which reveals that a complete reaction of the OTS molecules with the surface hydroxyl groups or trace water molecules has occurred. On both substrates the layers were approximately 30 Å thick as estimated by the attenuation of the substrate peak [14]. Figure 4b shows a series of NEXAFS spectra at the C(1s) edge of a monolayer of OTS (according to

a

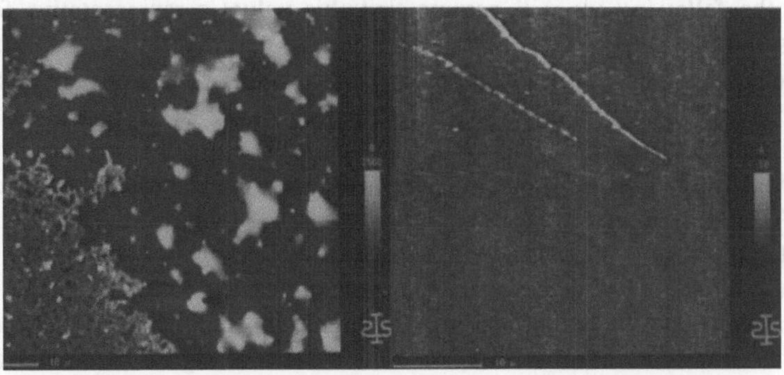

b

Fig. 5: AFM images of a Si(100) sample coated with ABDMS and a Si(100) sample coated with OTS. a) Si(100) sample coated with ABDMS. The scanned area is 100 μm x 100 μm with a maximum surface corrugation of 2500 Å. b) Si(100) sample coated with OTS. The scanned area is 40 μm x 40 μm with a maximum surface corrugation of 50 Å.

the XPS analysis). The intensities of the two main resonances vary strongly with the angle of the incident photons. The C-H* resonance is strongest for normal incidence while the C-C σ* resonance is strongest for grazing incidence. From this it can already be concluded that the chains are oriented predominantly normal to the surface plane. A more elaborate quantitative analysis yielded a tilting angle of 10° with respect to the surface normal [15,17,18]. The XPS and NEXAFS data therefore suggest a well ordered film.

Information about the lateral homogeneity of the OTS layers is obtained from AFM measurements. Contrary to the ABDMS films, AFM images of OTS films on Si(100) substrates show a flat and homogeneous surface topography (Fig. 5b). The maximum surface corrugation of 50 Å occurs only in two snake-shaped features, which could be caused by adsorbed oligomeric units of OTS molecules or multilayer adsorption on surface defects. Beside these defects the films are very smooth. This is in agreement with recent results on OTS monolayers on mica [19].

4. Conclusion

Acoustic plate mode devices offer the unique possibility of a strict separation between electrical contacts and the solution to be analysed. However, to develop a specific sensor for dissolved molecules using a selective adsorbent film, sensitivity and stability remain to be improved, in particular for the detection of proteins in a biological fluid. To answer the question whether an APM device can be developed into an immunosensor the detection limit of these devices has to be compared to the strength of the signal caused by an antigen-antibody reaction. At present, detailed data on noise level only exist for the 13 MHz device. If temperature is properly controlled and the immobilized films are stable, baseline variations of ± 20 Hz can be achieved.

In the following we will assume that this value is representative for our devices and that the limit of detection is twice the noise level. Thus, a minimum signal change of about 80 Hz is required. This corresponds to a change in surface mass density of approximately 120 ng/cm^2 for the 42 MHz device. A close-packed monolayer of IgG POD-conjugate - used as antigen component - has a mass density of about 250 ng/cm^2. However, independent immunological tests (ELISA) show that on average only 10 ng/cm^2 could be immobilized by using the present preparation methods. Ways of closing the gap between the mass sensitivity of the APM devices and the mass of antigen bound might be to use higher frequencies of operation and/or thinner crystal plates despite of the problems discussed above, or to operate the devices at higher harmonics and to reduce the noise level.

Besides the improvement of APM mass sensitivity, the derivatization procedures for the substrate surfaces need to be improved in order to get closer to the theoretical monolayer coverage of antigens. As concluded for our AFM and

NEXAFS results, short alkane chain silanes do not give homogenous covalent coupling layers assumed to be necessary for a high antibody coverage. Analogous to the alkanethiols, well ordered films are only obtained for long alkane chain silanes by self-assembly from solution. Whether well ordered covalent coupling layers also lead to an improved antibody binding capacity is a subject of present investigations. Clearly, the preparation and characterization of these complex surfaces are a challenge for the application of surface science concepts and techniques.

Acknowledgements

The authors are indebted to F. Josse for his valuable advice in sensor design and operation. We would also like to thank F. Bender for performing the experiments on mass loading and G. Hähner and C. Wöll for NEXAFS experiments and analysis. AFM measurements and analysis from A. Baski and H. Fuchs are gratefully acknowledged. The work has been funded in part by the Deutsche Forschungsgemeinschaft (SFB 247/E1) and the Max Buchner Stiftung.

References

[1] J.H. Peters, H. Baumgarten: Monoklonale Antikörper. Herstellung und Charakterisierung, 2. Auflage, Springer, Berlin 1990

[2] J. Ngeh-Ngwainbi, P.H. Foley, Shia S. Kuan, G.G. Guilbault: J. Am. Chem. Soc. **108**, 5444 (1986)

[3] S. Seeger, K. Bierbaum, R. Dahint, M. Mantar, M. Grunze: Synthetic Microstructures in Biological Research, edited by J.M. Schnur and M. Peckerar, Plenum Press, New York, 53 (1992)

[4] A.A. Oliner (ed.): Acoustic Surface Waves, Topics in Applied Physics vol. 24, Springer, Berlin 1978

[5] G.J. Bastiaans: chap. 14 in Chemical Sensors, ed. T.E. Edmonds, Blackie, Glasgow 1987

[6] G. Sauerbrey: Z. Phys. **155**, 206 (1959)

[7] R.M. White, F.W. Voltmer: Appl. Phys. Lett. **7**, 314 (1965)

[8] B.A. Auld: Acoustic Fields and Waves in Solids, John Wiley, New York 1973

[9] R. Dahint, M. Grunze, F. Josse, J.C. Andle: Sensors and Actuators B **9**, 155 (1992)

[10] S.J. Martin, A.J. Ricco, T.M. Niemczyk, G.C. Frye: Sensors and Actuators **20**, 253 (1989)

[11] F. Josse, Z.A. Shana, D.T. Haworth, S. Liew, M. Grunze: Sensors and Actuators B **9**, 97 (1992)

[12] Z.A. Shana, J.C. Andle, T. Zhou, F. Josse, J.F. Vetelino: IEEE Ultrason. Symp. Proc. **1**, 289 (1991)

[13] F. Bender, R. Dahint, F. Josse, M. Grunze, M. v. Schickfus: to be published
[14] K. Bierbaum, G. Hähner, Ch. Wöll, M. Grunze, A. Baski, H. Fuchs: to be published
[15] D.A. Outka, J. Stöhr, J.P. Rabe, J.D. Swalen: J. Chem. Phys. **88**, 4076 (1988)
[16] J. Stöhr, D.A. Outka: Phys. Rev. **B 36**, 7891 (1987)
[17] G. Hähner, M. Kinzler, C. Thümmler, Ch. Wöll, M. Grunze: J. Vac. Sci. Technol. A **10/4**, 2758 (1992)
[18] G. Hähner, M. Kinzler, Ch. Wöll, M. Grunze, M. Scheller, L.S. Cederbaum: Phys. Rev. Lett. **67**, 851 (1991)
see also: Erratum Phys. Rev. Lett. **69**, 694 (1992)
[19] D.K. Schwartz, S. Steinberg, J. Israelachvili, J.A.N. Zasadzinski: Phys. Rev. Lett. **69**, 3354 (1992)

Index of Contributors

Springer-Verlag and the Environment

We at Springer-Verlag firmly believe that an international science publisher has a special obligation to the environment, and our corporate policies consistently reflect this conviction.

We also expect our business partners – paper mills, printers, packaging manufacturers, etc. – to commit themselves to using environmentally friendly materials and production processes.

The paper in this book is made from low- or no-chlorine pulp and is acid free, in conformance with international standards for paper permanency.

MIX
Papier aus verantwortungsvollen Quellen
Paper from responsible sources
FSC® C105338
www.fsc.org

If you have any concerns about our products,
you can contact us on
ProductSafety@springernature.com

In case Publisher is established outside the EU,
the EU authorized representative is:
Springer Nature Customer Service Center GmbH
Europaplatz 3, 69115 Heidelberg, Germany

Printed by Libri Plureos GmbH
in Hamburg, Germany